IAASTD

International Assessment of Agricultural Knowledge, Science and Technology for Development

Central and West Asia and North Africa (CWANA) Report

IAASTD

International Assessment of Agricultural Knowledge, Science and Technology for Development

UNDP

UNEP

UNESCO

THE WORLD BANK

W.H.O. O.M.S.

WHO

GLOBAL ENVIRONMENT FACILITY

IAASTD

International Assessment of Agricultural Knowledge, Science and Technology for Development

Central and West Asia and North Africa (CWANA) Report

Edited by

Beverly D. McIntyre
IAASTD Secretariat

Hans R. Herren
Millennium Institute

Judi Wakhungu
African Centre for Technology Studies

Robert T. Watson
University of East Anglia

Library of Congress Cataloging-in-Publication data.

International assessment of agricultural knowledge, science and technology for development (IAASTD) : Central and West Asia and North Africa (CWANA) report / edited by Beverly D. McIntyre . . . [et al.].

p. cm.

Includes bibliographical references and index.

ISBN 978-1-59726-542-3 (cloth : alk. paper) —
ISBN 978-1-59726-543-0 (pbk. : alk. paper)

1. Agriculture—Asia, Central—International cooperation. 2. Agriculture—Middle East—International cooperation. 3. Agriculture—Africa, North—International cooperation. 4. Sustainable development—Asia, Central. 5. Sustainable development—Middle East. 6. Sustainable development—Africa, North. I. McIntyre, Beverly D. II. Title: Central and West Asia and North Africa (CWANA) report.

HD1428.I542 2008
338.96′07—dc22 2008046045

British Cataloguing-in-Publication data available.

Printed on recycled, acid-free paper

Interior and cover designs by Linda McKnight, McKnight Design, LLC.

Manufactured in the United States of America

10 9 8 7 6 5 4 3 2 1

Contents

Statement by Governments

All countries present at the final intergovernmental plenary session held in Johannesburg, South Africa, in April 2008 welcome the work of the IAASTD and the uniqueness of this independent multistakeholder and multidisciplinary process, and the scale of the challenge of covering a broad range of complex issues. The Governments present recognize that the Global and Sub-Global Reports are the conclusions of studies by a wide range of scientific authors, experts and development specialists and while presenting an overall consensus on the importance of agricultural knowledge, science and technology for development also provide a diversity of views on some issues.

All countries see these Reports as a valuable and important contribution to our understanding on agricultural knowledge, science and technology for development, recognizing the need to further deepen our understanding of the challenges ahead. This Assessment is a constructive initiative and important contribution that all governments need to take forward to ensure that agricultural knowledge, science and technology fulfill their potential to meet the development and sustainability goals of the reduction of hunger and poverty, the improvement of rural livelihoods and human health, and facilitating equitable, socially, environmentally and economically sustainable development.

In accordance with the above statement, the following governments accept the Central and West Asia and North Africa (CWANA) Report:

Azerbaijan, Bahrain, Iran, Kyrgyzstan, Lebanon, Libyan Arab Jamahiriya, Pakistan, Saudi Arabia, Tunisia, Turkey (10 countries).

Foreword

The objective of the International Assessment of Agricultural Knowledge, Science and Technology for Development (IAASTD) was to assess the impacts of past, present and future agricultural knowledge, science and technology on the:

- reduction of hunger and poverty,
- improvement of rural livelihoods and human health, and
- equitable, socially, environmentally and economically sustainable development.

The IAASTD was initiated in 2002 by the World Bank and the Food and Agriculture Organization of the United Nations (FAO) as a global consultative process to determine whether an international assessment of agricultural knowledge, science and technology was needed. Mr. Klaus Töepfer, Executive Director of the United Nations Environment Programme (UNEP), opened the first Intergovernmental Plenary (30 August – 3 September 2004) in Nairobi, Kenya, during which participants initiated a detailed scoping, preparation, drafting and peer review process.

The outputs from this assessment are a Global and five Sub-Global reports; a Global and five Sub-Global Summaries for Decision Makers; and a cross-cutting Synthesis Report with an Executive Summary. The Summaries for Decision Makers and the Synthesis Report specifically provide options for action to governments, international agencies, academia, research organizations and other decision makers around the world.

The reports draw on the work of hundreds of experts from all regions of the world who have participated in the preparation and peer review process. As has been customary in many such global assessments, success depended first and foremost on the dedication, enthusiasm and cooperation of these experts in many different but related disciplines. It is the synergy of these inter-related disciplines that permitted IAASTD to create a unique, interdisciplinary regional and global process.

We take this opportunity to express our deep gratitude to the authors and reviewers of all of the reports—their dedication and tireless efforts made the process a success. We thank the Steering Committee for distilling the outputs of the consultative process into recommendations to the Plenary, the IAASTD Bureau for their advisory role during the assessment and the work of those in the extended Secretariat. We would specifically like to thank the cosponsoring organizations of the Global Environment Facility (GEF) and the World Bank for their financial contributions as well as the FAO, UNEP, and the United Nations Educational, Scientific and Cultural Organization (UNESCO) for their continued support of this process through allocation of staff resources.

We acknowledge with gratitude the governments and organizations that contributed to the Multidonor Trust Fund (Australia, Canada, the European Commission, France, Ireland, Sweden, Switzerland, and the United Kingdom) and the United States Trust Fund. We also thank the governments who provided support to Bureau members, authors and reviewers in other ways. In addition, Finland provided direct support to the Secretariat. The IAASTD was especially successful in engaging a large number of experts from developing countries and countries with economies in transition in its work; the Trust Funds enabled financial assistance for their travel to the IAASTD meetings.

We would also like to make special mention of the organizations who hosted the regional coordinators and staff and provided assistance in management and time to ensure success of this enterprise: the African Center for Technology Studies (ACTS) in Kenya, the Inter-American Institute for Cooperation on Agriculture (IICA) in Costa Rica, the International Center for Agricultural Research in the Dry Areas (ICARDA) in Syria, and the WorldFish Center in Malaysia.

The final Intergovernmental Plenary in Johannesburg, South Africa, was opened on 7 April 2008 by Achim Steiner, Executive Director of UNEP. This Plenary saw the acceptance of the Reports and the approval of the Summaries for Decision Makers and the Executive Summary of the Synthesis Report by an overwhelming majority of governments.

Signed:

Co-chairs
Hans H. Herren,
Judi Wakhungu

Director
Robert T. Watson

Preface

In August 2002, the World Bank and the Food and Agriculture Organization (FAO) of the United Nations initiated a global consultative process to determine whether an international assessment of agricultural knowledge, science and technology (AKST) was needed. This was stimulated by discussions at the World Bank with the private sector and nongovernmental organizations (NGOs) on the state of scientific understanding of biotechnology and more specifically transgenics. During 2003, eleven consultations were held, overseen by an international multistakeholder steering committee and involving over 800 participants from all relevant stakeholder groups, e.g., governments, the private sector and civil society. Based on these consultations the steering committee recommended to an Intergovernmental Plenary meeting in Nairobi in September 2004 that an international assessment of the role of AKST in reducing hunger and poverty, improving rural livelihoods and facilitating environmentally, socially and economically sustainable development was needed. The concept of an International Assessment of Agricultural Knowledge, Science and Technology for Development (IAASTD) was endorsed as a multi-thematic, multi-spatial, multi-temporal intergovernmental process with a multistakeholder Bureau cosponsored by FAO, the Global Environment Facility (GEF), United Nations Development Programme (UNDP), United Nations Environment Programme (UNEP), United Nations Educational, Scientific and Cultural Organization (UNESCO), the World Bank and World Health Organization (WHO).

The IAASTD's governance structure is a unique hybrid of the Intergovernmental Panel on Climate Change (IPCC) and the nongovernmental Millennium Ecosystem Assessment (MEA). The stakeholder composition of the Bureau was agreed at the Intergovernmental Plenary meeting in Nairobi; it is geographically balanced and multistakeholder with 30 government and 30 civil society representatives (NGOs, producer and consumer groups, private sector entities and international organizations) in order to ensure ownership of the process and findings by a range of stakeholders.

About 400 of the world's experts were selected by the Bureau, following nominations by stakeholder groups, to prepare the IAASTD Report (comprised of a Global and 5 Sub-Global assessments). These experts worked in their own capacity and did not represent any particular stakeholder group. Additional individuals, organizations and governments were involved in the peer review process.

The IAASTD development and sustainability goals were endorsed at the first Intergovernmental Plenary and are consistent with a subset of the UN Millennium Development Goals (MDGs): the reduction of hunger and poverty, the improvement of rural livelihoods and human health, and facilitating equitable, socially, environmentally and economically sustainable development. Realizing these goals requires acknowledging the multifunctionality of agriculture: the challenge is to simultaneously meet development and sustainability goals while increasing agricultural production.

Meeting these goals has to be placed in the context of a rapidly changing world of urbanization, growing inequities, human migration, globalization, changing dietary preferences, climate change, environmental degradation, a trend toward biofuels and an increasing population. These conditions are affecting local and global food security and putting pressure on productive capacity and ecosystems. Hence there are unprecedented challenges ahead in providing food within a global trading system where there are other competing uses for agricultural and other natural resources. AKST alone cannot solve these problems, which are caused by complex political and social dynamics, but it can make a major contribution to meeting development and sustainability goals. Never before has it been more important for the world to generate and use AKST.

Given the focus on hunger, poverty and livelihoods, the IAASTD pays special attention to the current situation, issues and potential opportunities to redirect the current AKST system to improve the situation for poor rural people, especially small-scale farmers, rural laborers and others with limited resources. It addresses issues critical to formulating policy and provides information for decision makers confronting conflicting views on contentious issues such as the environmental consequences of productivity increases, environmental and human health impacts of transgenic crops, the consequences of bioenergy development on the environment and on the long-term availability and price of food, and the implications of climate change on agricultural production. The Bureau agreed that the scope of the assessment needed to go beyond the narrow confines of S&T and should encompass other types of relevant knowledge (e.g., knowledge held by agricultural producers, consumers and end users) and that it should also assess the role of institutions, organizations, governance, markets and trade.

The IAASTD is a multidisciplinary and multistakeholder enterprise requiring the use and integration of information, tools and models from different knowledge paradigms including local and traditional knowledge. The IAASTD does not advocate specific policies or practices; it assesses the major issues facing AKST and points towards a range of AKST options for action that meet development and sustainability goals. It is policy relevant, but not policy prescriptive. It integrates scientific information on a range of topics that are critically interlinked, but often addressed independently, i.e., agriculture, poverty, hunger, human health, natural resources, environment, development and innovation. It will enable decision makers to bring a richer base of knowledge to bear on policy and management decisions on issues previously viewed in isolation. Knowledge gained from historical analysis (typically the past 50 years) and an analysis of some future development alternatives to 2050 form the basis for assessing options for action on science and technology, capacity development, institutions and policies, and investments.

The IAASTD is conducted according to an open, transparent, representative and legitimate process; is evidence-based; presents options rather than recommendations; assesses different local, regional and global perspectives; presents different views, acknowledging that there can be more than one interpretation of the same evidence based on different worldviews; and identifies the key scientific uncertainties and areas on which research could be focused to advance development and sustainability goals.

The IAASTD is composed of a Global assessment and five Sub-Global assessments: Central and West Asia and North Africa—CWANA; East and South Asia and the Pacific—ESAP; Latin America and the Caribbean—LAC; North America and Europe—NAE; and Sub-Saharan Africa—SSA. It (1) assesses the generation, access, dissemination and use of public and private sector AKST in relation to the goals, using local, traditional and formal knowledge; (2) analyzes existing and emerging technologies, practices, policies and institutions and their impact on the goals; (3) provides information for decision makers in different civil society, private and public organizations on options for improving policies, practices, institutional and organizational arrangements to enable AKST to meet the goals; (4) brings together a range of stakeholders (consumers, governments, international agencies and research organizations, NGOs, private sector, producers, the scientific community) involved in the agricultural sector and rural development to share their experiences, views, understanding and vision for the future; and (5) identifies options for future public and private investments in AKST. In addition, the IAASTD will enhance local and regional capacity to design, implement and utilize similar assessments.

In this assessment agriculture is used to include production of food, feed, fuel, fiber and other products and to include all sectors from production of inputs (e.g. seeds and fertilizer) to consumption of products. However, as in all assessments, some topics were covered less extensively than others (e.g. livestock, forestry, fisheries and the agricultural sector of small island countries, and agricultural engineering), largely due to the expertise of the selected authors.

The IAASTD draft Report was subjected to two rounds of peer review by governments, organizations and individuals. These drafts were placed on an open access Web site and open to comments by anyone. The authors revised the drafts based on numerous peer review comments, with the assistance of review editors who were responsible for ensuring the comments were appropriately taken into account. One of the most difficult issues authors had to address was criticisms that the report was too negative. In a scientific review based on empirical evidence, this is always a difficult comment to handle, as criteria are needed in order to say whether something is negative or positive. Another difficulty was responding to the conflicting views expressed by reviewers. The difference in views was not surprising given the range of stakeholder interests and perspectives. Thus one of the key findings of the IAASTD is that there are diverse and conflicting interpretations of past and current events, which need to be acknowledged and respected.

The Global and Sub-Global Summaries for Decision Makers and the Executive Summary of the Synthesis Report were approved at an Intergovernmental Plenary in April 2008. The Synthesis Report integrates the key findings from the Global and Sub-Global assessments, and focuses on eight Bureau-approved topics: bioenergy; biotechnology; climate change; human health; natural resource management; traditional knowledge and community based innovation; trade and markets; and women in agriculture.

The IAASTD builds on and adds value to a number of recent assessments and reports that have provided valuable information relevant to the agricultural sector, but have not specifically focused on the future role of AKST, the institutional dimensions and the multifunctionality of agriculture. These include: FAO State of Food Insecurity in the World (yearly); InterAcademy Council Report: Realizing the Promise and Potential of African Agriculture (2004); UN Millennium Project Task Force on Hunger (2005); Millennium Ecosystem Assessment (2005); CGIAR Science Council Strategy and Priority Setting Exercise (2006); Comprehensive Assessment of Water Management in Agriculture: Guiding Policy Investments in Water, Food, Livelihoods and Environment (2007); Intergovernmental Panel on Climate Change Reports (2001 and 2007); UNEP Fourth Global Environmental Outlook (2007); World Bank World Development Report: Agriculture for Development (2008); IFPRI Global Hunger Indices (yearly); and World Bank Internal Report of Investments in SSA (2007).

Financial support was provided to the IAASTD by the cosponsoring agencies, the governments of Australia, Canada, Finland, France, Ireland, Sweden, Switzerland, US and UK, and the European Commission. In addition, many organizations have provided in-kind support. The authors and review editors have given freely of their time, largely without compensation.

The Global and Sub-Global Summaries for Decision Makers and the Synthesis Report are written for a range of stakeholders, i.e., government policy makers, private sector, NGOs, producer and consumer groups, international organizations and the scientific community. There are no recommendations, only options for action. The options for action are not prioritized because different options are actionable by different stakeholders, each of whom have a different set of priorities and responsibilities and operate in different socio-economic and political circumstances.

1

Setting the Scene

Coordinating Lead Authors:
Ali Taha Ayoub (Sudan), Rym Ben Zid (Tunisia)

Lead Authors:
Saad M. Alayyash (Jordan), Sonia Ali (Egypt), Yameen Memon (Pakistan)

Contributing Authors:
Jean Albergel (France), Nour Chachaty (Syria), Celine Dutilly-Diane (France), Mahmud Duwayri (Jordan), Ahmad El-Wakeel (Sudan), Alessandra Galié (Italy), Stefania Grando (Italy), Thameen Hijawi (Palestine), Ayfer Tan (Turkey), Selma Tozanli (Turkey), Lokman Zaibet (Tunisia)

Review Editors:
Azza Emara (Egypt) and Jad Isaac (Palestine)

Key Messages

Context

1. CWANA enjoys the presence of the largest oil reserves in the world, which makes it attractive to international interests and interventions. At the same time, CWANA has many people living below the poverty line.
2. Weak governance and poor human rights have exacerbated the wealth gap among societies and sparked ethnic conflict in many CWANA countries. Tribalism is still dominant, with conflicts, feuds and wars among tribes and ethnic groups, mostly over resources, becoming frequent.
3. About 85% of the region is dry, with highly variable rainfall accompanied by frequent drought: most of the deserts in the world are in this region. Natural resources are under high pressure from a high population growth rate and increased life expectancy, climate change and misuse.
4. Most of the current conflicts in the world are in this region. As a result, migration, population displacement, land degradation, water depletion, loss of biodiversity, disappearance of indigenous knowledge and degradation of livelihoods are becoming the norm in many CWANA countries.
5. Scarcity of water resources is compounded by inefficient water use. Most CWANA countries use more than 80% of their fresh water for agriculture and have inefficient irrigation systems. Most of the commercially produced desalinized water worldwide is in the Arabian Peninsula.
6. While most of the water resources are transboundary, there is weak cooperation to jointly manage the shared water. Current water allocations have not been arrived at through regional agreements, except in the Nile Basin and the Ganges Basin.
7. Most renewable groundwater in the region is already exploited and in some cases overexploited, leading to deteriorating water quality. This region has a large reservoir of fossil groundwater, used by some countries without any formal agreements among the sharing countries.

Well-being

1. High population growth and high rates of unemployment strongly drive increased urbanization and migration.
2. Improvement in living standards in some countries, changes in lifestyle, free access to market and development of tourism have increased demand for diverse agricultural commodities.
3. Uneven distribution of wealth within most CWANA countries leads to the formation of marginalized groups and poverty pockets and stems from several physical, social, economic, educational, ethnic and political factors. The middle class is shrinking and most of its individuals are getting poorer. More than 40% of the people live on less than US$1 a day in CWANA.
4. Access to microfinance to generate income is limited, especially for women.
5. Access to education is limited in many CWANA countries. Poverty and children involved in agricultural work, especially rearing livestock, limit access. In Sudan, where pastoralists move with their herds and families, mobile schools are provided.
6. Relief and emergency interventions are widespread in the region, especially in conflict zones. In conflicts, food for work or small income-generating projects proved to be better coping strategies than food aid, since food aid tended to perpetuate dependency.
7. In CWANA countries, animal products are major sources of protein, but their prices have increased significantly from increased feed cost. Decreased affordability and increase in poverty have shifted diets toward legumes and carbohydrates. This change in food habits may adversely affect human health.
8. Household farming using plant production and small ruminants is dominant in most CWANA countries, even in urban areas. The social and nutritional value of these systems in alleviating poverty and providing food outweighs its profitability.
9. Seasonal migration is common in many CWANA countries, which increases the pressure on natural resources and affects the social fabric. Most Arabian Peninsula countries have a high percentage of migratory workers in all fields. Jordan and Lebanon have many migratory workers in agriculture. High unemployment and low profitability in traditional farming make seasonal farming within and among states appealing; this depletes the indigenous skills to manage natural resources.

Agriculture

1. Large-scale farming, which relies on high investment and monocropping, is expanding at the expense of small-scale indigenous farming. It leads to loss of agrobiodiversity, land degradation, depletion of water, loss of livelihoods and conflicts over land and water rights.
2. Large areas of the agricultural land in CWANA are subjected to poor soil management, leading to wind erosion, water erosion, nutrient depletion and soil salinization.
3. As a result of globalization and free markets, small-scale farmers in CWANA are becoming less competitive because of low productivity and lack of subsidies, price systems and technical and marketing information.
4. Most CWANA countries are net importers of food, especially cereals. The Arab countries import US$20 billion each year. Imported food costs are rising rapidly and will probably rise further from world competition for food.
5. Agricultural imports in CWANA countries in 2004 reached US$41.8 billion, while agricultural exports did not exceed US$17 billion. CWANA exports fruits, vegetables, dates and olive oil and imports mainly grains. Considering the water scarcity, this appears to indicate unwise water-resource management. However, profitable grain production depends on large-scale land ownership and mechanization, while fruits and vegetables are less dependent on land and are labor intensive.

Key Issues

1. Land tenure, ownership rights and communal rangelands affect natural resource management and, consequently, agricultural development in CWANA. Social

and cultural values toward land ownership in most of the region go beyond economic profitability of agriculture. Communal land is overexploited and mismanaged.

2. Poor land-use planning, population growth and urbanization result in loss of agricultural land in most CWANA countries. Large-scale land reclamation projects have been implemented in many countries to overcome the loss of agriculture land.
3. Most CWANA countries are witnessing land fragmentation and small landholdings, which curtail investment and mechanization. This leads to reduced viability of agriculture and unsound land management. This also exists in tenured land systems.
4. Inefficient infrastructure and marketing and poor post-harvest management in CWANA lead to limited profitability and lack of competitiveness.
5. Women play a central role in agriculture in CWANA countries. Their role is not properly recognized; poor access to land tenure, unpaid family labor and gender issues are not high on national agendas.
6. Little cooperation between CWANA countries exists, although some success stories can be made use of: women's empowerment in Tunisia, water-harvesting systems in Palestine, agricultural input packages to small farmers in Egypt.
7. CWANA local genetic resources are disappearing fast, although this region is the origin of the world's domesticated agriculture.

Agricultural knowledge, science and technology

1. In general, little is allocated to research in CWANA countries, less than 0.2% of the GNP instead of the recommended 2%. Recently, some real progress has been made in promoting research and development, especially in Qatar and the Emirates.
2. Illiteracy is high within the farming community in CWANA, especially among women. This hinders technology adoption and advancement.
3. Most agricultural research does not reflect the real needs of farmers. However, in some countries, such as Morocco, new efforts have localized research and adapted it to meet the needs of stakeholders.
4. The gap between the results of national and international agricultural research and its usefulness to farmers comes from weak technology transfer, poor dissemination of information and ineffective extension services.
5. Brain drain is widespread in most CWANA countries. It is estimated that about 40% of migrating professionals to the developed countries come from CWANA countries. Lack of local opportunity, poor governance and conflicts drive the brain drain.
6. In most CWANA countries, civil institutions, including small farmer organizations and cooperatives in agriculture, are not promoting authentic agricultural development because of bureaucracy and centralization.
7. Community organizations started in developing household agricultural enterprises, community forests, water harvesting and environmental protection.
8. Most CWANA countries rank low in all good governance indices. Participation, rule of law, transparency, responsiveness, consensus orientation, equity and inclusiveness, effectiveness and efficiency, and accountability are essential for sustainable development in the region.
9. Many CWANA countries are deficient in local expertise in policy formulation, institutional development, research management and understanding of farming systems, knowledge systems and their dissemination.
10. Lack of regulation and enforcement results in overuse of pesticides, including banned ones, which pollutes water and creates health hazards.

1.1 Profile of CWANA

1.1.1 Overview

CWANA (Central and West Asia and North Africa) is complex and vast. In this report, CWANA has been divided into the following subregions: North Africa, Nile Valley and the Red Sea, Arabian Peninsula, West Asia, Central Asia, and the Caucasus, since these countries share similar agroecological characteristics (Figure 1-1). Where necessary other subdivisions are discussed, for example, the Euphrates riparian system.

CWANA extends from the Atlantic Ocean, Mauritania and Morocco, in the west to Pakistan and Kyrgyzstan in the east and from Turkey and Kyrgyzstan in the north to Somalia and Yemen in the south. It falls between longitudes 17°W and 80°E and latitudes 43°N and 2°S. It comprises 32 countries and occupies about 20.9 million km^2 or 15% of the world. Over half the world's dryland population lives in CWANA (Figure 1-1).

CWANA includes some of the most inhospitable places on earth. It has a great part of the Sahara, the Empty Quarter of Arabia and the Usturit and Kyzilkum deserts. CWANA countries are drylands susceptible to desertification and mostly drought prone (UNEP, 1997). They have the fastest growing food deficits in the world and could face catastrophe if their remaining natural resources are not properly managed and conserved. Drought management and mitigation are needed in these countries (Karrar, 2002). In CWANA, governments have to make difficult tradeoffs between short-term benefits and long-term solutions. Droughts always require immediate attention because they threaten human lives, but long-term solutions are necessary.

1.1.2 Ecosystems

1.1.2.1 Aridity zones

CWANA is characterized by low mean annual precipitation, high interannual variability and high potential evapotranspiration. In over 90% of the region annual rainfall is below 200 mm, but in a few areas rainfall can reach over 2000 mm. As precipitation decreases, annual rainfall variability increases.

Rainfall in CWANA is far from uniform. CWANA countries can be classified into four categories according to the average annual rainfall:

- Countries receiving more than 500 mm per year with 120 days of rain and with places surpassing 1500 mm: Kyrgyzstan (533 mm), Armenia (562 mm), Turkey (593 mm), Lebanon (661), Tajikistan (691 mm)
- Countries receiving 300 to 500 mm yearly with 60 to 100 days of rain: Afghanistan (327 mm), Morocco (346 mm), Sudan (416 mm), Azerbaijan (447 mm), Pakistan (494 mm)

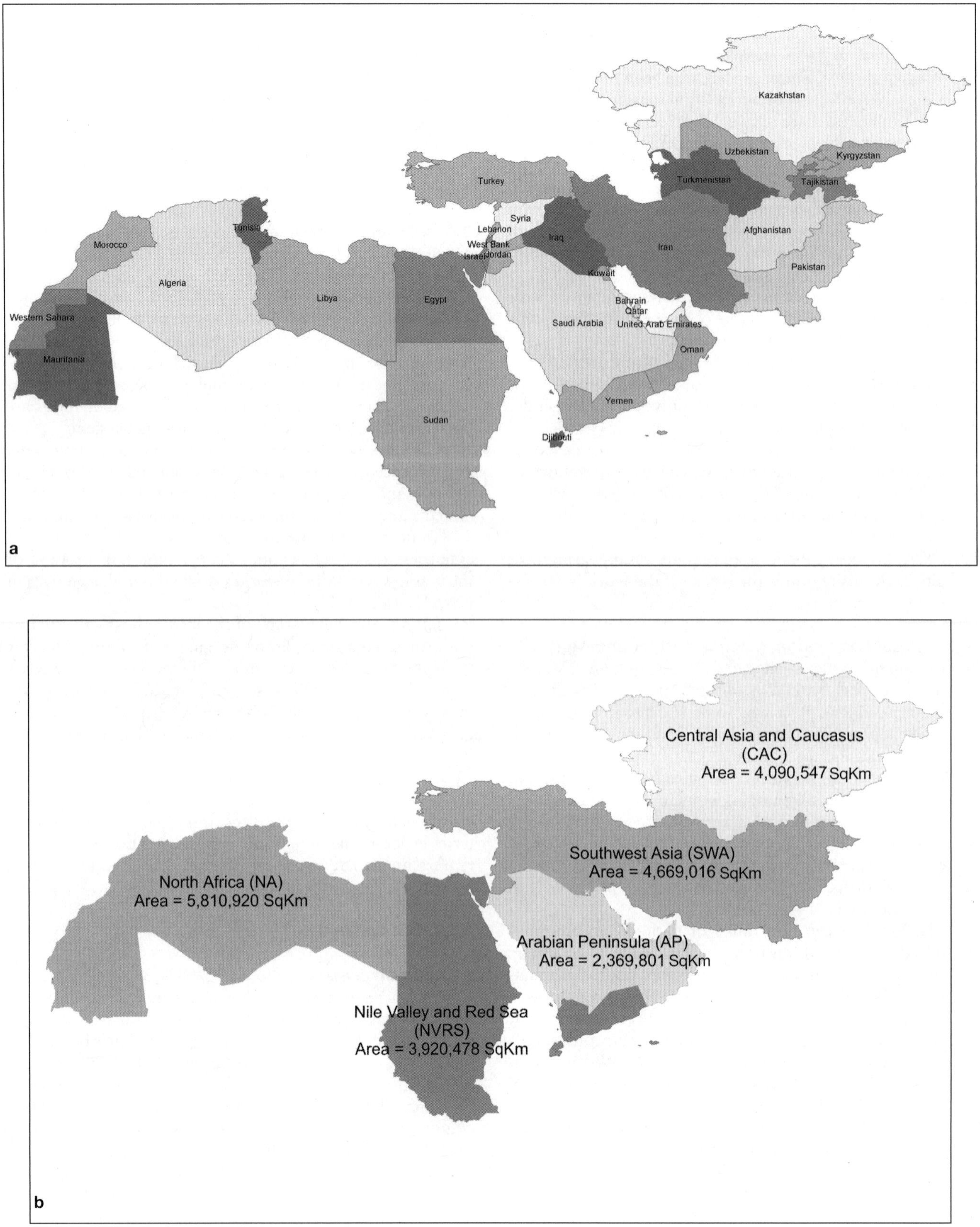

Figure 1-1. *The CWANA region: a) countries; b) subregions.* Source: ESRI's Arc View software supplemental map data, http://www.esri.com/.

- Countries receiving 100 to 300 mm annually with 40 to 70 days of rain, Jordan (111 mm), Kuwait (121 mm), Oman (125 mm), Turkmenistan (161 mm), Yemen (167 mm), Tunisia (207 mm), Iraq (216 mm), Djibouti (220 mm), Iran (228 mm), Syria (252 mm), Somalia (282 mm)
- Countries receiving less than 100 mm each year with fewer than 30 days of rain: Egypt (51 mm), Libyan Arab Jamahiriya (56 mm), Saudi Arabia (59 mm), Qatar (74 mm), United Arab Emirates (78 mm), Bahrain (83 mm), Algeria (89 mm), Mauritania (92 mm)

Rainfall in almost the entire region is intense, producing flash floods. Evaporation and transpiration cause much water loss from the surface. Potential evapotranspiration ranges from 1500 to 3000 mm per year. The Arabian Peninsula, which has almost one-fourth of CWANA's hyperarid zone, has annual precipitation deficits between 1700 to 2500 mm. A way to examine the consequences for agricultural systems in CWANA is to divide the region into aridity zones (Table 1-1):

- The hyperarid zone covers about 0.9 billion ha, over 86% in CWANA. Except for irrigated areas, such as in Egypt, few nomads live in this zone.
- In the arid zone the natural vegetation of short grass and sparse thorn scrub provides some grazing for sheep, goats, cattle and camels. Some wildlife, mostly gazelles and reptiles, live in this zone. Opportunistic rainfed cropping of sorghum and pearl millet is practiced in depressions and foothills. Nomadic groups live in this zone, which has a low carrying capacity.
- The semiarid zone has both animal and rainfed husbandry and therefore supports millions of rural people and major agricultural industries.
- The subhumid zone contains luxuriant vegetation, from savanna to broken woodland. Productive arable farming is the primary use.

Three more aridity zones occupy small areas:

- The coastal zones have various landscapes, coastal salty marshes, a series of medium-high hills lying a few kilometers away from the sea and open plains traversed by rocky sediments. While arid ecosystems dominate, fertile deltas of agricultural value exist.
- Swamps cover considerable areas in the region. In Sudan, the Sud swamps cover about 12 million ha. These swamps are inundated permanently or seasonally. The seasonally inundated parts are vast meadows of almost pure grass, which maintain much livestock.
- Mountain meadows occur on shallow soil at the higher altitudes and have high rainfall, high humidity and low temperatures. The meadows are rich in grasses, herbs and trees. Plants vary with the altitude, up to about 3500 m. Some of Central Asia and the Caucasus have 1-5 m snowfall. The climate has a short growing season and long, cold winters.

Table 1-1. **Characterization of major aridity zones in CWANA.**

Zone type	Land area (%)	Location	Characteristics	Type of agriculture
Hyper-arid	67.0	75% located in North Africa, Nile Valley and Red Sea subregions 25% in Arabian Peninsula and West and Central Asia	Low annual rainfall (0-100 mm) unequally distributed in time and space True desert Plain of loose red sand Scarce vegetation	Meager agricultural activities, except some irrigation on alluvial soils
Arid	7.5	West and Central Asia, Caucasus, Nile Valley and Red Sea	Total annual precipitation 100-300 mm Interannual variability rainfall range: 50-100% Short grass and sparse thorn scrub Wildlife—gazelles and reptiles Strong winds and summer thunderstorms causing sand storms	Nomadic grazing Opportunistic rain cropping and pearl millet in depressions and foothills Low carrying capacity Large irrigation schemes
Semiarid	17.3	North Africa Central Asia	Total annual precipitation 400-800 mm	Rainfed agriculture: cropland and permanent pastures Interannual variability in rainfall causes yield variation Cattle associated with rainfed agriculture Some supplementary irrigation schemes
Sub-humid	7.6	North Africa South Sudan Caucasus, Central Asia	Large variety of vegetation types Rainfall more than 800 mm a year Temperate climate to tropical climate	Productive rainfed farming Some plantation agriculture

Source: UNEP, 1997.

Rangelands occupy about half the area in West Asia. They have low carrying capacity.

Relative humidity is vital for human comfort. Areas close to the coast have high relative humidity in summer, while other parts of the region have high relative humidity only during the rainy season. The Arabian Peninsula is one of the hottest and driest regions of the world, with daytime temperatures often exceeding 50°C (De Pauw, 2002). Wind is more active in the hyperarid and arid zones than in the other zones. One reason is that these zones are relatively dry and have vast plains with little vegetation, over which the wind can blow unchecked. These powerful winds erode the soil and lift clouds of dust thousands of meters into the air. On many days in most of the region the atmosphere is hazy from dust.

1.1.2.2 Water resources

While CWANA covers 15% of the world area, it has only 2% of internal renewable water resources (WRI et al., 1998), making it the world's most water-stressed region. Water availability by CWANA subregion in the year 2000 was classified as follows: North Africa catastrophically low, Nile Valley and the Red Sea very low, Arabian Peninsula catastrophically low, West Asia very low, Central Asia and the Caucasus low. In the Arabian Peninsula, the water-stress index is 100%. The subregion has hyperarid and arid climates with an annual rainfall less than 100 mm.

The total available water is about 15 billion m^3 annually. Surface water comprises about 45%, groundwater about 41%, desalinization about 12% and agricultural drainage reuse about 2%. Agriculture uses about 86% of the available water in CWANA. Excessive use of groundwater has resulted in sharp decline in its amount and quality because seawater is intruding (UNEP, 2002b).

West Asia has a water stress index of 83%. The subregion is mostly arid and semiarid. Most countries in the subregion receive less than 250 mm of rain per year. Two major rivers, the Euphrates and Tigris, are shared by Iraq, Syria and Turkey. Total available water resources are about 80 billion m^3 each year, 85% surface water, 10% groundwater and 5% agricultural drainage reuse. Agriculture uses about 80% of the available water. Effluent, agrochemicals and industrial discharges have seriously affected water quality. CWANA's driest country is Mauritania, where the average annual renewable water resources are less than 0.5 billion m^3. Water resources are poorly managed and inefficient across CWANA. In some areas farmers over-irrigate wheat crops 20 to 60%.

Water scarcity must be dealt with before any development can be sustained. Any development in CWANA must recognize that vegetation and soils are sensitive to intensive use. Once degraded, many decades are needed to restore the production and hydrologic function of these ecosystems. Low and erratic precipitation prevents the rapid reestablishment of vegetation, leaving a degraded landscape exposed to water and wind erosion for a long time.

1.1.2.3 Status of soils

Data from the Global Assessment of Human Induced Soil Degradation (GLASOD) (UNEP/ISRIC, 1990) show that soil degradation is widespread in the whole CWANA region. In North Africa, 70% of the degraded soils are through wind erosion, 20% water erosion and 10% through nutrient depletion and soil salinization. Soil erosion is a widespread problem in North Africa, with severity highest at the western and eastern ends of the Atlas and Rif mountains. In the lower parts, centuries-old, traditional soil conservation techniques are breaking down. The Nile Valley of Egypt has substantial chemical degradation problems, involving both salinization and nutrient depletion. In many of the upland areas of the Nile Valley and the Red Sea, land degradation has reached critical levels (UNEP, 1997). About 64 million ha of soils are degraded to varying degrees in Sudan. Wind erosion is the dominant cause of soil degradation in the arid zone, while water erosion is dominant in the semiarid zone.

Nutrient loss affects all climatic zones of Sudan (Ayoub, 1998). Most of Somalia is affected by wind erosion, with a small area in the southern part affected by water erosion. In Yemen, 50% of the soils are affected by water erosion and 20% by wind erosion. High to very high severity of nutrient depletion is taking place in the sandy soils of Sudan, the Arabian Peninsula and Central Asia. The dominant cause of soil degradation is wind erosion in the Arabian Peninsula and West Asia, covering almost 60% of the degraded area, followed by soil salinization, 30%, and water erosion, mostly in Syria and Lebanon.

Inefficient use of irrigation water has resulted in salinization, alkalization and waterlogging. More than 50% of the irrigated lands in the Euphrates plains and in Pakistan have been badly affected by salinization and waterlogging (UNESCWA, 1997). Indeed, Pakistan is among the top five countries in the world with irrigated land damaged by salinization. About 300 million ha in CWANA are affected by soil salinity and alkalinity, nearly 30% of the world's saline and alkaline soils (Abrol et al., 1988). They occur in coastal areas and inland salt marshes. Soil salinity is severe in the Euphrates and Tigris valleys of Syria and Iraq. The southern part of Pakistan is affected by wind erosion and soil salinity, while the northern part is affected by water erosion. Afghanistan is mostly affected by water erosion, while the southwestern part is affected by wind erosion and soil salinity. The bulk of Iran is affected by wind erosion. Its northern and western parts are affected mainly by water erosion. Soil salinity is also widespread and most severe in areas bordering Iraq and Afghanistan. The total area of salt-affected soils in Iran is about 15% (Koocheki and Mohalati, 1992). Turkey is seriously affected by water erosion; a few areas have water erosion coupled with nutrient depletion and soil salinity.

The Caucasus is two-thirds desert, the Usturit and Kyzilkum deserts. Soils are commonly salinized. The lowlands are salt accumulation zones and saline soils dominate (Zhang et al., 1992). Turkmenistan lies within these deserts. There are two large seas, the Caspian and the Aral. The Aral Sea is dying because its input water is diverted for agriculture. It shows the most recent example of human-induced environmental degradation in the subregion (UNEP, 1997). Fortunately, a big portion of Tajikistan is stable under natural conditions. Azerbaijan has various degrees of water erosion and extensive soils affected by salinity. The whole of Kyrgyzstan suffers various degrees of water erosion, wind erosion and soil salinization.

Surface and gully erosion exacerbate problems of low productivity, further diminishing soil resources, threatening the future productivity of the land. Water quality can become impaired, which when coupled with high sediment levels, constrains developing sustainable water-resource management. Clearly, action must be taken on many fronts to develop sustainable solutions and improve management of land and water in CWANA.

1.1.2.4 Agrobiodiversity

Agrobiodiversity or agricultural biodiversity includes all components of biological diversity relevant to food and agriculture: "Agricultural biodiversity is a broad term that includes all components of biological diversity of relevance to food and agriculture. It encompasses the variety and variability of animals, plants and microorganisms, at the genetic, species and ecosystem levels. They are necessary to sustain key functions of the agroecosystems, their structures and processes for, and in support of, food production and food security. So it refers primarily to genetic variability in cultivated plants and domesticated animals together with their progenitors and closely related wild species growing and evolving under natural conditions. Plants and animals gathered from, and hunted in, the wild are also included in this term" (www.biodiv.org/convention).

CWANA has a wide range of diversity in climate, topography and soils; genetic diversity of many globally important crops and their wild relatives, such as cereals, food legumes, forages, industrial crops, fruit trees and vegetables and farm animals. Its cultural diversity affects agriculture. Agriculture began independently in different sites about 12,000 to 11,000 years ago. It spread along the East Mediterranean, North Africa and South Europe, southwards to Egypt and the Ethiopian Plateau and eastwards to Central Asia and Indus. Agriculture in CWANA centers upon sheep, goats, cattle, cereals (wheat, barley), and legumes (lentil, pea, vetch), flax, which are early domesticates of the Near East (Harlan, 1995). Historical evidence suggests that West Asia and North Africa (WANA) are among the most important domestication centers, from where small ruminant production expanded to other parts of the world. The small ruminants found in the region represent a rich source of the genetic diversity necessary for production, orientation and diversification efforts.

Most crops in the Near East were domesticated during the Neolithic period. CWANA is significant and unique for plant diversity. Vavilov's (1926) Near Eastern, Mediterranean and Central Asian centers of origin extend into the region. This, of course, indicates that CWANA is a center of origin or diversity of several crops and many other plant species. The wild plant relatives and landraces, with enormous genetic diversity, are still found for cereals, legumes, fiber and oil crops, pasture and forage plants, fruits, nuts and vegetables (Harlan, 1992).

Near Eastern centers of diversity, including Jordan, Lebanon, Palestine, Syria, southeast Turkey, southern Iran and Iraq, extend northeast to the Caucasus and south to Arab Peninsula. They encompass a megadiversity of important food crop and pasture species and have a nuclear center and center of origin (Vavilov, 1926). Here numerous species, notably wheat, barley, lentil, pea and vetch, of temperate zone agriculture originated 10,000 years ago. In Turkey, Harlan (1951) described microcenters for *Amygdalus* spp., *Cucumis melo, Cucumis sativus, Cucurbita moshata, Cucurbita pepo, Lens culinaris, Lupinus* spp., *Malus* spp., *Medicago sativa* and other *Medicago* spp., *Onobrychis viceaefolia, Phaseolus vulgaris, Pistachio* spp., *Prunus* spp., *Trifolium* spp., *Vicia faba, Vitis vinifera.* Almond, olive and pistachio trees also originated from this region and have dominated traditional agricultural systems. They have diverse wild relatives and local varieties. The biodiversity in this region is most outstanding for within-species genetic diversity and the many endemic species. Furthermore, the indigenous crop and food plants of the Near East are known for their resistance to disease and abiotic stress, making them valuable for germplasm enhancement, upon which global food security depends.

Central Asia is also an important center of diversity in cultivated plants. It has the richest species and intraspecies diversity for many globally important crops. Agriculture must have reached this center from the Near East about 5000 BCE (Zeven and de Wet, 1982). Major crops include apple, apricot, peach, pear, plum, grape, almond, pistachio, pomegranate, fig, wheat, barley, rice, maize, sorghum, bean, chickpea, tomato, potato, onion, garlic, coriander and melon. Industrial and stimulant crops cotton, sugar beet, groundnut, sesame and tobacco are also cultivated.

1.1.3 Climate change

The vulnerability of CWANA to variable rainfall and drought is compounded by the anticipated effects of global climate change.

Drylands, CWANA included, are net sources of CO_2 (Sombroek, 1995) from overuse of plant resources. Annual CO_2 emissions of CWANA total about 1.8 billion tonnes—about 7.8% of the world's emissions. Inventories of greenhouse gases show most emissions in most CWANA countries come from burning fossil fuel. In Sudan and Somalia, CO_2 emissions can be correlated to changes in land use, particularly deforestation. CO_2 can be sequestered through better soil management. In some Central Asian countries almost half the emitted CO_2 was absorbed, 90 million tonnes, by changing land use (UNDP et al., 2003). Some atmospheric pollution caused by human activity is related to oil production and export, fertilizers, cement factories and motor vehicles.

Seasonal sand and dust storms contribute to air pollution in the region. The US Environmental Protection Agency (US EPA, 1996) estimated annual dust fallout of about 1000 tonnes per km^2 along the coast of Kuwait. Dust storms absorb pollutants, such as pesticides, and transport them long distances. Transboundary air pollution is an emerging issue in the region. Climate change will potentially have major effects on the Nile Delta and Darfur (Box 1-1).

1.1.4 Macroeconomic indicators

1.1.4.1 Gross domestic product (GDP) and agriculture's share

The GDP varies among the countries from US$889 in Yemen to US$22,420 in United Arab Emirates (UAE). Most countries fall in the range of US$2,000-7,000; oil-producing countries have an average of US$16,657 (Table 1-2).

Though agriculture is a major employer in CWANA countries, with 50% of the labor force, its share in the GDP

Box 1-1. Climate Change: The Nile Delta and Darfur

The Nile Delta is one of the oldest intensely cultivated areas on earth. It covers an area of about 25,000 km^2. Almost 40 million inhabitants live in the delta. Deserts surround the low-lying fertile floodplains. The area is suitable for intensive agriculture. Most of a 50-km-wide land strip along the coast is less than 2 m above sea level and is protected from flooding only by a coastal sand belt 1 to 10 km wide, formed by discharge of the Rosetta and Damietta branches of the Nile. Erosion of the protective sand belt is a serious problem and has accelerated since the construction of the Aswan Dam.

Rising sea levels caused by expected global warming would destroy weak parts of the sand belt, essential to protect lagoons and the low-lying reclaimed lands. The impact would be serious. One-third of Egypt's fish catches are made in the lagoons. Sea-level rise would change the water quality and affect most freshwater fish. Valuable agricultural land would be inundated. Vital, low-lying installations in Alexandria and Port Said would be threatened. Recreational tourism beach facilities would be endangered and essential groundwater would be salinized. Dikes and protective measures would probably prevent the worst flooding up to a 50-cm sea-level rise. However, it would cause serious groundwater salinization and the impact of increasing wave action would be serious (UNEP, 2002a).

Darfur: Ecological degradation in the Sudano-Sahelian ecozone, and especially in Darfur, mainly caused by climate change, has been so severe that the traditional means for preventing and managing interethnic disputes have been rendered virtually unworkable. Indeed, many of the current disputes are not being fought along traditional political borders, but along ecological borders (in this case the borders of the semiarid plains roamed by "Arab" pastoralist nomads and those of the wet oases settled by "African" fur farmers) that divide richer and poorer ecozones. To continue to treat the conflict in Darfur, and many other parts of Africa, as purely ethnic, tribal, political or religious, and to ignore the growing impact of ecological degradation and depletion of the resource base, will ultimately lead to a distorted understanding of the real situation, and consequently limit the possibility for genuine conflict resolution.

In Darfur, as in most other parts of the continent, the balance of soil, climate, water and flora upon which human and animal life depend has been upset. In addition to persistent drought (six over the last hundred years, three occurring in the last 20 years), unsustainable methods of land use, such as large-scale mechanized rainfed farming and overgrazing in marginal lands, are destroying the ecosystem. As a result of ecological deterioration and armed conflicts, millions of people have been forced to abandon their homelands and have become displaced—so many in fact that Sudan has the world's highest proportion of internally and externally displaced people, one in every five (El-Nour, 1992).

Climate change can dramatically affect livelihoods and ecosystems. In these regions, lying between isohyets 100-600 mm, even the slightest decline in mean annual rainfall could bring people and livestock to the brink of disaster. Along with the general decline in rainfall, vegetation stripped from large areas has allowed sand dunes to move, which in turn has killed almost all remaining plant life. There has been an increase in incidents of conflict corresponding with decrease in rainfall (Suliman, 2000). Social phenomena in the region must be analyzed within the context of climatic and ecological transformations (Bachler and Spillmann, 1992; Suliman, 2000).

is only 13%. This share is lowest in the high-income countries, 3%, but can be as high as 60 to 80% in low-income countries (Rodriguez, 1997).

Countries with agriculture less than 10% in GDP, mostly the Gulf countries, have scarce natural resources and agriculture has not developed because of it. Countries, such as United Arab Emirates and Saudi Arabia, have to import staple products (Table 1-3). But, because they are rich from their oil income, they have no problem importing staple products. They do not really need to increase agricultural production, although in case of political crisis, the food weapon could be used against them. Jordan and Djibouti have scarce natural resources, nor do they have oil wealth. CWANA countries may be classified in three types.

Agriculture remains important in countries with agriculture in GDP between 10 and 20%. Nevertheless, most of these countries are dependent upon importing staple products. A large percentage of labor is employed in agriculture and the population density is high compared with agricultural production. This discrepancy is from natural resource scarcity, unequal access to resources and low productivity of labor. Countries such as Algeria and Iran can afford imported staple products because they have oil income. Egypt, Morocco, Tunisia and Yemen do not have oil. The share of agriculture in GDP in Lebanon is between 10 and 20% because agriculture is still important. Farmers have access to land and water and the farms are family owned. Lebanon is not dependent on imports for major staple products. Although population has increased in the last 30 years, migration has maintained the equilibrium between resources and population density. The Caucasus countries, although agriculture is more than 20% of the GDP, are still dependent on imports of major staple food because land tenure is uncertain and not secure (FAO, 2005). Nevertheless, in Tajikistan, agriculture has accounted for one-third of the economic growth since 1997 and has made a major contribution to the fall in rural poverty.

1.1.4.2 National saving and investment

The national saving rate induces level investment, otherwise the nation has to borrow (dissaving). National saving was high in Jordan, 24.4% in 2001, from external financial transfers. The saving rate was around 15.4% in Egypt in 2001 and the accumulation of capital as a percentage of

Table 1-2. **Total population, rural population, GDP PPP, human poverty index and human development index.**

Country	Total population in 1000 (2005)	Rural population (%)	GDP PPP in CI$ (2003)	Human poverty index (2002)[a]	Human development index (2004)[b]
Afghanistan	25,971	76	n/a	n/a	n/a
Algeria	32,877	40	6,248	21.9	0.728
Armenia	3,043	36	3,607	n/a	0.768
Azerbaijan	8,527	50	3,606	n/a	0.736
Bahrain	755	10	n/a	n/a	0.859
Djibouti	721	15	2,144	34.3	0.494
Egypt	76,117	58	3,950	30.9	0.702
Iran	70,675	32	7,145	16.4	0.746
Iraq	26,555	33	n/a	n/a	n/a
Jordan	5,750	21	4,319	7.2	0.760
Kazakhstan	15,365	44	6,556	n/a	0.774
Kuwait	2,671	4	n/a	n/a	0.871
Kyrgyzstan	5,278	66	1,714	n/a	0.705
Lebanon	3,777	12	n/a	9.5	0.774
Libya	5,631	11	n/a	n/a	0.798
Mauritania	3,068	36	1,896	48.3	0.486
Morocco	32,209	40	4,012	34.5	0.640
Oman	3,020	21	n/a	31.5	0.810
Pakistan	161,151	65	1,971	41.9	0.539
Palestine (West Bank, Gaza Strip)	3,901	35	n/a	n/a	0.736
Qatar	627	8	n/a	n/a	0.844
Saudi Arabia	25,795	12	n/a	15.8	0.777
Somalia	10,742	64	n/a	n/a	n/a
Sudan	35,040	59	2,046	31.6	0.516
Syria	18,651	50	3,575	13.7	0.716
Tajikistan	6,356	76	1,119	n/a	0.652
Tunisia	10,042	36	7,083	19.2	0.760
Turkey	73,301	33	6,749	12.0	0.757
Turkmenistan	5,014	54	5,884	n/a	0.724
United Arab Emirates	3,107	15	n/a	n/a	0.839
Uzbekistan	26,410	63	1,737	n/a	0.696
Yemen	21,481	74	889	40.0	0.492
Total	723,628				

PPP = purchasing power parity; CI$ = constant USD; n/a = not available

[a]scale of 1 to 100

[b]scale of 0 to 1

Source: World Bank, 2006.

Table 1-3. **Agricultural GDP as share of total GDP and major imports.**

Country	Agricultural GDP as share of total GDP (%)	Major imports
Afghanistan		–
		–
Algeria	10.0	Wheat, dry cow's milk, maize
Armenia	26.2	Wheat
Azerbaijan	–	–
Bahrain	0.7	Chicken
Djibouti	3.7	Wheat
Egypt	16.8	Wheat, maize, cake of soy beans
Iran	12.1	Wheat, oil of soybeans, rice milled
Jordan	2.2	Wheat, maize
Kuwait	–	–
Kyrgyzstan	38.6	Wheat
Lebanon	11.0	Cattle
Libya	–	Flour of wheat, paste of tomato, wheat
Mauritania	20.8	Sugar, oil of soybeans
Morocco	16.1	Wheat
Oman	3.2	Dry whole cow's milk, rice milled
Pakistan	23.0	Oil palm
Palestine	–	–
Qatar	–	–
Saudi Arabia	5.1	Barley, sheep, rice
Somalia	–	–
Sudan	39.2	Wheat
Syria	22.6	Sugar, maize
Tajikistan	24.3	Beef and veal, sugar, wheat
Tunisia	10.3	Wheat, maize, barley
Turkey	13.1	Cotton, skins, tobacco leaves
Turkmenistan	28.8	Sugar
United Arab Emirates	3.6	Rice milled
Yemen	15.2	Wheat, sugar, oil palm

Source: FAO, 2005.

GDP was about 17% in 2002. Syria achieved the highest rate of local saving, 30%, of all Arab countries in 2002 (AOAD, 2003). The percentage of saving of the GDP was the highest in Qatar (54.5%), Libya (46.1%), Emirates (38.4%) and Algeria (44.3%) in 2003. It was around 20% in Egypt, Jordan, Kuwait, Morocco and Tunisia and was less than 10% in Lebanon in 2003 (World Economic Forum, 2005).

Total Arab external investment is around US$1,400 billion. Almost half of this is Saudi Arabian investment. The other half comes from almost all the other OPEC Arab countries. This investment is either direct or in bonds and stocks. Foreign investment in Arab countries was about US$8,616 million in 2003. More than one-quarter of this foreign investment, 26.4%, is in Morocco and about 15.7% in Sudan. The remainder, 57.9%, is invested in the remaining Arab countries. Since Syria has achieved a high saving rate, it will be able to invest without borrowing or external investment, which will lead to a high economic growth rate.

1.1.5 Geopolitics

CWANA is subject to several geopolitical disputes. A key example is a shared sea among some Central Asian countries, which is becoming diminished (Box 1-2).

1.2 Well-being

1.2.1 Demography

The management strategy of a government toward its environment and to economic planning depends on its percep-

Box 1-2. Disappearing Aral Sea waters

The Aral Sea in Central Asia has been deprived of water sufficient to maintain its water levels since the 1960s. The fresh water that used to sustain the sea has been used by neighboring countries to produce export crops. Large amounts of water from the two main rivers feeding the Aral Sea were diverted into the desert to irrigate about 2.5 million hectares. It used to receive about 50 km^3 of fresh water per year in the 1960s; by the early 1980s it received none. By the 1990s, the surface area of the Aral Sea had shrunk by half and its volume had gone down by 75 percent. Its salinity had increased fourfold, preventing the survival of most of the sea's fish and wildlife. The negative environmental results include fisheries loss, water and soil contamination, and dangerous levels of polluted airborne sediments. Commercially useful fish catches of about 40,000 tonnes annually have ceased. Soil salinity has affected about 40% of the irrigated land. The regional water table has fallen. Many oases near the shore have been destroyed. Winds have picked up and scattered salt- and pesticide-laced particles, devastating surrounding regions. By 1990, more than 95% of the marshes and wetlands had given way to sand deserts. Communities face severe health problems. Drinking water is polluted, chronic bronchitis and kidney and liver diseases have increased by 3000%. The infant mortality rate is one of the world's highest.

Sources: UNEP, 1992, 1997, 2002b.

tion of the resources it has available and the likely future changes. As resource use links people, an understanding of population change is crucial to planning. The spread of health care and simple hygiene followed the end of World War II. Infant mortality went down and life expectancy went up. The effect was rapid population growth. By 1900, the population of CWANA was estimated at 710 million, about 12% of the world population.

Population distribution and population growth rates

The total population of CWANA as of 2005 was about 724 million (Table 1-2). Rural population, which is increasing, is about half of the total (World Bank, 2006).

Population growth rates vary among CWANA subregions and among countries. During the twentieth century, the populations of Egypt, Iran and Turkey more than quadrupled. Discrepancies in population growth rate among countries and subregions (UNDP et al., 2003) range from less than 2% in Algeria, Azerbaidjan, Caucasus, Iran, Kuwait, Kyrgyzstan, Morocco, Tajikistan, Tunisia, Turkey, Turkmenistan and United Arab Emirates; to 2.9% in the Nile Valley; 3.3% in Jordan, Libya and Saudi Arabia; 3.7% in Yemen and 4.2% in Oman.

Life expectancy

The highest life expectancy at birth is in the Arabian Peninsula, 73.3 years; West Asia, 67.7 years; and the Caucasus, 67.5 years. Afghans have the lowest life expectancy in CWANA—less than 50 years. Total fertility rates in CWANA vary from as low as 2.3 to 2.5 in Azerbaijan and Turkey to as high as 6.9 in Afghanistan, 7.2 in Oman and 7.6 in Yemen. The percentage of the population in the 15-65-year age group varies among member countries. In Ethiopia and Eritrea it is around 50%, in Turkey 66% and in the United Arab Emirates around 69.4% (UNDP et al., 2003).

Literacy

Central Asia and the Caucasus have the highest literacy rates in CWANA. Surprisingly, more than 98% of the population was literate in all countries in this subregion. Pakistan has the lowest literacy rate, 49%, in the subregion of Southwest Asia, with a significant difference between males, 61%, and females, 35%. Low literacy rates were observed for Yemen, 49%; Egypt, 55%; and Sudan, 59%, in Nile Valley and the Red Sea subregions. In North Africa, Morocco and Mauritania have the lowest literacy rates, about 50%. Comparatively better literacy rates were observed in the Arabian Peninsula—the lowest was Oman 74% and the highest, Qatar, 89%.

Access to potable water and sanitation

The percentage of the population with access to potable water varies from as low as 12% in Afghanistan to a high of 98% in Tunisia, with 94% in Lebanon and 90% in Iran. Variations among urban and rural areas in access to safe drinking water are great in many countries. In Morocco, 94% of the urban population has access to safe drinking water, while only 18% of the rural population enjoys it. In Tunisia the situation is much better. Safe drinking water is provided to all of the urban population and 95% of the rural population. Fairly high percentages of both urban and rural populations in Algeria, Iran, Lebanon, Pakistan, Saudi Arabia and Syria have access to safe drinking water.

CWANA countries have discrepancies in the percentage of populations with adequate sanitation. It is rated high for some countries: Algeria 91%, Iran 81%, Oman 78%, Syria 83%, Tunisia 80%; low for some: Egypt 32%, Morocco 41%, Pakistan 47%; and very low for others: Somalia 12%, Sudan 22%.

Employment and unemployment

Unemployment rates in 1999-2001 reached about 29.8% in Algeria, 14.9% in Tunisia, 15.5% in Jordan, 11.6% in Morocco, Syria and Yemen, and 9.2% in Egypt. An increase in unemployment in households will increase the dependency ratio. It may encourage the use of child labor, increase poverty and increase inequality of income distribution. In Arab countries, children under the age of 15 constitute about 40% or more of the population, which increases the dependency ratio and might speed the rate of population growth unless it is controlled. Landless and nearly landless people are not necessarily poor; off-farm income can compensate for daily needs. Nonfarm income, such as salaries and wages, may offset decreasing sources of income from the land. Internal and external migrations of unskilled laborers seeking work fluctuate according to oil prices. The lessening demand for unskilled labor in Arab countries particularly affects poor households. Also crises affect poor households, such as the tourism crisis in Egypt in 1997 (IMF, 2005).

In 1997, unemployment in Tajikistan was 2.4% for men and 2.9% for women. Unemployment was defined as persons aged 15 and older without paid jobs. In 2005, Afghanistan had unemployment rates of 7.6% for men and 9.5% for women; Azerbaijan 7.6% for men and 9.5% for women; Iran 10% for men and 17% for women; Pakistan 6.6% for men and 12.8% for women; and Turkey 10% for both men and women. No data were shown for Turkmenistan and Uzbekistan (Harvard College Library, 2007).

The lowest unemployment rates by far were in Tajikistan. Afghanistan and Azerbaijan had similar unemployment rates. In Iran and Pakistan, the difference in unemployment rates between men and women was wide; in Turkey, they were equal.

The underground economy as a component of informal work is big, but it is not counted in GDP. Also food and income aid from rich to needy persons through charity is large in the Arab world, but it is not counted either.

1.2.2 Poverty

Per capita income

Per capita income is an important economic indicator. Per capita income of the world was estimated at US$5,516, with an annual growth rate of 2.90%. If low per capita income is less than US$1500, medium between US$1,500 and US$6,000 and high income greater than US$6,000, CWANA countries can be classified accordingly. All eight Central Asian and Caucasus countries and all six Nile Valley and the Red Sea countries are low income. In North Africa four are low income and one, Libya, is high income. In West Asia, three countries are low income and three medium income. All six Arabian Peninsula countries are high income. Most high-income countries, 88%, produce and export oil; they represent about 4% of CWANA's population. Over half, 63%, of CWANA countries are low income, with 85% of the region's population.

Highest per capita income was recorded for the Arabian Peninsula countries. Much higher than the world average, per capita income of the United Arab Emirates is US$23,770. It also has one of the lowest per capita annual growth rates, 1.60%. Kuwait has the second largest per capita income, US$19,506. Bahrain follows with US$12,473; Saudi Arabia, US$9,608; and Oman, US$8,423. While these oil-rich Arab countries are classified high income according to the World Bank classification, the great majority of CWANA countries are considered lower middle income.

Countries belonging to Central Asia and the Caucasus have low incomes. The lowest income recorded was for Tajikistan, only US$118, with an annual growth rate of 9.40%. Kyrgyzstan also had a small per capita income, US$350. Kazakhstan had the largest in the subregion, US$1,966.

Pakistan had the lowest per capita income in Southwest Asia, US$556, with an annual growth rate of 4.2%. The second lowest, about double that of Pakistan, was for Syria, US$1,180. The largest per capita incomes in the subregion were for Iran, US$2,104, and Jordan, US$1,888.

In North Africa, Libya had the highest per capita income, US$4,965; Tunisia had US$2,366, followed by Algeria, US$2,062, and Morocco, US$1,394. The lowest was Mauritania, US$450 (World Bank, 2006).

Income distribution and poverty

Although average economic growth rates have been favorable, the high population growth rate of these last decades, 2.3%, has resulted in only a small net improvement. Moreover, there is now evidence of persistent income and nonincome inequality, which seriously limits prospects for pro-poor growth. The lack of data on income distribution over time limits considerably any research on the evolution of income distribution in this part of the world. Only a snapshot for the last decade can be given. The gap between the 10% of the population with the highest income and the 20% with the lowest income is tremendous. The greatest inequality is in Iran, where the highest 10% has 33.7% of the total national income and the lowest 20% has only 5.1%. Tunisia, Turkey and Turkmenistan have similar income distribution. In Algeria, Kazakhstan, Tajikistan and Uzbekistan this gap is less profound. According to the same data, Finland, Japan and Norway had smaller gaps for the same period, 22 to 23% for the highest 10% and 10 to 11% for the lowest 20 percent.

Poverty is a big problem in most CWANA countries. Linked to the incapacity of the urban economy to offer salaried work to most of the active population, 15-65 years of age, these countries face severe poverty problems. However, poverty statistics are not available for all CWANA countries and headcount indices have been estimated for only a few countries. The poverty line and survey year vary from country to country, so comparison among the countries should be treated with caution.

During the 1990s, half the populations in Mauritania (North Africa), Armenia, Azerbaijan and Kyrgyzstan (Central Asia and Caucasus) were under the national poverty line. In Pakistan, Turkey and Uzbekistan the share of the population under the poverty line was a little less than 30%, while in Algeria, Morocco, Jordan and Egypt, it was between 12 and 19 percent. Tunisia stands out as the country with the fewest poor; its national population under the poverty line was only 8 percent. In all the countries that delivered a poverty headcount ratio, except Armenia and Azerbaijan, poverty is more widespread in rural than in urban areas. The tertiary sector together with informal activities offers more opportunity for urban populations to be better off than rural populations. Most rural areas are becoming more isolated from cash-generating activities with a consequent negative impact on rural poverty. In the countries where undernourishment is high (more than 35%) the percentage of poor people is high and the GDP per capita very low.

Reduced demand for labor in the Gulf states has exacerbated the rising unemployment in the region. This has had a dramatic impact on remittances from migrant workers, especially in Egypt and Yemen, who provided much of the Arab casual labor. Unfortunately, data are not available for all CWANA countries, to assess the real gravity of the current situation.

In 2004, most inflation rates in Arab countries were moderate or low, except in Egypt (8.1%), and Yemen (12.5%). Inflation has a negative effect on stable incomes, like salaries, wages and pensions. Consequently, inflation will increase poverty and inequality in income in Arab countries. Inflation can be either cost-pull or demand-pull, but poor and lower middle class incomes will be more affected. Also, economic reforms in the region usually increase inflation, especially at the beginning. Data from the International Monetary Fund (IMF) and collected from Arab governments showed the inflation rate in Algeria was about 38% in 2002, about 19.5% in Morocco in 2001 and at least 10.4% in Egypt in 2003 (IMF, 2005).

Human poverty index and human development index

The data for 2002 indicate that eight countries ranged between 30 and 48 on the human poverty index (Table 1-2), on a scale of 1-100, where the higher the score the more severe the poverty. Data calculated for the human development index for 2004 ranged from a low of 0.492 for Yemen to a high of 0.871 for Kuwait, on a scale of 0 to 1. Two-thirds of CWANA countries have a human development index higher than 0.7. Oil-producing countries have on average a human development index of 0.82, considered high in CWANA (World Bank, 2006).

1.2.3 Food security

The definition of *food security*, according to the Food and Agriculture Organization of the United Nations (FAO), is the availability, accessibility, safety and sustainability of food. The self-sufficiency ratio is not a measure for food security if the country is capable of importing food. Most important is the availability of foreign currency to import food, in addition to foreign food aid. Grain storage capacity is important. Besides the purchasing power of per capita income—the real per capita income—farmers rely on the food they produce and factor in what it costs to produce it.

National food security
Agricultural imports constitute about one-fourth of CWANA's total merchandise imports, reflecting the region's vulnerability to food insecurity. Between 1991 and 1996, 15 countries had negative per capita food production growth rates (FAO, 2001). Oil-producing countries in CWANA are fast becoming the world's largest importers of food. In their quest to meet their priorities and obligations, people are driven to look at what is available today and disregard posterity. Extending cultivation to marginal and submarginal lands, overgrazing village common property and indiscriminately cutting vegetation for fodder and fuel are the easiest options. Farmers depend on subsistence farming to eke livelihoods on a fragile resource base, accelerating desertification in many countries. If the degradation continues, it may be irreversible.

Agricultural imports in CWANA countries reached US$41.8 billion, while agricultural exports did not exceed US$17 billion in 2004 (Figure 1-2) (FAO, 2006c). The main commodities exported were fruits and vegetables, dates and olive oil, while the main imports were grains. Considering the water scarcity in the region, this indicates unwise water-resource management. However, profitable grain production depends on large-scale land ownership and mechanization, while fruits and vegetables depend less on land and are labor intensive. Arab countries are net importers of food. In 2003 about 68.1% of imported food was grain, while other food groups were much less: milk and dairy products (15.0%), sugar (7.7%), and oil and fats (4.1%).

Because of the increase in population and other variables, the food gap between consumption and production increased and reached 40% in 2003. The Arab countries are relying more on other countries to feed their growing population.

In 2003, in the seven countries with high and high-moderate per capita GDP, food production was low or zero, especially in grains, legumes, sugar, oil and fats, and milk and dairy products in 2003. Other countries with low-moderate and low per capita GDP have better self-sufficiency in most foods, except Djibouti, which produces only some meat and fish and foods in no other groups (AOAD, 2003).

Sudan is almost self-sufficient in food. Sudan could be the food basket for all Arab countries, but there is a link between abundance of natural resources and poor growth. Also, it lacks institutions for marketing and financing, and it lacks infrastructure. These lead to low endowment and low performance. This situation also prevails in some other Arab countries. Sudan is reviewing laws of investment to encourage foreign direct investment.

Migration to Arab oil countries decreased poverty in Sudan. More than half the workers in rural areas migrated, so agricultural production declined. Also, government expenditure was mainly on defense, security and loan repayment. Government expenditure on health, education and other social services was limited.

Although it has natural feed sources, Libya has a protein gap. Because of drought and overgrazing, the government built 11 concentrated feed factories. Their capacity was about 963,600 tonnes per year. Actual production reached only about 367,000 tonnes in 2003. About 92% of the ingredients were imported. Since the factories were not working at full capacity, prices of local concentrated feed were higher and poorer quality than imported feed.

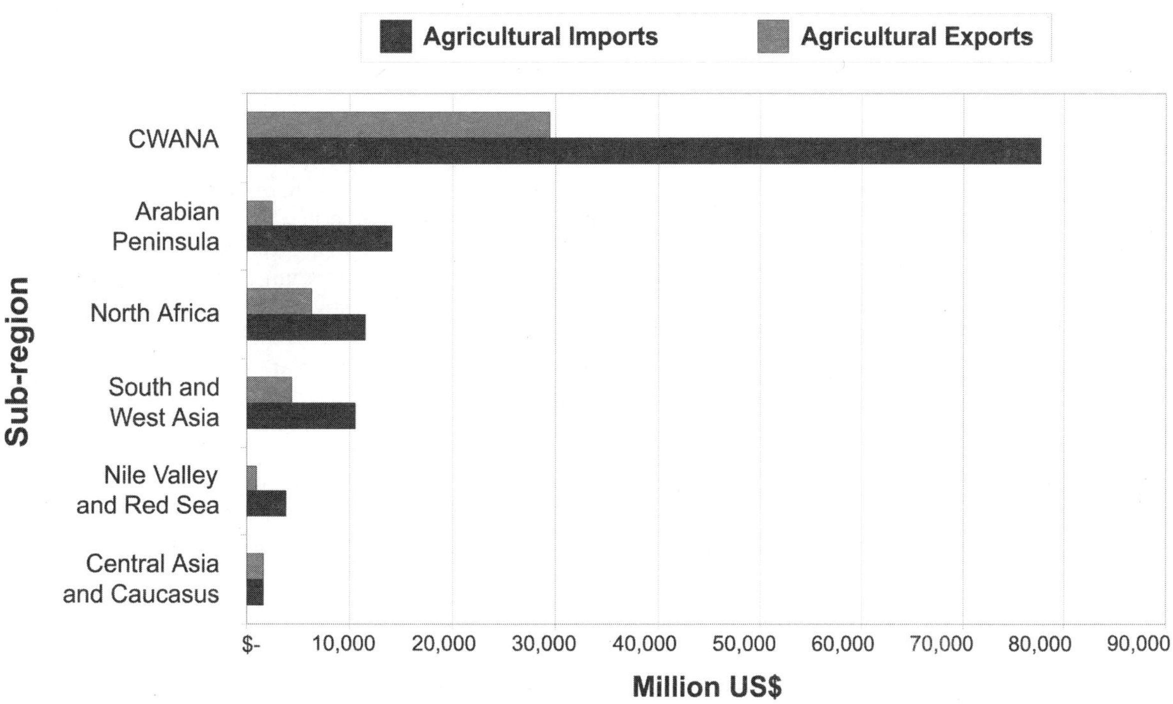

Figure 1-2. *Agricultural exports and imports in the CWANA region (2002–2004).* Source: FAO, 2006c

Food security in relation to health status

In 2003, 3 to 5% of the population was undernourished in nine CWANA countries, 6 to 10% in six, 23 to 37% in five; in Tajikistan, 61% of the population was undernourished (FAO, 2006d).

Infant birth weight provides an indicator of nutritional status. The percentage of low birth weight infants varies in CWANA. It ranges from less than 10% in Algeria, Iran, Jordan, Kuwait, Morocco, Oman, Saudi Arabia, Tunisia, Turkey, Turkmenistan and United Arab Emirates, 15 to 20% in Sudan and Afghanistan, to 25% in Pakistan. High infant mortality rates of 92 to 154 per 1000 live births are recorded in Mauritania and Afghanistan. These rates are almost twice the average world rate, 57 per 1000 live births. Most of the remaining CWANA countries had values lower than the world average.

Among countries of the region and excluding Afghanistan and Iraq, for which data are tenuous, only Tajikistan and Yemen recorded very high food insecurity in 2001-2003. About 61% of the population of Tajikistan suffered undernourishment and more than 33% of the Yemeni population was chronically undernourished. Pakistan, Sudan and Uzbekistan face serious food insecurity; the prevalence of undernourishment is about 25% in each of the three countries. Ten CWANA countries, such as Jordan and Morocco, show an increase in both prevalence and in absolute number of undernourished people between the baseline period and 2001-2003. The most alarming figures are in Kazakhstan and Uzbekistan, where the proportion of the undernourished has increased sevenfold and the number increased more than fourfold.

On the other hand, 12 other CWANA countries, such as Egypt and Syria, succeeded in reducing the prevalence of the undernourished in 2001-2003 compared with the baseline period. However, only four countries are on track toward achieving the Millennium Development Goal target. The most successful were Kyrgyzstan and Kuwait. Both started from a relatively high prevalence of hunger, but have cut the number of undernourished by at least two-thirds. In 1993-1995, Armenia had the highest prevalence of undernourishment (52%) in the region. It halved its number of hungry people, but at 29% the prevalence remains disturbingly high.

The lowest dietary energy supplies occur in countries with the lowest GDP per capita, such as Armenia, Pakistan, Sudan, Tajikistan and Yemen (FAO, 2006c). Capital stock in agriculture per worker is highly correlated to the agricultural value added per worker according to nourishment (Figure 1-3). It means that the value added per worker increases as the capital stock per worker increases. Countries where the percentage of undernourishment is more than 35% have the lowest value added per agricultural worker and the lowest capital stock per worker. Capital stock per worker and value added per worker are appropriate for assessing undernourishment.

1.3 Status of Agriculture

The proportion of arable land in CWANA is less than the world average and varies considerably among countries. Most, 62%, of the arable lands in CWANA are in Iran, Kazakhstan, Morocco, Pakistan, Sudan and Turkey. With Afghanistan, Algeria, Egypt, Iraq, Saudi Arabia, Syria, Tunisia and Uzbekistan, these countries have over 90% of the arable CWANA land. The remaining 18 countries have 10% of the arable land of the whole region.

1.3.1 Agricultural livelihood strategies

1.3.1.1 Production systems

A major characteristic of agriculture in CWANA is that it combines traditional subsistence farming with large-scale agrobusinesses. Traditional farming systems rely on labor, use few chemicals and pesticides, and use local landraces. Most traditional farming systems can be considered organic farming. However, production from of traditional systems is low compared with large-scale farming. Some small initiatives in Lebanon, Palestine and Tunisia focus on promoting organic farming, but labeling, certification and marketing need further development. In the future, demand for organic farm products is expected to increase. This might open a new window of opportunity for such products if they have proper labeling, certification and marketing. Indigenous farming relies on mixed systems at a small scale. The second farming system relies on big investment and monocropping. Corporate business farming is expanding at the expense of the small-scale family farm system. It is accompanied by a shift from traditional farming into business farming and uses intensive cropping and agrochemicals. This shift results in a loss of indigenous knowledge and biodiversity and increases water pollution, land degradation and loss of livelihood.

Most CWANA people live in hyperarid to semiarid zones. Population densities are generally less than 1 per km^2 in the hyperarid zone, except in Egypt, less than 5 per km^2 in the arid zone and about 10 per km^2 in the semiarid zone. About 72% of the population depend on agriculture, 7% on livestock and 21% on the urban areas for their livelihood. Rural people living in CWANA can be roughly classified into nomadic, seminomadic, transhumant and sedentary populations. Nomadic people are found in pastoral groups, which depend on livestock for subsistence and, whenever possible, farming as a supplement. Following the irregular rainfall, they migrate in search of pasture and water for their animals. Seminomadic people are also found in pastoral groups, which depend largely on livestock and agricultural cultivation at a base camp, where they return for varying periods. Transhumant people combine farming and livestock production during favorable seasons, but might migrate seasonally along regular routes when forage for grazing diminishes. Sedentary farmers practice rainfed or irrigated agriculture.

Land use is often a form of agroforestry or agrosilvopastoralism, as in western Sudan where the gum arabic tree, *Acacia senegal*, is cultivated along with sorghum or millet and raising small ruminants. Often, there is little distinction between a farmer and a pastoralist. People living in drylands take into account the local limitations and adapt to suitable and feasible land use in seeking a livelihood. However, the delicate balance achieved through traditional farming and livestock production is easily upset; this is shown by a general deterioration of grazing lands in drylands. A main cause of this deterioration—often referred to as desertification in its most severe form—is the overpopulation of

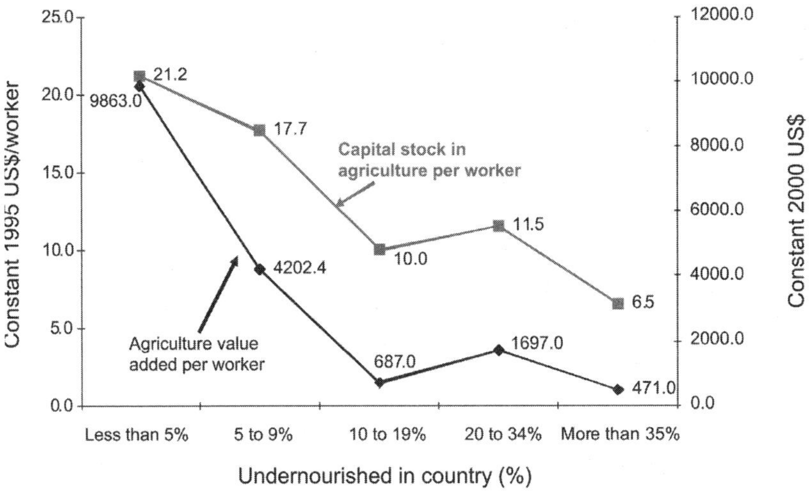

Figure 1-3. *Nutritional status in 33 CWANA countries per capital stock in agriculture and value added per worker.* Source: FAO, 2006c.

people and their livestock, coupled with the deregulation in access and use of rangelands, forests and land. Agricultural cropping and pastoralism become competitive, rather than complementary, forms of land use.

Major production systems in North Africa and the Middle East. The production systems in North Africa and the Middle East are diverse. The most productive systems are able to nourish or provide livelihoods for many people. The biggest portion (85%) of the area is covered by less productive systems: pastoral farming (3 people km^{-2}) and sparse farming (6 people km^{-2}). The most productive agriculture is concentrated in the remaining 15% of the land area because of greater land and water resources.

The population density in large-scale irrigated agriculture is the highest. The total cultivated area is irrigated. In the Middle East and North Africa, irrigation is a way to intensify agriculture. Highland mixed farming and rainfed mixed farming also have high population densities. These systems have family farms and integration between agriculture and livestock, essential for income diversification and managing soil fertility. They are diversified because they integrate cereal and legumes, fruit trees, fodder and livestock. These systems are productive because of the availability of good soils and rainfall.

Obviously, resource scarcity determines the productivity of each system and its natural resource management. In the arid zone reliable agriculture is possible through irrigation and nomadism. Rainfed agriculture is widespread in the semiarid zone. Two major elements should be looked at to intensify agriculture in this region:

- Availability of natural resources
- Access to these natural resources and control of them

The percentage of irrigated area, cultivated area and type of land and water management varies with production system (Table 1-4). As water resources are scarce, irrigation efficiency is important, but it is low and rarely exceeds 50 percent. Production might be intensified by improving irrigation efficiency through better resource management and introduction of technical innovations, especially irrigation water-saving techniques.

Major production systems in Central Asia. In Central Asia, many countries from the former Soviet Union have a lot of land issues. Land allocation, land reform and restructuring and the transition to open markets have not yet been achieved. Land issues have negative effects on production. The most productive in central Asia are the highland mixed rice and wheat farming systems. Those systems, set in Pakistan and Afghanistan, have benefited from the Green Revolution, especially the rice or wheat farming system. High-yielding varieties, mechanization and agrochemicals boosted agricultural production. These two systems are highly intensified and the ratio of irrigated area to the total cultivated area is large: 86% of the cultivated area is irrigated, with a population density of about 40 people per km^2.

Similar to the Middle East and North Africa, two major elements determine agricultural production and food security: land and water. There is scope for extending the irrigated area, and therefore agricultural production, especially in Turkey. All conditions are set, socially and economically, to boost agricultural production in Turkey. It has the political will, water resources, water-management policies and irrigation projects, such as the South Eastern Anatolian Project (GAP). Pakistan and Afghanistan have only limited potential to extend irrigated area because water resources are decreasing, especially groundwater. Intensification of agricultural production may occur by introducing new water-storage and water-saving practices. Although, the share of irrigated areas in cultivated areas is very low (10%) in the countries of the former Soviet Union, there is no scope for implementing large-scale irrigation schemes as

Table 1-4. **Percentage of irrigated, cultivated area and type of management for major production systems in North Africa and the Middle East.**

Production system	Cultivated area irrigated (%)	Type of land and water management
Irrigated farming system	–	–
Large-scale irrigated subsystem	100.0	Large-scale irrigation Intensive year-round cropping: cropping intensity 120-160% Large-scale centralized management of water access and distribution Water access and distribution managed centrally but land attributed to many tenants (0.5-5 ha) organized in water-user associations Large-scale fully irrigated individual schemes
Small-scale irrigated subsystem	Low	Traditional irrigation practices Small units (0.02-1 ha)
Highland mixed farming system	23.0	Supplementary irrigation in summer for vegetables or high-value fruits (source of water)
Rainfed mixed farming system	4.3	Supplementary irrigation in summer for vegetables or flowers
Dryland mixed farming systems	18.0	Small irrigated areas grown in vegetables
Pastoral farming system	1.0	Small-scale irrigation (1-2 ha)
Sparse (arid) farming system	0.1	Irrigation schemes set up in oases
Coastal artisanal fishing system	–	–
Urban-based farming systems	High	Family gardens

Source: FAO/World Bank, 2001.

in the past. Nevertheless, cultivated areas can be extended mostly through rainfed and small-scale irrigation. To have sustainable agricultural production, land tenancy should be secured; farmers should access and control their land.

1.3.1.2 Role of women in agriculture

In CWANA, agricultural work is mostly performed in small-scale households and often involves all family members. The work of women, however, is little reported and that of children is even more rarely noted. Analysis of the statistics on agriculture in CWANA suffers from the great diversity of situations that characterize male and female farm work. Among the factors that contribute to agriculture are production systems, some labor intensive; social and marital status, age, household composition and economic status, availability of male or female labor; mechanization of the work and farm size; and ethnic, religious, cultural and social norms. Despite these variations some common trends in CWANA can be detected. Involvement of a woman in agriculture may put strains on her domestic duties, including child rearing, housekeeping, cleaning, cooking, and fetching water and fuelwood. Women contribute 28 to 70% of agricultural labor. This can be from, among other things, the growing number of female-headed households because of male migration or war. Together with performing domestic and agricultural work of the household, in some countries these female farmers have started looking for off-farm work. Revenue from migrated relatives is often not sufficient for survival. In Syria, women, particularly from low-income households, constitute a large share of seasonal agricultural workers. The highest rates of participation are in the 15-24-year age bracket (Ramsis Farah, 1999). In other CWANA countries, women over 40 are often more involved in agriculture than younger ones. In Syria, 44% of women work for a wage, while 56% work as unpaid farm labor. If paid, women usually receive lower salaries than men. In Egypt and Yemen, women earn roughly two-thirds of men's wages. In Iran, they earn 46% of male salaries, while in Lebanon they earn 50% (FAO, 1995). Women from poorer and smaller households are usually more involved in agriculture and are more likely to work off-farm for daily wages than those from richer households.

Women mainly perform manual, time-consuming and labor-intensive work on the farm; mechanized work is generally a male task. Women are usually responsible for horticultural crops and agroprocessing. They are involved mainly in planting seeds or transplanting seedlings, harvesting, picking fruit and vegetables, and postharvest threshing, selecting and storing. Men mainly prepare the land, irrigate, spray, mechanically harvest and market the produce. Duties connected to livestock rearing differ by animal. Mainly, women take care of small livestock. Herding and marketing are generally male duties. Fishery and agroforestry are also mainly male tasks. In Egypt fisheries and fish marketing are primarily men's tasks, while more than half of the labor in fish processing is performed by women, who also contribute in making, maintaining and repairing nets (FAO, 1996). Despite their substantial agricultural work, women have limited control or ownership of resources and revenues. Percentage of land area owned by women ranges from none in Oman, about 5% in Syria and Lebanon, 11% in Jordan and 24% in Egypt. Statistics about asset entitlement and access are scanty and rarely separated by gender. Generally, women own plots smaller than men's. In Syria, 7% of women own animals and 1% own agricultural machinery. According to the report of the Convention on the Elimination of All Forms of Discrimination Against Women in Yemen, female farmers do not control land, water, agricultural equipment, credit or capital (CEDAW, 2001).

Some critics have highlighted the reasons for the underrepresentation of women's involvement in agricultural work, particularly in Muslim countries (El-Fattal, 1996). These are the association of women with domestic spaces to the exclusion of outdoor activities, such as work in fields; association of agricultural work with wage labor, while women are mainly unpaid; and association of farmers with plot holders—women working in the fields are rarely landowners. Many land and agrarian reforms have increased concentration of property entitlements, access and control in the hands of the male heads of households and assigned access to men for basic agricultural resources, such as water, seed and fertilizer, distributed by government agencies. Lacking control over, and entitlement to, production often implies restricted access to loans and social security, limited autonomy and decision-making power, and, eventually, curtailed ability to achieve food security. Women's limited access to markets also affects their control of revenue and decisions.

The increasing number of female-headed households, visible in many countries of CWANA, corresponds to an increase in women's workload and a decrease in their independence. In Pakistan and Sudan, 25% of households are headed by women, 16% in Egypt and Morocco, 13.6% in Yemen, 11% in Lebanon, Oman and Tunisia and 6% in Iran, Jordan, Syria and Turkey (FAO, 1995; Hartl, 2003).

In some cases, women become empowered because men are absent. They participate in decision making by managing small budgets and their mobility is increased because they sometimes go to the market to sell their produce, even if they still must let male relatives make major decisions, such as selling a cow (CNEA, 1996). The feminization of agricultural labor increased the rates of women's work in the unpaid and informal systems. Their employment in wage labor is still characterized by gender wage differentials, precariousness, lack of social services and vulnerability. The increase in household work involves children, who contribute their share of work, to the detriment of their school attendance, free time, health and other children's rights.

Finally, despite women's increasing involvement in the fields, agricultural machinery is still usually designed for men, limiting, together with social biases, women's and children's access to technological improvements. In Syria, the introduction of agricultural machinery from the 1960s often increased women's and children's drudgery by strengthening the gender division of labor. Men were assigned mechanized work, leaving manual work to women and children.

1.3.2 Agricultural land use

A detailed examination of the aridity zones shows that over 4 million km^2 of land in CWANA is available for good cropping and animal husbandry (Table 1-5). The greatest land use is permanent pasture, 550 million ha. Cropland and forests are 141 million ha and woodlands 124 million ha. Most of the permanent pastures and forests and woodlands are in the Nile Valley, Sudan and Somalia. Central Asia and North Africa are also rich in croplands and permanent pastures. Among the low-income countries, per capita accessibility to arable land is about 1 ha, of which 40% is irrigated and 60% rain-fed. The estimated daily income of an average farming family is about US$2.82 (Rodriguez and Thomas, 1998). By 2025, the lowest per capita cropland—less than one-tenth of a hectare—will be in the Arabian Peninsula and the Nile Valley. There is great disparity in cropland per capita among countries within a subregion. For example, cropland per capita in Sudan, 0.448 ha, is eight times higher than in Egypt, 0.053 ha. These are expected to be reduced to 0.0313 ha for Egypt and to 0.277 ha for Sudan by 2025, if current trends of population growth and land degradation continue.

There has been no appreciable overall increase in cropland in the last three decades. In 2005, the area of arable and permanent crops is 5.6% in North Africa, 10% in Central Asia, 1.9% in the Arabian Peninsula and 16.3% in the Middle East (FAO, 2006b). Cereal production increased slowly in North Africa, about 50% between 1975, 18.5 million tonnes, and 2005, 28 million tonnes. Central Asia had an

Table 1-5. **Croplands per capita and CWANA land use, 1994.**

Subregion	Croplands per capita (ha)	Cropland (million ha)	Permanent pasture (million ha)	Forests and woodlands (million ha)
North Africa	0.330	25.0	108.0	18.0
Caucasus	0.264	6.0	48.0	6.0
West Asia	0.260	12.0	13.0	0.8
Central Asia	0.244	75.0	91.0	37.0
Nile Valley	0.156	19.0	169.0	60.0

Source: WRI et al., 1998.

appreciable increase in cereal production up to 1990, 31.5 million tonnes, but since then production has fallen sharply, dropping back by 2005 to the 1975 figure of 19 million. In the Arabian Peninsula cereal production increased steadily and peaked in 1990 at 4.9 million tonnes, three times that of 1975, 1.5 million tonnes; it has now dropped to twice the figure of 1975. Cereal production in Mashreq (Iraq, Jordan, Lebanon and Syria) performed well between 1980 and 1995: 9 million tonnes, but dropped to about 130% of the 1975 figure, 5 million tonnes (FAO, 2006a).

Sustainable land use is a challenge to all people living in drylands. Problems in CWANA include desertification, inadequate knowledge of more productive land-use practices, political marginalization and low investment. People also confront major problems in attaining sustainable land use because of inadequate knowledge of alternative land-use practices. Many people observe a "tradition" in agriculture not always matched by similar, traditional, approaches toward forestry, wildlife ranching and ecotourism, all of which have become profitable enterprises in many dryland regions of the world. This lack of appreciation can be a barrier to innovation in land use, especially on marginal agricultural land. Such barriers are often overcome through farmer education, extension services, and, most of all, through demonstrating the benefits of more diversified land use.

The drylands of the region suffer from the vicious cycle of low productivity, low investment and, as a result, poverty. Investments, apart from those made for irrigated agriculture, are relatively small. Low productivity, low investment and land degradation often lead to desertification and are responsible for regional poverty and income disparity. The poverty and hunger prevalent in some CWANA countries, like Sudan, are poignant examples of this situation. Other critical problems include the inherent problems of water scarcity, tenure considerations and ineffective development policies. Improving this situation requires that a variety of technical and institutional problems be solved. An example would be increasing the investment in appropriate agriculture, alternative land-use practices and other appropriate, income-generating interventions. Other solutions include designing strategies for risk management and implementing programs for more equitable land distribution and income.

As noted, CWANA is climatically diverse, mainly with hyperarid, arid and semiarid zones. The region has highly variable and uncertain climates. Climatic variability and associated floods and droughts result in increased risks of crop failure and reduced food security. While precipitation in many CWANA countries averages 200-500 mm annually, suitable for some crops, the extreme fluctuation in precipitation from year to year make such averages irrelevant. We cannot depend upon average precipitation to plan agricultural and natural resource development or urban expansion. As a result of low and erratic precipitation in the drylands, ephemeral or intermittent streams are the norm. Flash floods from intense rainfall are highly variable and are common in many CWANA countries. Dry stream channels may become torrents within hours of convective storms that occurred several kilometers upstream.

Land use and water are inextricably linked but often are not managed in concert. Watershed management offers the framework for achieving integrated management to increase or sustain food and natural resource production while protecting the soil and water (Brooks et al., 1997). This approach recognizes that land use in uplands affects the flow and quality of water downstream. Water resource development in the form of reservoirs and canals can affect the type and intensity of land use throughout a watershed. Transforming this recognition into effective solutions, however, is currently hampered by inadequate policies and an absence of institutions and arrangements needed to integrate sectoral cooperation, upon which implementation depends (Kundzewicz, 1997). Understanding and coping with the links between land use and water is critical. Water should be viewed as the most valuable product of the land and the one resource upon which all other production depends. In discussing needs and opportunities for enhanced management of CWANA, we must recognize and be able to work with variable and often unpredictable precipitation and water yield.

Soil erosion reduces the productivity of upland watersheds. The sediment that reaches stream channels can adversely affect reservoirs, water systems and water quality in downstream riparian corridors. This is a typical environmental problem in downstream countries of the Nile Valley. However, soil erosion in the highlands could be interpreted as a "good thing," a source of nutrients for lowland agriculture. For example, the annual flooding of the Nile in Egypt was the basis for productive agriculture, now disrupted by the Aswan High dam.

In all drylands under intensive use, common issues and problems arise that need to be addressed through management. Foremost, drylands have commonly been viewed as wastelands, not worthy of economic concern or political attention. Global concerns about desertification over the past few decades have, however, focused more attention on dryland issues and the need for land-use reform. "Desertification" often describes areas that have become desert-like from human-induced degradation. To some extent, emerging programs to combat desertification have helped generate the political, economic and technical support needed to reverse land degradation. However, certain inherent characteristics of drylands place limits on the potential for agricultural, natural resource and urban development. Conditions also make watersheds vulnerable to degradation and curtail our ability to restore or rehabilitate the land.

1.3.3 Livestock and fisheries

For many livestock owners in CWANA, livestock is a sign of wealth and social prestige. Herd numbers count for more than herd quality. Husbandry is mainly on natural pastures and extensive nomadic grazing. Most of these pastures are poor from frequent droughts and overgrazing. The number of cattle in the region goes up to 124 million head, about 9% of the world's cattle. Sheep number about 320 million head, 30% of the world total; goats about 189 million (23%); and camels about 14 million head (68%). Cattle are most abundant in Sudan (38 million), Pakistan (24 million), Turkey (10 million) and Iran (9 million). Small ruminants abound in Sudan (90 million head), Pakistan (83 million) and Iran (81 million). Camels are mostly in Somalia (7 million head), Sudan (3 million) and Mauritania (1.4 million).

Buffalo total about 31 million head, 18% of the world total. Over 80% of CWANA's buffaloes are in Pakistan and 13% are in Egypt (FAO, 2006a).

Annual meat production from livestock in CWANA is estimated at 6.5 million tonnes, 55% from cattle and buffalo, and the rest from small ruminants. CWANA consumes about 24% of the world small ruminant meat. Additional 5.5 million tonnes of meat came from fish, poultry, and game, making total meat production 12 million tonnes in 2003, about 5% of world meat production (FAO, 2006a). The biggest producer of meat in the region in 2003 was Pakistan with 1.89 million tonnes; followed by Iran (1.6 million tonnes); Egypt (1.45 million tonnes); and Turkey (1.35 million tonnes).

Fish catch in CWANA was 1.3 million tonnes in 2002. Egypt caught half of that with Pakistan and Iran almost equally catching about 17% of the CWANA catch. Egypt and Pakistan got their fish mostly from fresh water; Iran got its fish equally from marine and fresh water (FAO, 2006a).

1.3.4 Policies and interventions in rangeland management

The primary concern for governments of the region is to develop policies to check overgrazing, a problem recognized by all. Developing water resources was thought to spread the burden of livestock over a wider area and reduce overgrazing. This could work only if livestock populations were stabilized. But, uncontrolled by governments, the livestock population is rising steadily in most CWANA countries. The inevitable result is more overgrazing. Land tenure was the next to be tried. Three land-tenure systems were progressively established in the region following independence from foreign rulers:

- government lands not subject to any public use
- government lands subject to use by a tribe or village or group of individuals
- private lands registered to individuals

Most CWANA countries established state ownership of rangelands during the twentieth century. The rangelands of Algeria, Iran, Jordan Sudan and Syria, to mention a few, were considered government property, with tribal rights to use these rangelands recognized. Tunisia and Morocco recognized and established collective properties of tribes to the land as early as 1918. But soon after its independence, Tunisia chose to promote privatization of common land. Some of these reforms were accompanied by measures to promote settlement of nomadic pastoralists, improve rangeland management by limiting stocking rate, establish reserves, ban cultivation and ban uprooting of shrubs in rangelands. As these measures led to a clash of interests, they could not be applied.

The attempt was to organize pastoralists to sustainably use common rangelands through state ownership and state cooperatives, herder cooperatives, community cooperatives and comanagement of community rangelands. Governments also built roads to facilitate moving herds and access to markets. Expanded road network and improved transportation subjected areas high in biodiversity and in good range condition to grazing pressure never before experienced; pastures near the most popular routes were overgrazed.

To help herders reduce drought losses, governments throughout CWANA introduced drought-management policies, such as feed subsidies and credit rescheduling. While these interventions succeeded in protecting incomes in drought years, they introduced a bias to keep livestock numbers high, which accelerated rangeland degradation. This protection undermined adjusting herd size to annual climatic variation and increased herd size. In parallel, policies such as subsidizing agricultural inputs, like fuel or tractors, were not restricted geographically, and they favored crop encroachment in pastoral areas.

After several decades of rangeland management through promoting rehabilitation measures, planting shrubs and cactus, and preventing grazing, most experts agree that rangelands are still degrading and solutions should rely on institutional change and tenure reform. Approaches promoting natural resource management in local communities or "co-management" of resources under the regime of common property rights are relatively new in the region. Implications of the initiatives have not been discussed extensively (Dutilly-Diane, 2006). In Sudan, it was recommended in the mid-1950s that rangelands be allocated and registered to tribal owners. This was considered crucial because until individuals or groups knew the benefits of new or improved ranges would be theirs, all efforts to develop rotational grazing would fail (Wallach, 1989).

1.4 Key Issues

Ecosystems

Some 85% of the CWANA land area is considered desert and dryland susceptible to desertification; 70% of the region's agricultural areas are arid or semiarid, and only 35% is cultivable. During the last 20 years many CWANA countries have suffered long-term droughts, with various degrees of severity: Afghanistan, Iraq, Iran, Jordan, Morocco, Oman, Pakistan, Sudan, Syria, Tajikistan, Tunisia and Turkmenistan (FAO, 2001). The successive droughts that hit CWANA countries have devastated plant, animal and human lives alike. Livestock herders suffered most as incomes fell sharply and vulnerability to food shortages increased dramatically.

Desertification continues to be the most significant environmental issue in most of CWANA. It has affected wide areas of rangeland. Soil erosion in excess of 20 tonnes ha^{-1} per year is common in many areas. There is a close correlation between drylands and the location of areas that are likely to be affected by desertification in the future. This correlation may be explained by the peculiar vulnerability of fragile dryland environments to wind and water erosion, soil salinization, and loss of vegetation by overgrazing by livestock, overcutting of fuelwood and trees, and other excessive uses of the land and natural resources, and also by the deregulation of natural resource management. The prevailing climate also exerts persistent stress on both soil and vegetative resources. Relatively little disturbance can cause instability and imbalance, leading to desertification. Drought, overgrazing, clearance of woody species and tillage are the principal causes of rangeland degradation. In North Africa up to 90% of the area is affected by desertification (UNEP, 1997).

Key issues pertaining to soils are degradation through water and wind erosion and through nutrient loss. Salinization, waterlogging and alkalization are major issues in irrigated areas, especially in large-scale irrigated schemes.

Population growth and demography

Population in CWANA will be about 1.2 billion by 2025; it is expected to reach 1.6 billion (17% of the world population) by 2050. The population is rapidly becoming urbanized in many countries of the region. By 2020 the percentage of the urban population to the total population will be 93% in Oman, 91% in Libya, 90% in Saudi Arabia and UAE, 85% in Turkey and 75% in Tunisia. In some countries of the region such as Afghanistan the urban population will remain under 30% (UNDP et al., 2003).

Water scarcity and its implication for agricultural production systems

Scarcity of water is the chief challenge to agricultural development in almost all CWANA countries. Freshwater scarcity threatens CWANA's ability to achieve food security, alleviate poverty and improve human health. Land scarcity compounds the problems of water scarcity, making people more vulnerable to the extremes of droughts and floods and leading to widespread exploitation of natural resources.

Access and use of agrobiodiversity

Due to the megadiversity of crop species in CWANA, major crops, wild relatives and landraces vary significantly in the region. They are important sources of useful genes for several characters such as adaptation to extreme conditions of abiotic stress, resistance to pests and diseases, high quality, and factors affecting productivity. They also play an important cultural and social role, and they secure evolutionary continuity.

The richness of agrobiodiversity and the existence of low-input farming systems in the region is key to food security and sustainable agricultural production in the region as well as outside. Many farmers in CWANA countries cannot afford expensive external inputs such as fertilizers, pesticides or seed of improved varieties adapted to the particular ecological and economic situation. Plant genetic diversity, at both intra- and interspecific levels, is crucially important in CWANA farming systems. So this existing diversity helps stabilize farming systems by maintaining the wide range of crop diversity.

Domestic animals contribute to food and agriculture in many ways, providing meat, milk and milk products, eggs, fiber, and fertilizer for crops, manure for fuel, and essential draft power. They are an extremely important economic resource, reducing risk, generating employment, and evening out seasonal farm labor demands. Some 4,000 breeds of cattle, horse, donkey, pig, sheep, buffalo, goat, chicken and duck have been developed and used worldwide. Animal genetic resources of breeds and strains and wild or semi-domesticated relatives are important, their diversity being insurance against future adverse conditions. Farmers and breeders have successfully selected animals for a variety of traits and production environments, using their genetic diversity extremely effectively.

But because of the Green Revolution that took place in Pakistan and the adoption of high-yield varieties, genetic resources in the CWANA region are being degraded. The shift from small-scale to large-scale farming systems using technical packages that feature mechanization and chemicals is contributing to biodiversity loss. Many governments are unaware that they should establish gene banks to conserve local varieties. The situation with livestock seems to be better. Foreign breeds, especially of cattle, have been imported in many countries but their adoption has been slow because they need large amounts of feed that farmers cannot afford. So farmers are still rearing local breeds of cattle, sheep and goats.

Climate change

According to the Intergovernmental Panel on Climate Change (IPCC, 2001), CWANA is among the most vulnerable regions in the world in predicted decreases in water and food security. Changes in rainfall and temperature patterns could also alter biodiversity, with many species not being able to adapt or migrate. However, the consequences of global climate change on agriculture and ecosystems are uncertain. The most likely impact on CWANA, based on various simulation models, is adverse consequences for its semiarid zone. These models show that doubling the CO_2 concentration in the atmosphere will induce the following:

- In West Asia and the Arabian Peninsula, temperatures are projected to increase by approximately 4°C. Rainfall and soil moisture will decrease.
- In Central Asia temperatures will increase substantially in winter. More rainfall and a slight increase in soil moisture levels will result. Summer changes include a 6°C temperature increase, greatly varied changes in precipitation and a general decrease in soil moisture (Williams and Balling, 1994).
- In North Africa, the Nile Valley and the Red Sea, grain yields are projected to decrease, further diminishing food security. Desertification will be exacerbated by reductions in average annual rainfall and increased evapotranspiration. Significant extinctions in plant and animal species are projected, importantly affecting rural livelihoods (UNEP, 2002b).

The type of climate change expected will have a major impact on dryland soils, where most of the salinization will occur. These soils are inherently vulnerable to degradation, since they have low biological activity, organic matter and aggregate stability. The resilience of the dryland ecosystems to deficits in moisture, temperature extremes and salinity is still inadequately known.

Land and water access, tenure and management

Land and water access is greatly differentiated in CWANA and generally unequal. Few countries have conducted real land reform, but Turkey did in 1945. Major measures were applied under the rule of Mustafa Kemal in 1926: the tithe to the landholder was abolished in 1925 and the former tenures located on state-owned land were given to farmers (Dufumier, 2004). Land reform was conducted in irrigated areas in the governorates of Deir Ezzor and Rakka in northeastern Syria. Although many problems arose after the government distributed land to farmers, access to it was more or less equal. It created the basis for agricultural production.

In Algeria and the former Soviet Union republics, the transition to a market economy has not yet been accomplished and land regime is still uncertain as the former state-owned farms have completely disappeared and conditions for gaining access to land are not clear.

Many countries such as Jordan, Morocco and Tunisia have adopted a capital-intensive model of agricultural development at the expense of small-scale farming systems. This model is capitalistic and export oriented and based on private property rights of water and land.

The Green Revolution increased agricultural production, as, for example, in Pakistan: "The Green Revolution generated tremendous increases in yields, particularly in large agricultural irrigated plains where cropping intensity was high because of efficient water management. But even in the regions where the Green Revolution occurred, small-scale farmers could not invest to develop their production systems and to progress. Although the Green Revolution can be extended in terms of yield and production to other areas where natural resources are available . . . it will not alleviate poverty neither provide food for hundreds of millions of small scale farmers" (Mazoyer, 2001).

Infrastructure and financing for agricultural development

The capitalistic model mentioned above has spread in many countries thanks to infrastructural development. There are discrepancies in infrastructure within CWANA. In countries where market-oriented agriculture has been adopted, infrastructure has been improved although there are still pockets of subsistence farming. In countries like Syria, where agricultural policies have led to self-sufficiency, roads and highways link production areas to major marketing centers.

Governance

Good governance is characterized by participation, rule of law, transparency, responsiveness, consensus orientation, equity and inclusiveness, and effectiveness, efficiency and accountability. Several indices are used worldwide to measure governance, and aid has been tied to good governance indices. Most CWANA countries rank low in all these indices. For most CWANA countries, the environmental governance corruption index is below 4, in a scale from 0 to 10, where 0 is most corrupt. Most CWANA countries rank below average in the environmental policy and freedom index (Kaufmann et al., 2003).

Local knowledge

Local knowledge has been generated for centuries; it is empirical, based on farmer experience. According to the ecosystems they live in, communities have developed knowledge that is quite diverse. It varies according to physical conditions such as climate, soil and vegetal cover but also social and economic conditions. Local knowledge encompasses agricultural practices and techniques concerning cropping patterns and animal husbandry, and also resource management systems like water-harvesting, water-management and rangeland-management systems. Small-scale farmers have also manufactured locally used tools such as plows. Managing biodiversity and conservation is also considered part of local knowledge.

Some research has been done on biodiversity management and conservation, water management systems (especially in arid areas) and rangeland management systems but very little on local agricultural techniques and practices such as cropping patterns. Local knowledge can easily be transferred from farmer to farmer as it has been generated at a small scale. Generally, farmers do not need major investments to adopt it.

As aridity is widespread in the region, local knowledge about water management and conservation is quite developed. Community-managed irrigation networks, water- and land-conservation systems such as *tabias* and *jsour* in Tunisia are mostly located in arid areas in southern and central Tunisia where rainfall does not exceed 200 mm. Water- and land-conservation systems are mostly small-scale catchments built manually to harvest rainwater.

Those systems are no longer maintained and are disintegrating as other job opportunities with higher opportunity costs are available in nearby regions. Community technologies no longer play their traditional role in managing resources because most present-day activities are large scale, like constructing dams or reclaiming land, and are carried out by government agencies. Village or community water-management systems have almost disappeared, but individual farmers still maintain their own small water works.

Highly sophisticated irrigation networks have been set up by communities in areas where the main constraint has been water scarcity. Irrigation systems were based on community organization; village dwellers contributed to their maintenance (*foggaras*) by cleaning up drainage and irrigation canals. Local grassroots organizations were in charge of water management and distribution. In the beginning of the 1970s, because of the evolution of technology, major water-harvesting works of dams and drilling were carried out, and new irrigation systems have been adopted.

In southern Morocco, water for irrigation came from the Atlas Mountains and downstream communities in the Draa Valley and Tafilaelt developed irrigation areas. They used their own techniques for capturing, conveying and managing water—techniques adapted to the local conditions of labor available for digging and maintaining the canals, water flows, and social organization. In the 1970s, the government built two dams upstream and created huge irrigation schemes downstream. Communities could no longer manage irrigation, the amount of water available per hectare decreased, and profitability was not as high as before. The combination of these factors plus Bayoudh disease among date palms caused a decrease in date fruit production (Ben Zid, 2002).

Moroccan date palm production has declined by 80% since the 1920s (Ben Zid, 2002). This decline is because the production system changed. The former system was viable because cheap labor was used to maintain the irrigation system. In 1920, laborers started migrating to France and northern Morocco, so less labor was available, affecting the whole system.

Because Bayoudh disease had decreased date production, farmers tried to keep date palm biodiversity by growing and multiplying indigenous cultivars that bore disease-resistant genes.

Farmers have their own way to distinguish date palm varieties, which is quite different from the researchers' method. They characterize leaves as well as fruits. Researchers use only one criterion: the fruit. A farmer in Algeria has written an index classifying various parts of the date palm tree (Bakkay and Tirichine, 2005). Written in Amazigh and Arabic and published by the High Secretariat of Amazigh in Algiers, it will be translated into French. Such an index will be helpful in identifying the cultivars most resistant to diseases and in combining farmers' and researchers' knowledge. It will be a tool for setting up future biodiversity management programs in the Maghreb.

All parts of the date palm tree are used: leaves are used for building materials and sometimes to make tools for fishing (IPGRI, 2005). Fruits are classified according to their ripeness and are processed accordingly: some are dried and stored and some are eaten fresh (IPGRI, 2005). Date fruits are processed into different highly nutritive products (ICRA, 2003).

In arid areas and rangelands, farmers and village dwellers know the biology of many range species as well as their location according to topography and soil features. They know therefore where to take their flock to pasture according to the season.

Because the vegetal cover is degraded from the heavy pressure on range resources combined with climate change and other external factors—development of agriculture and introduction of new techniques like drilling—some species are threatened and pastoralism itself is declining. Local knowledge related to pastoralism is disappearing; young men no longer work as shepherds in rangeland areas but seek job opportunities in other areas or regions. Local knowledge on range species is not documented and will thus be lost.

In North Africa, farmers were growing local cultivars of cereals. The French occupiers developed national research systems at the beginning of the twentieth century to breed varieties or cultivars adapted to their needs. National research centers worked mostly on wheat that would produce flour suitable for bread making.

Cereal breeding went on even after the North African countries got their independence. At that time, research centers focused on producing high-yield varieties adapted to mechanized techniques, which have gradually replaced locally bred cultivars. Indigenous wheat cultivars are still grown in remote hilly areas where there is no mechanization. The plants are usually short and easy to mow manually. They are also resistant to fungi and disease. They are usually grown by resource-poor farmers.

Farmers over centuries have developed different cropping patterns for wheat to optimize production and manage risk. In area where rainfall is more than 500 mm, wheat density is quite high (more than 100 kg ha^{-1}). In arid areas, farmers usually grow barley, not wheat, as it is more resistant to drought. Barley is grown for two purposes: farmers can get grain and it can be pastured—so they limit losses if it does not rain enough. Barley harvest is low (40 to 50 kg ha^{-1}). If rainfall is not as high as expected, livestock pasture the barley crop.

Social equity and gender

New processes that are transforming the rural areas of CWANA and bringing women to the forefront of agricultural work are not directly reflected in adjustments to either the legal system or prevailing social habits. On the contrary, discriminatory gender practices in agriculture persist to date.

Capacity development

The region needs to develop its capacity in a number of areas. Higher agricultural education, irrigation water management and conservation and use of plant genetic resources can all play major roles in increasing food security, alleviating poverty and meeting the Millennium Development Goals in the region if capacity is adequately built.

The new challenge for universities in developing capacity is in agricultural education for sustainable rural development and for strengthening rural communities. Higher agricultural education has contributed to the growth and modernization of agricultural production. It has focused on professional development of those responsible for agriculture and rural development. Curriculum and management adjustments have not paralleled this growth (Atchoarena, 2006).

Interaction with the farmer and farmer organizations or with the private sector has not been a university priority. Participatory research has been negligible and therefore the effect of research results has been moderate. To maximize the benefits from research at institutions of higher agriculture education, agricultural information systems and transfer technology units need to be put in place and links made to national, regional and international systems.

Many research results can be transferred or extended, but the community is not presently benefiting from them because the resulting technology is not being transferred or made available to the beneficiaries and the stakeholders.

Agriculture is central to rural development and poverty alleviation. Unfortunately the ministries of agriculture in many regions have not been able to take an active role in developing national strategies for generating jobs, improving livelihoods and alleviating poverty. Human resources in this area are lacking, and neither the universities nor the national agricultural research centers have provided the expertise or initiated effective programs in this aspect.

The region lacks expertise in impact studies, and in monitoring and evaluation. Lack of this expertise has kept important emerging technologies or practices from being disseminated in the field.

Expertise on risk analysis and assessment and on national commitments and benefits from international agreements and conventions is also lacking. Cooperation among institutions of higher education in agriculture with international and regional agriculture centers and organizations will reduce the gap in these areas. There is a lack in institutionalization of the participatory and community-based research and technology dissemination approaches in universities and research institutions. There are no policies for strengthening public–private partnerships.

Large areas in the CWANA region suffer severely from poor water management; inefficient irrigation and drainage practices and technologies; lack of knowledge and know-how on the part of farmers, farmer associations, and service providers; and institutional weaknesses. The problems in meeting growing water needs stem not only from water scarcity, but also from weak water-management capacity. With agriculture using a high percentage of the world's available water, improvement in capacity building at every level—from farmers to government—is required.

Genetic resources are important for food security. Since CWANA harbors a wealth of plant genetic resources, we are giving them special attention. Capacity development in this area focuses too heavily on technical training. Management and strategic planning, fund raising, public awareness and policy have not been considered in the past. What the capacity-building needs are of the plant genetic resource centers or the farmers has not been identified. Initiatives are not assessed and there is the risk of overemphasizing capacity needed in some areas and underestimating that in others.

Marketing and policies

Marketing is seen as a main element (or a main constraint) in developing the agricultural sector in the region. It affects the improvement of livelihoods of the rural population and rural development in general. Marketing opportunities and market participation are related to the sustainability of the farming and production systems, mainly for the poor. Markets in CWANA lack facilities—they are sometimes gathering places rather than markets; they lack regulations—or at least the existing regulations are not enforced; and conditions are not competitive—the marketplace suffers from collusion and bilateral negotiation. Importers and exporters have high transaction costs. Some markets are extremely protected from foreign competition, as in the Gulf region.

Locally, production is scattered and there is no marketing chain. Quality control is not well developed and the consumer has no role. Transport costs are high for small-scale farmers; few farmer organizations are able to transport and market agricultural products in bulk. Some commodities, like citrus, tomatoes, peppers, dates, have seen a major increase in yields—sometimes overproduction. Because there are no local storage and processing facilities and because farmer organizations are not as structured and strong as is needed, middlemen control farm-gate prices. Hence small-scale farmers do not get good prices and cannot invest in their system and repeat it the following year.

1.5 Status of Agricultural Knowledge, Science and Technology

1.5.1 Knowledge

Since the beginning of human settlement, major civilizations have started at the shores of rivers and lakes. Water is an important factor in initiating a human settlement. At the edge of desert and in valley areas where water is only plentiful during the winter, how to manage water resources became a big challenge to the builders of settlements. How to collect and store winter floodwater and use it during dry seasons was extremely important and valued knowledge. Ancient peoples had simple tools and techniques to build water-collecting systems adequate to support their demands during the dry months. This knowledge and associated techniques were developed over time, and efficient water-harvesting systems were installed in some parts of the desert. Indeed, CWANA contains the ruins of many civilizations who were pioneers in water harvesting such as the Maareb civilization in Yemen and the Nabateans in Jordan. These civilizations have left a heritage of rainwater-harvesting knowledge that helped people sustain themselves in the harsh environment of the drylands. Over time water-harvesting techniques have been developed and modified to adapt to different geomorphologic and climatological situations. Some are discussed here.

In areas where the catchments are significantly large, macroharvesting systems are implemented that have large-scale collecting and distributing schemes. For local and small catchments, smaller water-harvesting systems are more suitable.

One technique of micro water harvesting developed and used for rangelands enhancement is to build a series of check dams and contour lines to concentrate runoff water for wild vegetation. This technique results in effective rainfall many times actual rainfall. In Jordan, such techniques have been adopted in various parts of the country. Research is being conducted to see their efficiency and sustainability, mainly in enhancing rangeland. One popular technique is earth bunds (or *hafira*) (10,000-50,000 m^3). Water collected in the earth bund is used for watering livestock and sometimes for domestic use in remote areas, as in Sudan. The bunds were dug and managed by local communities in the past, but now the government uses earth-moving machines to dig them. The bunds are maintained by governments and some NGOs.

For crop production and in wet areas, terraces are built to serve two purposes: to stabilize soil from erosion and to harvest runoff. These techniques have been practiced for thousands of years in Yemen.

1.5.2 Science and technology

Investments in agricultural science and technology have expanded rapidly during the last four decades. Major technical and institutional reforms have occurred, which have shaped the pattern of developing and disseminating technology. In the early 1970s, the Consultative Group on International Agricultural Research (CGIAR) was established and national agricultural research systems (NARS) were greatly strengthened. During the 1980s and 1990s, partnerships among CGIAR centers and NARS were established, including the ecoregional consortia.

Historically, research has been conducted on the role of organic matter in soils, the development of reduced tillage systems, the use of on-farm organic resources in combination with inorganic fertilizers and the role of legumes in biological nitrogen fixation. Similarly, there has been research in integrated pest management (IPM) and in weed and pest control. These topics are of little interest to the private sector and are in danger of neglect by public research institutions. In most CWANA countries, agricultural research is not a priority. Levels of funding do not meet international requirements. Internationally, 2% of the GDP is allocated to research, but not in CWANA. NARS are short of financial support and personnel, which has emigrated from some countries. Some particular features of the international agenda are these:

- The global research agenda is gradually moving from a focus on individual crop performance to a growing acceptance of the importance of increased system productivity. This is viewed largely in terms of better-managed interactions among diversified farm enterprises, sustainable resource management, and improved targeting of technologies toward women farmers and poorer households.
- Perhaps even more importantly in the long term, institutional modalities are now shifting. From a public sector focus, largely led by the international system, more emphasis is now being given to public–private partner-

ships driven mainly by client demands. These changes are being accompanied by a growing understanding of farmers' problems and opportunities and a greater willingness to blend indigenous knowledge and modern information.

- Growing investments in biotechnology aim to increase agricultural research productivity and have the potential to revolutionize production practices by generating customized crop varieties. While national and international public funding available for agricultural research and extension systems has gradually decreased, private sector biotechnology research has attracted considerable support. Most of this research is likely to focus on profit-generating inputs, export crops and agroprocessing.
- Research on water resources has mainly focused on water management, water saving, and new sustainable processes to reuse wastewater and desalinize salty water.

Some examples of research conducted in CWANA are outlined here:

- One of the first research institutions in CWANA was in Sudan. Started in 1907, it concentrated only on cotton research during the British administration. After independence, it focused on diversification and intensification; food crops were introduced as part of agriculture research programs. The fields covered were soil management, crop husbandry, crop protection, and plant and animal breeding. In the 1940s, 1950s and 1960s a lot of research was conducted and published in international journals. Now little is published. Research concentrates only on irrigated crops, neglecting rainfed crops, although the rainfed area (mechanized and traditional) is ten times the size of irrigated areas in Sudan. Research on livestock is meager, concentrating mainly on veterinary issues as opposed to increasing production. Nowadays, the agricultural research situation in Sudan is bleak because of the lack of funds, the brain drain and partisan issues. Agricultural research once was the responsibility of the Ministry of Agriculture, but it has now been transferred to the Ministry of Science and Technology. The implications are not positive; the connection between farmers and extension agents has been considerably weakened.
- In Jordan, one of the pioneer specialized research institutes in agriculture is the National Center for Agriculture Research and Technology Transfer, whose field stations serve as research and demonstration farms. Research is mainly to develop drought-resistant varieties of cereals and legumes and to breed livestock. Field stations disseminate findings. Seeds of improved varieties are produced in large scale for farmer supply. The Ministry of Agriculture has an extension division whose agents (animal production engineers, plant production engineers, soil engineers) provide advice to farmers and consult with them. The main constraint is financial; all research depends on international funds. Most work has been on developing and selecting varieties for rainfed farming adapted to arid lands, mainly barley and wheat. Government funds mainly concentrate on extension.
- Irrigation systems have been developed to encourage efficient irrigation. Most developments have focused on adapting and transferring new irrigation techniques, such as drip and sprinkler irrigation, which during the 1980s took over from traditional practices using basin irrigation. These new techniques were first introduced by the Jordanian Ministry of Agriculture and its extension system, with institutions and regulations set up to discourage farmers from using more water. Another incentive for adopting irrigation techniques that use less water is it saves pumping, as most farming activities depend on groundwater pumped from deep aquifers. For surface-water users, the main reason to consume less water is that they must pay for the water they use. To encourage take-up of new technologies, new irrigation techniques are introduced into research stations. Over time, people start seeing the benefits of saving water and the ease and practicality of operating these techniques versus the traditional basin and channeling system. Where costs are saved by using less water, the incentive to use water-saving techniques is strong. However, when water is free or the cost is not tied to the amount of water used, farmers prefer the easier traditional basin techniques. This is typical for countries with abundant water or government-subsidized irrigation water.
- Egyptian farming systems represent all the different situations: Nile River water is channeled to the farming areas where farmers get it for free and most irrigation systems are basin. In the farming areas using groundwater two systems are in operation. One is that the government digs wells and pumps the water into channels to the farming areas at no cost to the farmers. In this system mostly basin irrigation is used. The other is that farmers have their own wells and pump their water; almost all use efficient irrigation systems that save pumping costs. These technologies are associated with added costs to the farming system. Most of the time, the savings in water pay this cost, but sometimes farmers do not have the initial start-up costs. In such cases government may intervene to help farmers adopt these irrigation water-saving techniques. Indeed, one of the major tasks of the agricultural credit fund in Jordan is to provide soft loans, with a subsidized interest rate, to farmers to adopt the new technology. In Tunisia, government greatly subsidizes adoption of new irrigation techniques.
- With increased domestic water and sanitation requirements, more treated wastewater is reused for agricultural production. Research is carried out on reusing reclaimed, treated wastewater for restricted irrigation (forage, wood trees, etc.). This reuse of treated wastewater differs from country to country and is mostly carried out where water resources are limited and in high agricultural demand. Some countries, such as Tunisia, have had good experience and use treated wastewater in farming on a large scale; others, like Jordan, are now testing; but the experience of most CWANA countries is limited.
- With the evolution of structural engineering, humankind is able to build massive structures that can dam tremendous amounts of water. In the last century, a number of mega dams were built on major rivers, allowing people to regulate water flow and farm all year round with water for dry cycles. Examples are the Aswan High dam

in Egypt and the Ataturk dam in Turkey. There are also several medium-size dams designed for agricultural and electricity purposes, and small-size dams (as in Tunisia) for agriculture.

1.5.3 Extension and systems of technology transfers

Agricultural research and extension services are not playing their important roles adequately in many CWANA countries. A prominent reason is that in many countries funds have not been adequate to maintain salaries at reasonable levels, leading to an increase in absenteeism as scientists take up other jobs to supplement their income. As a result, extension services have suffered. Both research and extension services in many CWANA countries depend heavily on donor funding. Given the fragile economies and extensive demands on the public sector countries, donor support for research and extension will continue to be important for some time to come.

Links between farmers, extension agents and research systems in many CWANA countries are weak. Often researchers interact little with extension services and farmers, and do not reflect their priorities in the research agenda. In some cases the national research program is defined by donors or individual researchers and may have little relation to farmer needs.

In other cases, farmers never learn about new technologies the research systems develop because effective mechanisms to transfer innovations from research to the extension system do not exist. Finally, the extension services have often failed to reach farmers because their communication strategies are not effective. Thus, extension services often miss the farmers who would benefit the most from good advice—the women farmers who are responsible for the great majority of agricultural output in most CWANA countries.

Even when farmers recognize that new technologies will raise productivity, they are often reluctant to bear the risks associated with a new approach. Methods are needed to reduce the risk that farmers face when adopting new technologies and to increase their access to sound rural financial services, including savings, credit and insurance.

Finally, research and extension systems must be opened to more providers, strengthening links between universities, NGOs, private firms, and others. Presently they lack coordination.

Present evaluation of research activities is not appropriate. Applicability of research findings should be evaluated. Presently, evaluation is only on the number of papers published and not on how research findings have reached farmers. In some places, research is farmer driven but its results do not go back to farmers because extension is weak.

Some messages for technology transfer are sent through the print media, radio and TV, but the messages are too technical as sent and cannot meet a wide range of farmer needs. Demonstration plots are rare and it is seldom feasible to scale up the new techniques. Farmers have no money for hiring labor, integrating new techniques, developing a marketing chain for new products.

Technology transfer occurs mostly through agricultural development projects as government agencies do not have financial or human resources to carry out adequate extension programs. There are some success stories about adopting new techniques, but they are rare and the impact of such projects is still limited. The Watershed Management Project, an FAO project implemented in three Tunisian governorates (Kairouan, Siliana and Zaghouan), introduced a type of water reservoir in the project area (FAO, 2004). Farmers settled in other regions adopted the same way of building their water reservoirs. This new technique has not been scaled up because capital is usually lacking within the small-scale farmer community. Sometimes there is farmer-to-farmer technology transfer through group farmer visits to groups of farmers in other regions, but more often transfer occurs in an agricultural development project. Resources are available while the project is ongoing, but after it closes, there is no way to upscale widely the new techniques the project has promoted.

Interest in agricultural education is decreasing. Education is becoming more knowledge based than skilled based. Little critical thinking is required, and the curriculum lacks multidisciplinary aspects. There is no holistic vision. Governments are not hiring people anymore. The main problem in the region that made past extension initiatives fail is that too many people are working in the ministries of agriculture, resulting in inefficiency and poor allocation of human resources.

References

Abrol, I.P., J.S.P. Yadav, and F.I. Massoud. 1988. Salt-affected soils and their management. FAO Soils Bull. 39. FAO, Rome.

AOAD (Arab Organization for Agric. Dev.). 2003. Agricultural statistical publication. Arab League, Cairo.

Atchoarena, D. 2006. The contribution of higher agricultural education to learning and development in rural areas: Experiences and policy implications. Int. Working Group on Education (IWGE) meeting, FAO, Rome, 13 Jun 2006. Available at http://www.unesco.org/iiep/eng/networks/iwge/2006/I_5b.pdf. UNESCO, Paris.

Ayoub, A.T. 1998. Extent, severity and causative factors of land degradation in the Sudan. J. Arid Environ. 138:397-409.

Bachler, G., and K. Spillmann. 1992. How to cope with environmental conflicts. Environment and Conflicts Project (ENCOP), Swiss Fed. Inst. Tech., Zurich.

Bakkay, S., and A. Tirichine. 2005. Index of Arabic and Mozabite terms for the description of the date palm tree. Government of Algeria, Algiers.

Ben Zid, R. 2002. Agriculture in the elaboration of development strategy of southern areas of Morocco. Dirasset and Sec. Environ., Rabat.

Brooks, K.N., P.F. Folliott, H.M. Gregersen, and L.F. De Bano. 1997. Hydrology and the management of watersheds. 2d ed. Iowa State Univ. Press, Ames.

CEDAW (Convention on the Elimination of All Forms of Discrimination Against Women). 2001. Consideration of reports submitted by states parties under Article 18 of CEDAW. CEDAW/C/YEM/5.

CNEA. 1996. Impact assessment of the northwestern integrated development project on women. World Bank, Washington, DC.

De Pauw, E. 2002. An agroecological exploration of the Arabian Peninsula. ICARDA, Aleppo.

Dufumier, M. 2004. Agriculture et paysanneries des tiers mondes. Karthala, Paris.
Dutilly-Diane, C. 2006. Gestion collective des parcours en zone agro-pastorale: Le cas de Ait Amar. Afrique Contemporaine 219(3):103-117.
El-Fattal, L. 1996. Women in agriculture in West Asia and North Africa: A review of literature. World Bank, Washington, DC.
El-Nour, I. 1992. Displaced and refugee studies in the Sudan, an annotated bibliography. Univ. Juba, Sudan.
FAO. 1995. A synthesis report of the Near East Region: Women, agriculture and rural development. FAO, Rome.
FAO. 1996. Fact sheet: Women, agriculture and rural development—Egypt. Available at http://www.fao.org/waicent/faoinfo/sustdev/WPdirect/WPre0016.htm. FAO, Rome.
FAO. 2001. Millions of people in East Asia hit by heavy monsoon rains as Central Asia is gripped by drought. Near East reels from three consecutive years of drought. Press release 01/63. FAO, Rome.
FAO. 2004. Mid-term evaluation of the Watershed Management Project in three governorates, Kairouan, Zaghouan and Siliana. GCP/028/ITA/TUN, Tunis.
FAO. 2005. FAO country profiles and mapping information system. Compendium of food and agriculture indicators. Available at http://www.fao.org/es/ess/compendium_2005.
FAO. 2006a. FAOSTAT database: agricultural production, primary crops. Available at www.fao.org/faostat/servlet. FAO, Rome.
FAO. 2006b. Land use. Available at www.fao.org/faostat/servlet. FAO, Rome.
FAO. 2006c. The state of food and agriculture: Food aid for food security? FAO, Rome.
FAO. 2006d. The state of food insecurity in the world 2006. FAO, Rome.
FAO/World Bank. 2001. Farming systems and poverty: Improving farmers' livelihoods in a changing world. FAO, Rome, and World Bank, Washington, DC.
Harlan, J.R. 1951. Anatomy of gene centers of origins. Am. Naturalist 85:97-103.
Harlan, J.R. 1992. The dynamics of domestication. p. 117-133. *In* Crops and man. ASA, Madison, WI.
Harlan, J.R. 1995. The living fields. Cambridge Univ. Press, Cambridge.
Hartl, M. 2003. Rural women's access to land and property in selected countries [Online]. Available at http://www.fao.org/sd/2003/PE07033a2_en.htm. FAO/IFAF/Int. Land Coalition, Rome.
Harvard College Library. 2007. International labor. Available at http://hcl.harvard.edu/research/guides. Harvard, Cambridge, MA.
ICRA (Int. Center for Development-Oriented Res. Agric.). 2003. Knowledge conservation: Stakeholders and women's involvement in grass root conservation of the biodiversity of palm date trees in the oasis of Djerid, Tunisia. ICRA.
IMF (Int. Monetary Fund). 2005. World economic outlook, April 2005. IMF, Washington, DC.
IPCC (Intergovernmental Panel on Climate Change). 2001. The scientific basis: Contribution of Working Group I to the Third Assessment Report of the Intergovernmental Panel on Climate Change. Cambridge Univ. Press, UK and NY.
IPGRI (International Plant Genetics Research Institute). 2005. Analysis of the genetic diversity of the date palm tree in Kerkenah Islands (Tunisia). IPGRI, Rome.
Karrar, G. 2002. Long-term drought mitigation and management in the Near East. Policy paper prepared for the FAO Near East Reg. Off., Cairo.
Kaufmann, D., A. Kraay, and M. Mastruzzi. 2003. Governance matters III: Governance indicators 1996-2002. Policy Res. Working Pap. 3106. Available at http://www.worldbank.org/wbi/governance/pubs/govmatters3.html. World Bank, Washington, DC.
Koocheki, A., and M.N. Mohalati. 1992. Feed value of some halophytic range plants of arid regions of Iran. p. 249-253. *In* V. Squires and A.T. Ayoub (ed) Halophytes as a resource for livestock and for rehabilitation of degraded lands. Kluwer Acad. Publ., Dordrecht.
Kundzewicz, Z.W. 1997. Water resources for sustainable development: Hydrological sciences. J. Sci. Hydrolog. 42(4):467-480.
Mazoyer, M. 2001. Protéger la paysannerie pauvre dans un contexte de mondialisation. Protecting small-scale farmers in the context of globalization. FAO, Rome.
Ramsis Farah, N. 1999. Situation analysis: Syrian women in agriculture. FAO, Rome.
Rodriguez, A. 1997. Rural poverty and natural resources in the dry areas: The context of ICARDA's research. Working Pap. ICARDA, Aleppo.
Rodriguez, A., and N. Thomas. 1998. Mapping rural poverty and natural resource constraints in dry areas. ICARDA Social Sci. Pap. 6. ICARDA, Aleppo.
Sombroek, W.G. 1995. Aspects of soil organic matter and nutrient recycling in relation to climate change and agricultural sustainability. *In* Nuclear techniques in soil–plant studies for sustainable agriculture and environmental preservation. Proc. Symposium. Int. Atomic Energy Agency, Vienna.
Suliman, M. 2000. Sudan civil wars: New perspective. In Arabic. Cambridge Acad. Press, UK.
UNDP/UNEP/World Bank/WRI. 2003. World resources 2002-2004. Decisions for the earth: Balance, voice, and power. World Resources Inst., Washington, DC.
UNEP. 1992. The Aral Sea: Diagnostic study for the development of an action plan for the conservation of the Aral Sea. UNEP, Nairobi.
UNEP. 1997. World atlas of desertification, 2nd ed. John Wiley, NY.
UNEP. 2002a. The Nile Delta: Potential impact of sea level rise. Vital climate graphics—Africa. Available at www.grida.no/climate/vitalafrica. UNEP/GRID-Arendal.
UNEP. 2002b. Global environmental outlook 3. Past, present and future perspectives. Earthscan, London.
UNEP/ISRIC (Int. Soil Reference and Information Centre). 1990. World map of the status of human-induced soil degradation. UNEP, Nairobi.
UNESCWA (UN Econ. Social Commission for West Asia). 1997. Regional report on implementation of Agenda 21. UNESCWA, Beirut.
US EPA (Environ. Protection Agency). 1996. Annual report. EPA, Washington, DC.
Vavilov, N. 1926. Centres of origin of cultivated plants. In Russian. Tr. po Prikl. Bot. Genet. Sel. [Bull. Appl. Bot. & Genet. Sel.] 16(2):139-248.
Wallach, B. 1989. Improving traditional grassland agriculture in Sudan. Geograph. Rev. 79(2):143-160.
Williams, M.A.J., and R.C. Balling, Jr. 1994. Interactions of desertification and climate. Prepared for WMO andUNEP. Arnoldy, London.
World Bank. 2006. World development indicators 2006. World Bank, Washington, DC.
World Economic Forum. 2005. The Arab World competitiveness report. World Econ. Forum, Geneva.
WRI/UNEP/UNDP/World Bank. 1998. World resources: A guide to the global environment. Oxford Univ. Press, Oxford.
Zeven, A.C., and J.M.J. de Wet. 1982. Dictionary of cultivated plants and their regions of diversity. Centre for Agricultural Publication and Documentation, Wageningen.
Zhang, L., Y. Xia, and Y. Zou. 1992. Halophytes and halophytic plant communities in Inner Asia. p. 115-122. *In* V. Squires and A.T. Ayoub (ed) Halophytes as a resource for livestock and for rehabilitation of degraded lands. Kluwer Acad. Publ., Dordrecht.

2

Historical and Current Perspectives of AKST

Coordinating Lead Authors:
Kawther Latiri (Tunisia), Alia Gana (Tunisia), Kamel Shideed (Iraq)

Lead Authors:
Jean Albergel (France), Stefania Grando (Italy), Yalcin Kaya (Turkey), Farzana Panhwar (Pakistan), Manzoor Qadir (Pakistan), Ayfer Tan (Turkey), Selma Tozanli (Turkey)

Contributing Authors:
Mohamed Annabi (Tunisia), Celine Dutilly-Diane (France), Gulcan Eraktan (Turkey), Alessandra Galie (Italy), Lubna Qaryouti (Jordan), Lokman Zaibet (Tunisia)

Review Editors:
Iftikhar Ahmad (Pakistan), Muna Hindiyeh Kazaleh (Jordan)

Key Messages

1. In the CWANA region, diversity in history and policies and natural resources has resulted in diversity in production, needs and trends. Development of AKST is uneven and differs from one country to another. AKST has promoted the initiation of a green revolution in a few countries. However, AKST has been ineffective in avoiding overexploitation of natural resources (water, soil and biodiversity), providing enough food, reducing poverty and removing social inequity. More attention needs to be given to the multiple functions and sustainability of agriculture.

2. The major factors that have negatively affected agriculture in the last 50 years are (1) the degradation of already limited natural resources (land, water, biodiversity); (2) extreme climate variability—recurrent and severe drought episodes aggravated by the severe effects of climate change; and (3) recent and rapid changes in organizing input and output markets (privatization, trade liberalization). In some countries conflicts, political instability and poor governance have hampered agricultural development.

3. During the last 50 years and despite these constraints, agricultural production and yields increased, mainly in the irrigated systems. In the rainfed systems, yields remained below the world average. High interannual climate variability and climatic hazards affect yields. Insufficient consideration of rainfed agriculture in AKST has been associated with major environmental and natural resource degradation in marginal lands. Despite important investment in AKST for irrigated areas, environmental problems expressed in overexploitation of water resources, soil degradation and pollution have occurred. Opportunities for organic farming are appearing, giving prospects for better incomes to farmers as well as protecting the environment. Organic farming is generally more environmentally friendly than conventional agriculture but may require well-planned policies and regulations to meet international standards.

4. Most countries in CWANA became water scarce during the last 50 years. As this trend will intensify in the future, problems of water quality are also expected to increase. Agriculture uses from 85 to 97% of water resources depending on the country (global average = 70%). It is also the primary reason for water pollution by pesticides and nutrients in groundwater, waterways, wetlands and coastal waters. In recent years these trends have begun to reverse through improved techniques, policies and strategies that encourage less exploitative water use, quality resilience, water protection, water harvesting, recharging and reuse, improved water use efficiency, water saving crops, drought-resistant varieties and better hydraulic infrastructure management.

5. Arable land resources in CWANA have been pressured by expansion of cropped area, overgrazing, loss of soil organic matter and depletion of nutrients and by the salinization caused by irrigation. Agricultural land is also being lost due to urbanization. This situation is aggravated by the lack of appropriate regulation aimed at protecting farmland. The following priorities for AKST are essential:

- Improvement and reinforcement of land-use regulation and soil-protection policies;
- Implementation of good practice guides for agriculture;
- Rehabilitation and agricultural intensification by promoting local and regional associations to facilitate community-based soil management and restoration;
- Restoration of the vegetal cover through agroforestry projects; and
- Rehabilitation of salt-affected soils in irrigated areas.

6. A high rate of deforestation has contributed to dust blows, atmospheric pollution and carbon loss and to climate change. AKST is helping to promote new productive land-use systems that could reduce greenhouse gas emissions and act as sinks for CO_2, CH_4, NO and N_2O. Additionally, new approaches for producing clean energy are being encouraged.

7. CWANA, cradle of the main cultivated crops in the world, is rich in unique agrobiodiversity in cultivated plants and their wild relatives, domestic animals and other species such as medicinal plants. However, as a consequence of the extension of the agricultural area and of some agricultural practices, this agrobiodiversity is in danger. The development of monocropping systems has led to a reduced number of crops and species cultivated in the area and a significant loss of diversity, thus threatening the environmental and social sustainability of farming systems. According to the IUCN Red List, more than 1600 plant and animal species are threatened. In CWANA 27 countries are parties to the Convention on Biological Diversity (CBD) and 17 are parties of the Cartagena Protocol on Biosafety (CPB). Many actions have been taken to implement these agreements and joint projects were initiated to maintain biodiversity. However, some policy actions and technology transfer such as application of early warning systems and capacity building are still needed.

8. Agriculture originated in CWANA and the region is rich in traditional knowledge on water harvesting, cultural practices and animal breeding. In the last decade initiatives were developed to recognize, validate and maintain traditional knowledge. However, complete coverage is still lacking and there is danger that with increased urbanization this knowledge will be lost. AKST could benefit from projects that encourage its retention.

9. The countries of CWANA have made significant progress in raising per capita food consumption. They vary significantly in per capita income and living standards, and hunger and malnutrition still prevail in some regions, including rural areas. This appears to be from insufficient attention given to food security at farm household levels.

10. Agricultural risk management policies in CWANA have mainly consisted of emergency measures, especially to cope with drought consequences and epidemics and programs to improve farming techniques. Most countries of the region need to design and implement a comprehensive and active risk policy. This would include the establishment of early warning systems, development of crop insurance and improvement in infrastructure, water management, agriculture and extension. Policies that protect human health, the environment and discourage cultivation of marginal land should be implemented. Marketing systems must be reinforced and farmer organizations promoted.

11. Agricultural activities in CWANA are undergoing major changes; policies have been reoriented to reduce public investment and support mechanisms favoring farm production. These changes negatively affect small- and medium-size farms, which play a major role in agricultural production and rural employment, including female employment. They reduce urban drift. Agricultural development strategies in CWANA are faced with major challenges: reducing poverty and securing food self-sufficiency and a better position on international markets, while protecting the environment. The contribution of the private sector to agricultural research and development is still limited in CWANA compared to the large contributions in industrialized countries.

12. In most countries of the region, rural migration and urbanization have been a major trend. However, while the share of the rural population has declined, the rural population has dramatically increased. Persistent high demographic growth amplified the pressure on the labor market and on natural resources. Food insecurity, aggravated by drought and climate change and unemployment, could intensify migration pressure.

13. Changes in farm structures in most CWANA countries have been characterized by two major trends: a movement toward the concentration of farmland within a minority of private and public farmers and a movement toward the fractioning of farmland, mostly through inheritance and demographic growth, which constrains consolidation and intensification of family farms. This is in contrast to industrialized countries, where the intensification of farming systems has eliminated farms and enlarged the average farm size.

14. In recent years, employment dynamics of the agricultural sector in most countries of CWANA have been characterized by a significant decline in the share of the active agricultural population, which went in average from over two-thirds of the active population in the 1960s to less than one-third, with increasing participation of women in agricultural production. Despite their major and increasing contribution to agricultural production and rural livelihoods, women's activities have remained unrecorded and undervalued, their role mainly restricted to unpaid family labor, as well as cheap and seasonal wage labor. While the illiteracy rate of rural women has remained very high in some countries (80% in Morocco), agricultural extension has continued to target mainly male heads of households. Agricultural development programs have frequently failed to integrate women's needs and priorities, gender equity objectives and, instead, frequently contributed to increasing women's workloads.

15. With few exceptions, farmers associations have remained very weak because of insufficient public policies. These weaknesses represent major constraints for consolidating the agriculture sector. Countries with strong farmer associations have a stronger agriculture sector and have successfully promoted more decentralized and participatory development as well as reinforced the professional organization of farm producers. Strong farmer associations will likely promote the participation of farmers in technology development, transfer and adoption.

2.1 Natural Resources, Agricultural Production and Infrastructure

2.1.1 Land use and land cover

Land use and land cover characteristics are affected by a changing climate and increasing climate variability. Both land use and climate affect the biogeochemical cycles and properties of ecosystems, altering the supply of goods and services to society, including carbon sequestration. A slight increase in agricultural land has been observed over the last 50 years, but has reached a plateau in most countries and decreased in others. At present, agricultural land covers from 3 to 80% of a country. The highest proportion of use of agricultural land is observed in Central Asia and Caucasus (CAC) countries.

Agricultural land is mainly devoted to permanent pasture and rangeland. For the whole CWANA region the proportion is 83%, the lowest proportion in Southwest Asia (55%) and the highest proportion in the Arabian Peninsula (98%). In the Arabian Peninsula, an almost twofold increase in permanent pasture was observed in the last 50 years; permanent pastures increased from $86x10^6$ ha to about $171x10^6$ ha. In the other subregions, permanent pastures increased only slightly.

Arable land area in CWANA increased the last 50 years, but differences were observed among the subregions. In some, arable land increased significantly and is still increasing (Nile Valley and Red Sea: from $15.5x10^6$ ha in 1961 to $22.5x10^6$ ha in 2002; Arabian Peninsula: from 1.1 to $3.7x10^6$ ha for the same period). In North Africa and in Southwest Asia, arable land increased in the 1970s, but is now stable (Skouri and Latiri, 2001). In the Central Asian countries, arable land area decreased in the beginning of the 1990s but is now increasing again.

During the last 50 years, irrigated land area increased in all the subregions except in the CAC countries. This increase and its proportion in irrigated land worldwide varied only slightly, except in the Arabian Peninsula. There irrigated land almost doubled due to considerable investment that supported dynamic development in the Arabian Peninsula. The share of irrigated cropland increased from about 30% in the early 1960s to more than 50% in 2000. The increase in irrigated land in the other CWANA subregions was much lower. At present, the proportion of irrigated land varies. The largest concentration of irrigated land is in Southwest Asia, around 23% of the agricultural area, corresponding to 14% of the world irrigated area. In the other regions, irrigated land varies between 2 and 5%.

Forests and woodlands are only around $136x10^6$ ha, 3.3% of the world area in forest. This proportion has not changed significantly in the last 50 years.

2.1.2 Agricultural production, cropping patterns and productivity

While West Asian and North African countries have important advantages in agriculture due to natural endowments, countries in the Nile Valley, Arabian Peninsula and near the Red Sea have natural constraints to overcome to satisfy their food needs. Since the mid-1950s this has driven them to import food.

Crops grown are a function of climate and soil, country priorities and needs, irrigation water and profitability. Cropping patterns changed distinctly from 1961 to 2005. CWANA depends increasingly on expanding yields of rainfed crops and cash crops in irrigated areas.

2.1.2.1 Cereals

Among the agricultural products produced in the CWANA region, cereals continue to be important. Grains are the essential resource for human nutrition. Wheat is the most important grain in the Mediterranean, rice in Pakistan. Cereals represent over 35% in crop rotation systems with wheat as the most widespread crop (27%), followed by barley. In dry areas, sorghum and millet are important as well.

In the last 50 years, cereal production increased from $51x10^6$ tonnes in 1961 (without CAC countries) to about $173x10^6$ tonnes in 2005. Yet production is not stable and yearly variations are high partly due to variation in the timing and amount of rainfall. Because of high demographic pressure, only South and West Asian countries had stable per capita production in the four decades. The Nile Valley, Red Sea and North African countries saw per capita production fall quite sharply, requiring significant grain imports. Compared to world averages, grain production in the CWANA region is lagging behind.

The increase in production is associated with an increase in yield during the last 50 years (1960-1965: 1 tonne ha^{-1} to 2000-2005: 1.9 tonnes ha^{-1}). Overall yields are still low, compared to the world average, even when annual rainfall is good. The world average increased from 1.4 tonnes grain ha^{-1} in 1960-1965 to 3.2 tonnes ha^{-1} in 2000-2005. Only in countries where cereals are irrigated is the yield high and stable (Figure 2-1).

Southwest Asia produces about 60% of the cereals grown in CWANA, about 4.6% of world production. The rapid production increase in Southwest Asia, $35x10^6$ tonnes in the 1960s to more than $100x10^6$ tonnes in 2005, can be attributed to both increased yield (1.0 to 2.2 tonnes ha^{-1}) and area harvested (35 to $44x10^6$ ha). Some countries, e.g., Turkey had high yield increases, while others did not.

In North Africa, production increased mainly from yield increases since land under cultivation had reached its maximum ($9.4x10^6$ ha). Annual yield variations are high and yields are still extremely low during dry years. These countries have high annual variation in rainfall; and drought is considered a permanent risk and the main factor affecting

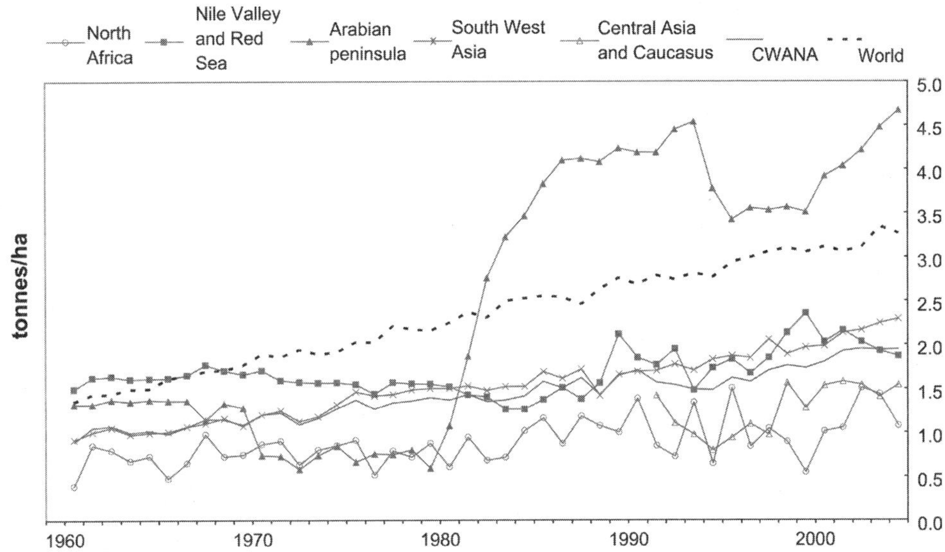

Figure 2-1. *Cereal yield in CWANA, 1961-2005.* Source: FAOSTAT database, data for CAC countries are not available for the whole period

yield. When rainfall is adequate, these countries still have low yields. Increases the last 50 years have been limited (1960-1965: 0.7 tonnes ha^{-1} to 2000-2005: 1.2 tonnes ha^{-1}) (Latiri, 2005).

In some Mediterranean countries, irrigation allows a more stable yield and increases have been observed since the early 1970s. In the Nile Valley and Red Sea countries, cereal production increased due to area and yield increases from $8.9x10^6$ tonnes in 1960-1965 to $26.3x10^6$ tonnes in 2000-2005. In the Arabian Peninsula, the area under cereals increased, but is still minimal: $0.7x10^6$ ha. Yields are above the world average because of irrigation. In CAC countries yields are about 1.9 tonnes ha^{-1}, but some countries, like Uzbekistan, have experienced regular increases the last few years. Like animal products, grains production fluctuated greatly in Central Asia and the Caucasus. Economic liberalization had a destabilizing effect in the early 1990s but this situation has shown recovery since the early 2000s.

2.1.2.2 Maize, legumes and industrial crops

Maize and rice are the most important irrigated crops. Oil crops, such as groundnut, sunflower and sesame are significant. Cotton is an important irrigated crop in most of Central Asia, covering over $3.8x10^6$ ha.

The most important legumes in the region are chickpea, lentil and dry beans. West Asian countries have been known for their pulses for a long time. Most pulses originate from Mesopotamia. The total share of South Asia and West Asia in the average pulse production in CWANA was 73% in the 2000s, with Turkey, Pakistan and Iran standing out as the most important producers. Egypt and Sudan in the Nile Valley and Red Sea subregion and Morocco in the North African subregion are also important producers. International demand for pulses has had a positive effect on volume and productivity Their area increased by about 50% and yield increased from 0.62 to 0.87 tonnes ha^{-1} during the last 50 years. This production increase was mainly in Southwest Asia.

Legumes could play a major role in rainfed systems. About 30 million ha of land is left fallow in CWANA every year. If 70% of this land could be sown with forage legumes, it could produce enough to feed 80 million sheep. Moreover, there would be $1.4x10^6$ tonnes of nitrogen added from symbiotic nitrogen fixed each year.

Sugar beet and sugar cane are mainly produced in Southwest Asia, comprising 8.5% of world production. In the other subregions, production increased but is still low. North Africa increased from 0.1 to 1.5% and Nile Valley from 0.1 to 1.2% of world production from the 1980s to today.

Sugar beet is an industrial crop highly supported by government policies in CWANA and in other parts of the world. Subsidies, until the 1990s, increased volume and productivity. Facilities offered to producers for storing and transporting their crops had significant results on the production and marketing of the sugar beet in West Asian and North African countries. Since the mid-1990s production volumes fell because of agricultural policy changes. Most of the sugar refining units were public enterprises that have gone private. Tunisia gave up its sugar industry, judging it more expensive to produce its own sugar than to import it. Most of the sugar beet farms converted to raising livestock. In Turkey, the government continues to subsidize beet production, but the changes in transportation patterns, including the switch from rail to truck transportation, diminished the willingness of the farmers to cultivate sugar beet. Many switched to other crops. Nevertheless, Turkey stands out as the largest sugar beet producer in CWANA, followed by Iran, Morocco, Egypt and Syria. Production in CWANA has increased, while the overall world production of sugar beet has decreased since 2000.

2.1.2.3 Horticultural and vegetable crops

Horticulture and vegetables also exist in the cropping systems in the region. Potatoes, tomatoes, grapes, almonds, apples, olives and oranges are the most important crops in CWANA countries. Permanent crops under irrigation (dates, citrus) are in almost all the countries of North Africa, the Arabian Peninsula and Southwest Asia. These regions have very diverse fruit and vegetable species. The high demand for them in international markets pushes most of these countries to invest heavily in their horticultural sector to increase productivity.

Vegetable production has increased from about 20x10^6 tonnes in the 1960s to 96x10^6 tonnes in 2005 (representing 11% of world vegetable production). It is mainly concentrated in Southwest Asia (6% of world production) and the Nile Valley and Red Sea subregions. Due to increases in harvested area and in yields production rose from 10.6 tonnes ha^{-1} in 1960-1965 to 19.5 tonnes ha^{-1} in 2000-2005 (world average in 1960-1965: 9.7 tonnes ha^{-1} and in 2000-2005: 16.8 tonnes ha^{-1}). While production in the Nile Valley and Red Sea regions continues to climb, production in South Asia and West Asia is stable and even slightly decreasing.

The same growth trends can be observed in overall fruit production, which took off in the early 1980s. Countries of the Arabian Peninsula have followed this trend in fruit production, while their horticultural production has lagged behind. The North African countries have fluctuating horticultural production. Fresh fruits and vegetables are becoming important export products for exporting countries and should be an important component of protected agriculture and crop diversification programs in CWANA countries.

CWANA produces about 11.4% of world fruit, with yields slightly below the world average. While yields in the 1960s were 5.1 tonnes ha^{-1} (world average: 7.5 tonnes ha^{-1}), they reached 8.9 tonnes ha^{-1} (world average in 2000-2005: 9.7 tonnes ha^{-1}). Yield did not increase evenly. In the Nile Valley and the Red Sea regions, fruit yields are the highest, with an average of about 13 tonnes ha^{-1}, while the world average yield is about 10 tonnes ha^{-1}. Southwest Asia demonstrated the highest increase in yield, from 4.3 tonnes ha^{-1} in 1961 to near 10 tonnes ha^{-1} in 2005. In North Africa, fruit yield slightly decreased during the last 50 years. Increased production in the region is related to an increase in the area harvested, which doubled the last 50 years. The main increase came from the Nile Valley and Red Sea regions.

2.1.2.4 Oil crops

Important public investments in the agricultural sector in Turkey and Pakistan in West Asia and South Asia, in Egypt and Somalia in the Nile Valley and the Red Sea and in Morocco and Tunisia in North Africa caused important increases in oil crop production of the CWANA region. The production increased from about 700,000 tonnes in 1961-1965 to more than 2.3x10^6 tonnes in 2001-2005 in South Asia and West Asia, accounting for 56% of the total production in CWANA in the 2000s. Nile Valley, Red Sea and North Africa subregions also had important increases in oil crop production.

2.1.2.5 Crop–livestock–range systems

Crop–livestock–range systems are the most widespread form of agricultural production in CWANA. They are the least intensive in land and water use. Their distinguishing features are the use of arable crops and natural rangeland to feed small ruminants, the principal economic output of the system. Although cropland and animals are generally associated with strong systems of private property and smallholders, the predominant production unit, property regimes associated with rangelands are less well defined. Open and uncontrolled access to rangeland often results in extreme overgrazing.

Crop–livestock–range systems are primarily on the margins of major rainfed and irrigated cropping zones and are often associated with seasonal movement of animals and households. The integration of crops and livestock is promising for low-income small-scale farmers. A great advantage of crop and livestock integration is that it uses diverse resources, such as fodder legumes, crop residues and livestock manure in a system of nutrient recycling. Livestock provide high profit per unit of labor, plus valuable manure for a soil improver. Livestock also have the advantage of being relatively easy to market, compared with harvested crops. There is a steady demand for livestock products, with relatively high and stable prices.

Increasing population, urbanization and incomes in CWANA are leading to a growth in demand for animal products, which opens opportunities for resource-poor farmers in domestic and export markets. However, these farmers face the challenge of producing for a competitive market. The role of technology is to promote the adoption of improved animal health and nutrition practices, genetic enhancement and better handling to achieve higher productivity. Plant species adapted to these dry areas can increase feed supply. During dry years barley, vetches, oats or other forage crops can improve the supply and quality of feed and prevent soil erosion, especially on hillsides. Spineless cactus or shrubs (like *Atriplex halimus*) alone or intercropped with other forage crops can also improve the supply and quality of feed.

Multinutrient feed blocks made from agroindustrial by-products and other ingredients are a low-cost source of supplements that can increase animal productivity. In addition, addressing and preventing endemic diseases leads to improved livestock productivity. Early weaning of lambs is another way to increase milk production. Improved rams can be distributed to producers to improve flock performance. Lamb fattening and dairy processing into high-value commodities and targeting niche markets can help increase earnings from small ruminant enterprises (Aw-Hassan et al., 2005). A holistic program that includes production, management and health of both small and large ruminants would increase productivity and animal health.

Irrigated fodder crops are important in Egypt. Berseem (trifolium forage legume) represents over 20% of the cropped area. These fodder crops are present in each country of the Arabian Peninsula, from 12% in Saudi Arabia to 32% in Qatar. In Kyrgyzstan, fodder crops represent 37% of the irrigated cropped area.

2.1.2.6 Use of inputs

Use of agricultural inputs varies in different CWANA subregions. The countries in the Arabian Peninsula, Nile Valley and Red Sea regions and North Africa practice extensive agriculture, while South Asian and West Asian countries,

Pakistan, Syria and particularly Turkey seem to have opted for intensive agriculture since the 1960s. Data obtained from the FAO time series show the total number of agricultural tractors increased more than 20-fold between 1961 and 2002 in South Asia and West Asia. It increased fourfold in Northern Africa, Nile Valley and Red Sea countries.

During the last 50 years, fertilizer use has increased significantly across CWANA. Turkey invested in huge fertilizer plants in the 1960s and 1970s. In Egypt, Iran, Iraq, Morocco, Syria and Tunisia, governments also invested in fertilizer production, although these investments were less than those in Turkey. Fertilizer use is currently about 7.5% of world consumption ($10.46x10^6$ tonnes); before it was 1.7% ($0.53x10^6$ tonnes). This increase is still occurring in Southwest Asia, although in North Africa and the Arabian Peninsula fertilizer use has slowed significantly or fallen.

Profiting from important government subsidies, insecticides, pesticides and herbicides for pest and disease management were introduced. Their production and marketing stayed in the hands of large multinational firms that created local subsidiaries. Turkey, Morocco and Egypt host multinational affiliates, while other CWANA countries import inputs. The fall in government subsidies to agriculture, since the 1980s in Turkey and Egypt and since the 1990s in Morocco, has had a negative effect on the use of these products. Another reason for the decline in use are changes in international rules, e.g., banning the use of pesticides in fresh produce for general food safety and environmental concerns. North African countries, Egypt and Turkey limit use of these inputs on agricultural products for export, while Syria started to increase their use in the second half of the 1990s.

Use of agricultural chemicals is uneven among farmers. Small-scale landholders all over CWANA continue to practice traditional agriculture, while capital-intensive agriculture is practiced by only a small portion of wealthy landholders. However, there are no statistical data to show this, and real field research is drastically lacking in this area.

The increased use of certified seeds and the development of variety protection in CWANA subregions are governmental concerns and have been included in international grant programs since the beginning of the 2000s. Most of the countries, like Afghanistan, Algeria, Iran and Syria, intensified state initiatives to create seed trade associations to improve relationships among breeders, producers and farmers. Afghanistan is profiting from a USAID financial grant that helped it establish 20 village enterprises in five target provinces to produce and distribute certified seeds to farmers. Syria received a grant from the Japanese government that helped the Syrian General Organization for Seed Multiplication found a culture laboratory. Turkey stands out because it started seed improvement programs in the 1980s. The Turkish seed industry has shown remarkable progress the last 25 years. In Turkey, nearly 150 private seed companies deal with hybrids, vegetables and forage crops; foreign investments, mainly from the Netherlands, Israel and the United States of America, account for approximately one-quarter of them. A new law enacted in 2005 gathers dominant seed companies under one roof to better coordinate efforts.

Since liberalization policies were put in place, state control is loosening in the agricultural inputs market. Marketing channels are controlled more and more by NGOs (national associations or international organizations) or private enterprises. Governments still undertake important infrastructure investment in irrigation systems, but agricultural chemicals are now marketed mainly by private enterprises. State subsidies have been greatly reduced, affecting the use of these chemicals by small landholders. The decreased use of agricultural chemicals is in line with new international measures to reduce or eradicate pesticide residues or nitrates in water. However, lack of integrated agriculture has significantly reduced agricultural yields.

2.1.2.7 Animal products

The 1990s saw important structural changes in product markets, when many countries of West Asia, South Asia and some countries in North Africa invested in raising livestock and developed milk-processing industries. In Pakistan, Turkey, Egypt, Tunisia and Morocco, heavy public investment resulted in an increase in milk production between 1961 and 2005. Milk became a commodity with high value. In all, CWANA milk production per capita increased considerably. For Nile Valley and Red Sea countries this sector really took off in the early 1980s and sharply increased from the mid-1980s to the present. Accordingly, there has been an increase in the number and capacity of milk-processing industries in these countries, creating an attraction for foreign investment.

Meat production did not show the same increase as milk production between 1961 and 2005. The South Asian and West Asian countries increased meat production from about 110,000 tonnes (1961-1966 average) to nearly 6,000,000 tonnes (2001-2005) the last 40 years. But their high population increase somewhat saps this positive trend, so the yearly meat production per capita stays relatively low, when compared with industrialized Western countries. In North African countries and in the Arabian Peninsula, yearly meat production per capita increased considerably between 1961 and 2005, while in Central Asian and Caucasus countries a fall occurred, from 34 kg per capita annually during 1986-1990 to 27 kg per capita annually from 2001-2005. The poultry sector showed the most important increase with the spread of intensive production.

2.1.2.8 CWANA case studies

To understand how cropping patterns changed, five countries have been selected for detailed examination: Turkey, Pakistan, Iran, Morocco, Egypt.

Turkey. Turkey is self-sufficient in food and is also the largest exporter of agricultural products in CWANA. Turkey has the most arable land in CWANA, 11th in the world, and its climate gives it the highest yield capacity. All field crops, including aromatic plants and different fruits and vegetables, can easily be produced.

During the last 50 years, arable land increased with only small modifications in the cropping systems. The main crop is wheat; it has had the highest value of all agricultural products for 50 years (FAO, 2007). Its area increased, but its percentage in the cropping system stayed stable, between 43 and 45%. Oats, rye and forage declined in importance and chickpea and lentils have replaced fallow areas. Grape decreased, but it is still among the top-valued crops. Cash crops, such as maize and hazelnuts, are grown in both rainfed and irrigated areas. The amount of irrigated land increased. Cow's milk is also important. Chicken meat production increased and this meat is now among the ten top products in value. Beef is also in the top ten, while mutton decreased in value.

Pakistan. Pakistan has the second most arable land, 22%, in the region. The climate is arid with low rainfall and humidity and high solar radiation over most of the country. Most areas receive less than 200 mm annual rainfall. The high-altitude northern mountains receive more than 500 mm. Similar cropping patterns have existed in Pakistan for the last 40 years, with some modification, showing a large increase in permanent crops (1961, $150x10^6$ ha; 2005, $700x10^6$ ha) and irrigated land (1961, $10x10^6$ ha; 20005, $18x10^6$ ha). Cereals yield showed regular increases, going from 0.8 tonnes ha^{-1} to 2.5 tonnes ha^{-1}.

Iran. Iran is one of the largest countries in CWANA, half of which is unproductive land. The diversity of climate in Iran provides a good opportunity for varied agriculture and horticulture. Water shortages are compounded by unequal distribution of rainfall. Similar to Turkey, wheat and barley are the two main crops. Chickpea and lentils are planted in late winter and early spring. Dates and watermelon replaced pistachios. Irrigated area increased from 5 to $8x10^6$ ha and permanent crops increased from 5 to $20x10^6$ ha. During the same period, yields in cereals, fruits and vegetables showed regular increases, cereal yield from 1 to 2.5 tonnes ha^{-1}, fruits from 4 to 11.3 tonnes ha^{-1} and vegetables 6.8 to 22.6 tonnes ha^{-1}.

Morocco. Morocco is extremely dependent on its agricultural sector. Over 90% of the country's agriculture is rainfed; output is unreliable and only 19% of the land is cultivated. Morocco produces wheat, barley, citrus, wine, vegetables and livestock. Agricultural land increased from 4.5 to $7.5x10^6$ ha without major changes in the cropping systems. However, the land area devoted to permanent crops and olives has increased in recent years. Vegetable yield increased threefold. Irrigated areas almost doubled. Like other Mediterranean countries, yields in cereals show huge annual variability.

Egypt. The total agricultural land of Egypt is about $3.28x10^6$ ha, of which about 92% is in the Nile Valley. Almost all agricultural lands are entirely dependent on irrigation. In the last century, the arable area increased by about $1x10^6$ ha, while the country's population increased nearly sixfold, from 11.2 to 65 million. While cotton was the main crop (26% of arable land), its area has been reduced from 834,000 ha in the 1960s to 315,000 ha in 2005. Grains, maize, wheat and rice occupy about 50% in the cropping systems, with large increases in wheat and rice over the last 50 years. Wheat is the most cultivated crop. Clover for animal feed—almost nonexistent in 1960s—increased along with the rotation system and now represents between 20 to 25% of the cropping area. Tomato production currently occupies 195,000 ha and is the highest grossing crop in Egypt.

Excessive use of irrigated agriculture has had adverse effects on the environment. Around 840,000 ha in Egypt, mostly cultivated lands in the North Delta, are salt affected.

2.1.3 Water resource development and management

2.1.3.1 Trends in potential water resources in CWANA

Economic growth remains dependent on water. Most areas of CWANA have scarce water and pronounced variable climate. CWANA's exploitable water resources have undergone a considerable evolution since the 1970s, particularly in the 1990s. Countries mobilized more of their water resources. New infrastructures were built; for example, water that Tunisia can use increased from $2.6x10^9$ m^3 in 1990, 57% of the resource, to $4.1x10^9$ m^3 in 2004, 90% of the resource. New hydraulic structures will be completed during the next decade, including 11 large dams and 50 retention dams or lakes.

Despite this, CWANA water resources per capita have decreased, due to the increase in population. About half the CWANA countries are below the threshold of 500 m^3 per person annually (Figure 2-2), This threshold was once considered a minimum for easy development for a country, but now it is thought the average needs are around 1000 m^3 (Falkenmark, 1997; Tropp et al., 2006).

2.1.3.2 Evolution in water uses and water demand

Agriculture uses 98% of the water in Afghanistan and Turkmenistan and about 65% or less in Algeria, Lebanon, Armenia and Bahrain. Agricultural water withdrawal has evolved rapidly the past few years. Agriculture seems to be the sector where water must be saved (Table 2-1). Just a few countries, for example, Tunisia and Uzbekistan, have decreased water consumption in agriculture through the introduction of water saving policies.

For some countries, total water withdrawal is greater than the amount that can be renewed. Supplementary water comes from depletion of renewable groundwater, abstraction of fossil groundwater and nonconventional resources, desalinated water, treated wastewater and saline drainage water. Water withdrawal, expressed as a percentage of internal renewable water, indicates the capacity to rely on renewable water sources. Values above 100% indicate that water flows in from outside the subregion or country, or that fossil or nonconventional water sources are used. The CWANA region has the highest percentage of water withdrawal worldwide and it is expected to intensify (Table 2-1). The withdrawal rates are already above the 40% threshold for water scarcity, beyond which cost and groundwater depletion increase dramatically. For some countries these percentages are extremely high, for example, the Kingdom of Bahrain.

2.1.3.3 Trends in water strategies and policies

Water conservation remains the main component for water resource management. The loss between producing the water and its use was estimated to be more than 25% in most countries. If one adds the waste from poor or underutilization of water to the loss between producing and using it, the amount wasted may be 40 to 50% of the global volume. Methods to conserve water in irrigated agriculture include modernization of irrigation networks, improvement of plot

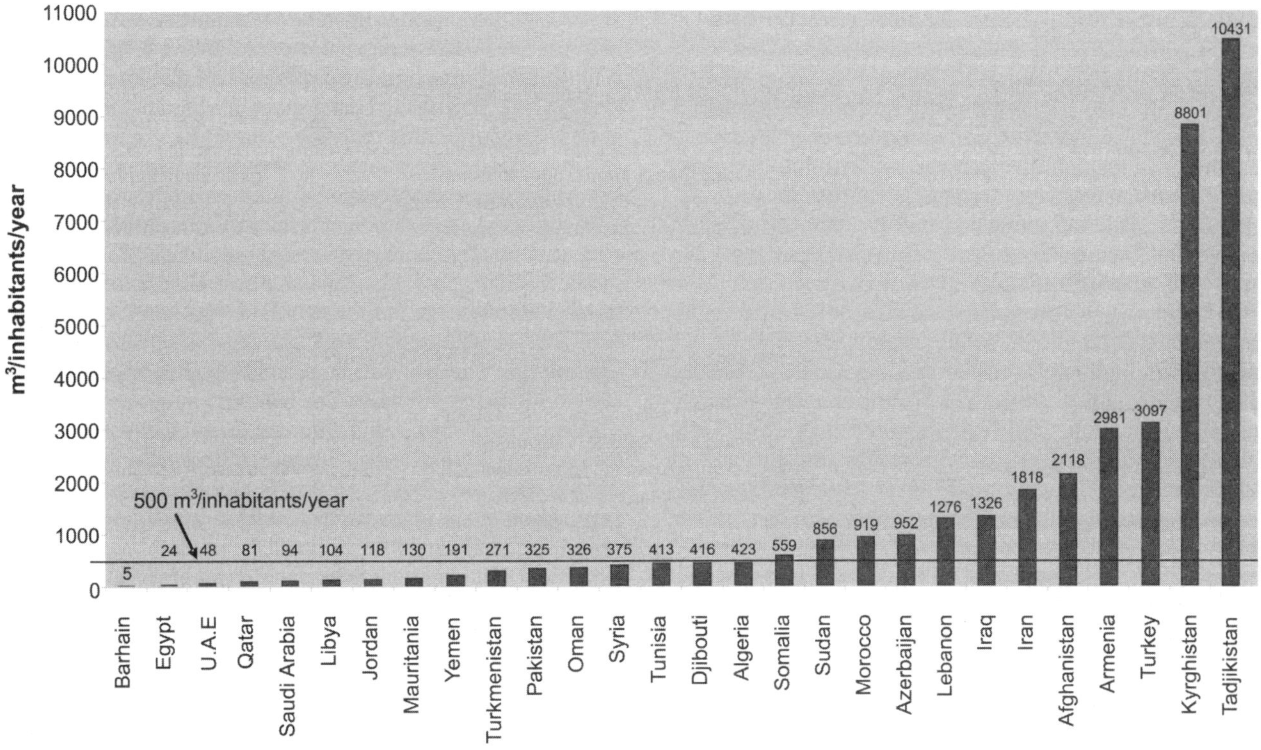

Figure 2-2. *Water resources per inhabitant in the CWANA region.* Source: AQUASTAT, FAO database

water management and irrigation techniques, introduction of new techniques to regulate irrigation and encouragement of farmer associations to share and manage water. In Tunisia, drip irrigation is partly subsidized.

In rainfed agriculture water-harvesting techniques help supply crops with water. In CWANA arid and semi-arid areas, more than one million hectares are managed to harvest water for agriculture, mainly in West Asia and North Africa.

Several countries in North Africa, the Arabian Peninsula and the Nile Valley and Red Sea regions that have few renewable water resources overlie important nonrenewable fossil groundwater basins. These basins are partly shared with neighboring countries. For Algeria, Tunisia, Libya, Egypt, Saudi Arabia and the United Arab Emirates fossil water is an important resource. Libya exploits 3.7×10^9 m^3 per year of its fossil groundwater. Although groundwater reservoirs may allow storage of huge quantities of water accumulated during the pluvial periods of the Quaternary, their development cannot be considered sustainable. The lack of recharge results in the slow depletion of the aquifers. Moreover, as the water level declines, it increases the cost of pumping, and the water quality deteriorates in some areas. This may make the extraction of fossil water less attractive.

Urbanization, tourism and industry pushed authorities in some countries to undertake adequate measures to protect wastewater collection sites from likely medium- to long-term pollution. The creation of treatment plants and the adoption of policies to reuse treated wastewater have contributed to increases in wastewater produced and treated.

Desalinated water is important in the Arabian Peninsula and Egypt. Saudi Arabia, the United Arab Emirates and Kuwait are, by far, the largest users of desalinated water. They use 71% of the total for CWANA, with Saudi Arabia alone accounting for 42%.

Groundwater is a significant resource. Irrigated areas in CWANA are estimated at 30×10^6 ha, which is 7% of the agricultural area, using nearly 130×10^9 m^3 of irrigation water per year. Roughly 13×10^9 m^3 of irrigation water could be recuperated, monitored for quality and used.

Many countries have sustained agricultural development in diverse regions and urbanization of neighboring zones has accelerated. Water transfers can be used to more equitably distribute water and reduce the gap between demand and supply by according priority to potable water needs and getting the most benefit from each unit of water. Internationally, the debate about water diversion has become heated. Many water-transfer projects have already been made in many CWANA countries. Resistance is growing to further developments, such as diverting water from water-abundant to water-scarce regions, even when the projects promote economic development, poverty alleviation and environmental protection. Examples of huge water-transfer projects are the Southeastern Anatolia Project in Turkey, the Nile River initiative and water management in the Jordan River region. The technical issues of these different water-management strategies are not the main problem. Environmental and social issues have become the main constraints.

Taking turns among water users started about 1760 BCE in Palmyra with Hammurabi's Code. Water pricing was used in 137 CE (Sartre, 2001). Today water pricing is increasingly accepted as public policy throughout CWANA.

Table 2-1. **Water withdrawal (ww) and gap in CWANA region.**

		Agricultural ww	Total ww	Total ww per capita	Total ww
		(10^9 m^3 year^{-1})		(m^3 inhabitant^{-1} year^{-1})	(% of internal renewable water resources)
North Africa	Algeria	3.94	6.07	194	52
	Libyan A.J.	3.54	4.27	784	711
	Mauritania	1.5	1.70	606	15
	Morocco	11.01	12.60	419	43
	Tunisia	2.17	2.64	271	57
Nile Valley and Red Sea	Djibouti	0.00	0.02	27	6
	Egypt	59.00	68.30	969	117
	Somalia	3.28	3.29	347	22
	Sudan	36.07	37.32	1135	58
	R.Yemen	6.32	6.63	343	259
Arabian Peninsula	K.Bahrain	0.23	0.44	180	2200
	Kuwait	1.23	1.36	491	138
	S.Oman	0.21	0.29	483	547
	Qatar	15.42	17.32	736	722
	Saudi Arabia	1.57	2.30	783	1533
	U.A. Emirates	6.32	6.63	343	162
South and West Asia	Afghanistan	22.84	23.26	1014	36
	R.I. Iran	66.23	72.88	1071	53
	Iraq	39.38	42.70	1742	57
	Jordan	0.76	1.01	190	115
	Lebanon	0.92	1.38	384	31
	Pakistan	162.65	169.39	1130	76
	Palestine	–	–	–	–
	Syria	18.93	19.95	1148	76
	Turkey	27.86	37.53	534	18
Central Asia and Caucasus	Armenia	1.94	2.95	960	28
	Azerbaijan	11.65	17.25	2079	57
	Kazakhstan	28.63	35.00	2263	32
	R Kyrgyz	9.45	10.08	1989	49
	Tajikistan	10.96	11.96	1931	75
	Turkmenistan	24.04	24.65	5142	100
	Uzbekistan	54.37	58.34	2270	116

Source: AQUASTAT, 2002; FAO database.

Heavy governmental investments have been made in water cost-recovery in the whole CWANA region and high crop yields have been recorded in North Africa, West Asia and Central Asia. Reasonable and effective water pricing systems provide incentives for efficient water use and water quality protection (Easter and Liu, 2005). Water pricing should cover the costs for infrastructures and their depreciation and operations and maintenance. Metering water consumption is necessary for efficient water-pricing policies. Water metering has three major approaches: area, volume and market equilibrium, which is related to the crop and its profit. The predominance of gravity fed irrigation means water consumption must be measured on irrigation time and pipe capacity. Meters are used for systems under pressure. In Jordan, Al Hadidi (2002) describes the widespread installation of meters at private wells; billing can discourage overexploitation of the aquifers. Weak cost recovery translates into financial resources inadequate to maintain minimum operations and maintenance, causing services to deteriorate (Baroudy et al., 2005).

Most CWANA countries have adopted water pricing for irrigation and drainage. Water tariffs are increasingly important to manage demand. In Morocco and Tunisia water bills cover operating costs but exclude infrastructure costs. Water pricing is moving from reimbursement of maintenance costs to protecting the natural resource. However, some disadvantaged areas still benefit from government support. Generally, governments continue to support investment and fill the gap left by payment arrears (Baroudy et al., 2005).

2.1.3.4 Effect of global changes on water resources

Climate change will affect precipitation, evapotranspiration, runoff—and ultimately, water resources. Changes in the water cycle likely will affect the magnitude, frequency and cost of extreme weather and the water available to meet growing demand. Recent reports (IPCC, 2007) show that climate change is likely to increase the days of intense precipitation and the frequency of floods in northern latitudes and snowmelt basins. Severe droughts could also increase because of a decrease in rainfall, more frequent dry spells and greater evapotranspiration. In the arid and semiarid areas of CWANA modest changes in precipitation can have a large effect on water supply. In mountain watersheds, higher temperatures will increase the ratio of rain to snow, accelerate the spring snowmelt and shorten the snowfall season, leading to a faster, earlier and greater spring runoff.

Temperature projections of climate are less speculative than projections of precipitation. The scenario problem lies in the scale mismatch between global climate models, with monthly data over several tens of thousands of square kilometers and catchment hydrological models, which require daily data and at a resolution of a few square kilometers. In CWANA few studies on climatic change scenarios are available, CWANA is not in the GEWEX experiment (Global Energy and Water Experiment) while the HYMEX (Hydrological Cycle in Mediterranean Experiment) will cover part of CWANA.

2.1.4 Water management infrastructure for agriculture

The control of water always has been critically important in the development of civilizations (Job, 1992). Since ancient times, different hydraulic systems have been built to augment rain and collect, store and transport water (El Amami, 1983; Khouri et al., 1995; Prinz, 1996). From simple cisterns in the ground, vital for the survival of families in arid zones, to large dams that are part of national policies to guarantee water to the greatest possible number of people, water storage always has been a primary preoccupation of CWANA governments, which face recurring droughts and limited water (Hamdi and Lacirignola, 1994; Jaber, 1997; ESCWA, 1998a).

The earliest recorded dam was on the Nile River at Kosheish, where a masonry structure 15 m high was built about 2900 BCE to supply water to King Menes' capital at Memphis. Evidence exists of a masonry-faced earthen dam built about 2700 BCE at Sadd-el-Kafara, about 30 km south of Cairo; this dam failed shortly after completion when, in the absence of a spillway, it was overtopped by a flood. The oldest dam in use is a rock-fill structure about 6 m high, built around 1300 BCE, on the Orontes River in Syria. By 1950, governments were building more dams as populations increased and national economies grew. Today nearly half of the world's rivers have at least one large dam. However, the last 50 years have also highlighted the performance and the social and environmental effects of large dams. Dams have fragmented and transformed the world's rivers and perhaps 40-80 million people have been displaced by reservoirs.

2.1.4.1 Large dams

Since the 1950s, CWANA has built many dams and some are among the largest in the world. In November 2000, the World Commission on Dams published an overview of dams and development (Clarke, 2000). It is based on 150 case studies, 13 in CWANA: Iran, Morocco, Pakistan, Tunisia and Turkey. While the large dams vary greatly in performance, most dams in CWANA underperform in achieving the intended benefits and services. In some instances, though, benefits occurred for much longer than predicted and still continue. Adverse effects on ecosystems occur frequently and many were unanticipated.

Irrigation components fell well short of targets in the irrigation command area developed, irrigated area achieved and, to a lesser extent, cropping intensity. It is difficult to find data on predicted and actual crop values. A high variability is observed among projects, data on smaller dams suggested greater consistency in performing closer to targets than larger ones. In contrast to irrigation, hydropower performance of large dams was, on average, closer to the goal, although performance varied.

Larger than necessary reservoirs may reflect overestimates of water demand or high reserves for drought. Multipurpose projects had higher variability and lower average performance than single-purpose projects. When a new dam is planned, the number of families and people that will be displaced and involuntarily resettled is routinely underestimated. The lack of records reporting these aspects also remains contested and continues to fuel controversy in the large-dam debate. Many positive and negative ecosystem effects from large dams were unanticipated, even in the 1990s. Mitigation was the most practiced response to ecosystem effects, but it has mostly failed or worked only sporadically.

Participation and transparency in decision making were neither open nor inclusive through to the 1980s. Emphasis has grown on transparency and participation in decision making involving large dams, especially in the 1990s. Many dam projects still do not plan for public participation of affected people. Participation of NGOs has increased in the projects since the 1990s.

2.1.4.2 Small dams

CWANA countries have had, over varying lengths of time, policies of constructing small dams designed to use surface water and control erosion. The dikes of these dams are between 5 m and 15 m high (lower limit for large dams established by the International Commission on Large Dams). They are constructed of rubble in small rural catchments in areas of moderate relief. They have rustically designed lateral spillways with a discharge capacity of tens of cubic meters per second or, in some cases, just over 100 m^3 per second. Some, but not all, have a sluice gate. The dams cost around 500,000 euros, sometimes far less. Their reservoirs are relatively small, a few hectares and have a holding capacity from a few tens of thousands to 10^6 m^3 (Albergel and Rejeb, 1997). Their main objectives are to reduce losses in agricultural land (estimated at 10,000 ha per year) and dam siltation, to increase water table recharge and to provide water for irrigated cropping (Albergel and Nasri, 2001).

Small dams have been known in West Asia since ancient times. The dam on the Nahr El Asi, near Homs, in Syria, was built in the reign of Seti 1 (1319-1304 BCE). Many were built at the start of the Christian era, one of which was Badieh Dam on the road to Palmyra. Numerous ruins testify to their presence in the dry steppes. Some still exist, but are completely filled with sediment. The first small dams built using modern techniques were those constructed during the 1960s in the province of Swaida to supply drinking water to villages on a basalt plateau that has no underground water. These reservoirs are usually stocked with fish. The lakes are used for fishing and for leisure activities, as with the small dam at Al Corane, not far from Damascus, in a small high valley. In the Middle East, the idea of a hill reservoir is not as well defined as in the Maghreb, but numerous small reservoirs have been constructed to create water reserves for the cattle of the nomadic Bedouin tribes.

In North Africa, the drought in the early 1980s, considered to be the longest ever experienced, marked the start of a policy of labor-intensive, small dam and hill reservoir construction. Small dams were primarily designed for irrigation, livestock watering, flood protection or supplying drinking water to rural areas that had no readily exploitable underground water. North African countries built a large hydraulic infrastructure in the 1970s and 1980s. Almost all the large dams are affected by significant sediment. Numerous small dams have been built to slow down siltation. For example, the largest dam in Morocco, the Al Wahda Dam on the Ouergha River in the province of Sidi Kacem (88 m high, with a capacity of 3.4 10^9 m^3), is protected by many small dams in the upstream catchment designed to hold the erosion coming from the steep marly slopes of the Rif. Some have already been constructed, while others are planned. Erosion from the Ouergha catchment, estimated at 98 tonnes ha^{-1} annually over 6150 km^2, causes the dam to lose 60 10^6 m^3 volume each year (Anonymous, 2001).

2.1.4.3 Infrastructures for water harvesting

Water harvesting is a way to increase water available for crops. Section 2.5.2 on traditional knowledge presents the history of water harvesting, describing modern or rehabilitated infrastructures for water harvesting. Water harvesting is a proven technology to increase food security in drought-prone areas. Erosion control and recharging groundwater are additional advantages. Water harvesting can also be considered for rudimentary irrigation or storage of drinking water for animals. The farmers and pastoralists have no control over timing. Runoff can be harvested only when it rains. Although it exists in extremely old cultures (Nasri et al., 2004) and has been observed in most countries, extension and irrigation staff often have quite limited knowledge about various water-harvesting techniques and the associated socioeconomic implications. A growing awareness of the potential of water harvesting for improved crop production arose in the 1970s and 1980s. The stimulus was the well-documented work on water harvesting in the Negev Desert of Israel (Evanari et al., 1971). ICARDA is helping disseminate information on water technologies, promoting adoption and providing training for field staff in CWANA (Oweis et al., 1998, 1999).

Three main kinds of infrastructure for water harvesting are distinguished (Prinz, 1996; Prinz and Prinz et al., 1998).

- *Rainwater harvesting:* These infrastructures induce, collect, store and conserve surface runoff in arid and semiarid regions (Boers and Ben-Asher, 1982). This type of harvesting covers water collected from rooftops, courtyards and treated surfaces. The water is used for domestic purposes or garden crops. Microcatchments collect runoff from a small catchment, to feed the root zone of a tree, bush or annual crops. Macrocatchments convey runoff from hill-slope catchments to a cropping area at the hill foot or to a tank or artificial pond for watering cattle.
- *Floodwater harvesting,* also called "large catchment water harvesting" or "Spate Irrigation," is collecting and storing of flow from floods in intermittent wadis for irrigation. Floodwater harvesting has two forms: (1) *Floodwater harvesting within the stream bed.* The water flow is dammed and stored in reservoirs or is allowed to inundate the flood plain; the wetted area can be used for agriculture or pasture improvement. (2) *Floodwater diversion.* The wadi water is forced to leave its natural course to nearby cropping fields. Various technologies and different names exist for these two floodwater harvesting types (see section 2.5.2 on traditional knowledge). Pakistan has more than 1.5 10^6 ha under floodwater harvesting. The irrigated area under floodwater harvesting in North Africa and the Middle East is increasing.
- *Groundwater harvesting* covers different infrastructures that concentrate and extract groundwater using little energy or only gravity. Qanat systems, underground dams and special wells are a few examples of groundwater-harvesting infrastructures. *Qanats,* widely used in Iran, Pakistan, North Africa and Spain, consist of a horizontal tunnel that taps underground water in an alluvial fan and brings it to the surface due to gravitational effect. Qanat tunnels have an inclination of 1 to 2% and a length of up to 30 kilometers. Many are still maintained and steadily deliver water to fields for agriculture and villages for drinking water. *Groundwater dams* are subsurface dams built in the wadi beds. They obstruct the underground flow of ephemeral streams. The water is stored in the sediment below ground and can be used after floods. A more sophisticated infra-

structure is the sand storage dam, which is a sand-filled reservoir watered by the wadi flow.

2.1.4.4 Irrigation infrastructures

Irrigation covers about 48 million hectares in CWANA. Central Asia represents 59% of this total, although it covers only 21% of the total area. Pakistan alone, covering a little over 4% of the region, accounts for 33% of the irrigated areas. By adding Egypt, Iran, Iraq and Turkey, 72% of the land under irrigation is controlled by five countries in CWANA. Surface irrigation is by far the most widely used technique, practiced on 88% of the area. Sprinkler irrigation is practiced on 11% and microirrigation on 1.4% of the total area. In Libya and Saudi Arabia, sprinkler irrigation is clearly predominant, while in Jordan, Oman and the United Arab Emirates, microirrigation is most widely used. It is practiced on over half of the full and partial control irrigation areas (AQUASTAT, 2002). In Kuwait and Lebanon, sprinkler irrigation is used on more than 37% and microirrigation 39% of their full and partial control irrigation areas. The arid countries without large rivers choose to develop more intensively microirrigation and sprinkler irrigation to save water (AQUASTAT, 2002).

CWANA is subject to salinization because the volume of rainwater dissolving the salts generated by the soil is low. The problem has long been recognized; however, national assessment of salinization is difficult and little information on it can be found. Furthermore, no commonly agreed upon methods exist to assess the degree of irrigation-induced salinization. More information on salinization will probably become available. Strategies to improve the situation, recognized as a priority by most CWANA countries, should be defined. All countries reported having salinization problems, varying from 3.5% in Jordan to over 85% in Kuwait.

A measure necessary to prevent irrigation-induced waterlogging and salinization in arid and semiarid regions is installing drainage facilities. Drainage, in combination with adequate irrigation scheduling, allows leaching of excess salts from the plant root zone. Figures on drained areas are available for 13 of the CWANA countries in the FAO AQUASTAT database (2002). About 34% of these irrigated areas have been provided with drainage facilities, varying from 0.6% in Iran to over 90% in Egypt.

2.1.5 Land degradation and water quality deterioration

The degradation of arable land resources is as old as the history of irrigation and human settlement. A well-documented case occurred between 4000 and 2000 BCE, within the boundaries of CWANA, where secondary salinity affected the land and water in the Tigris and Euphrates valleys in Mesopotamia. History has repeated itself in the last century in many other countries of the region.

Water-quality deterioration, recognized as a global problem, has been exacerbated in dry regions. Research on water-quality deterioration began later than research on land degradation in CWANA, where problems from water-quality deterioration are expected to intensify, due to anthropogenic interventions and the increasing possibility of extreme events of climate change (IPCC, 1998, 2007). In addition, saline water intrusion is expected to increase, from sea level rise and overexploitation of groundwater in coastal zones. Salt-prone water and land problems are expected to increase in arid and semiarid regions (Qadir, Wichelns et al., 2007). Declining water quality has also increased water supply problems, especially in drier climates. Pollutants have a greater opportunity to accumulate during dry periods. A heavy reliance on irrigation in some areas has compounded the problem (UNEP, 2002). Research long focused more on water quantity and use than on its quality. Since the 1980s, there has been increased attention on the productivity of irrigated agriculture and conservation of the environment.

2.1.5.1 Land degradation

Processes and causes of land degradation have been studied since the beginning of the twentieth century. Land degradation remains important for the twenty-first century because it adversely affects agronomic productivity, the environment, food security and quality of life. Land quality declines. According to the Global Assessment of Human Induced Soil Degradation (GLASOD) (Oldeman, 1988), land degradation is linked to climate variation and unsustainable human activities, such as overgrazing, deforestation and poor agricultural practices (Bossio et al., 2007). Land degradation is also linked to population density (Roose, 1996); in an agrarian system, if the population passes a certain threshold, land starts to run short and soil restoration mechanisms seize up. In some dry zones, when the population reaches about 100 inhabitants per square kilometer, the zone is rated as a densely populated, degraded area. A rapidly growing population and limited land resources mean that combating desertification will be difficult for developing and poor countries.

The Global Assessment of Soil Degradation is the first baseline study using a consistent methodology to estimate global soil degradation (Oldeman, 1988). Its estimates of the extent of affected areas are rough and should not be used as precise data (Table 2-2). However, they provide a good overview of the situation. National authorities concerned with land degradation need to update and refine present estimates and mapping. Major soil constraints and vulnerability for CWANA countries are related to sodicity, shallowness and erosion risk. Erosion risk is major, affecting 5% of Central Asia and the Caucasus and 20% of South Asia and West Asia (Table 2-3) (Nasr, 1999). Mechanisms that start land degradation are numerous (Bossio et al., 2007). A decline in soil structure causes crusting and compaction, leading to decreased infiltration rates and increased erosion. Significant chemical processes include acidification, salinization, pollution and fertility depletion. These processes can operate individually, simultaneously or in combination and threaten the sustainability of natural resources.

The salinization of land in CWANA is the consequence of both naturally occurring phenomena and human activities. Anthropogenic activities contribute more to salinization of land and water resources than primary salinity. Excessive irrigation, accompanied by inefficient drainage (Qadir and Schubert, 2002) caused by either lack of information on crop water requirements or a need to apply leaching water,

Table 2-2. **Land degradation: Severity of human-induced degradation in CWANA.**

	Total area (1000 km²)	None	Light	Moderate	Severe	Very severe	Cause	Type
				(%)				
North Africa	5,552	50.6	12.8	10.5	25.4	4.6	A, O, D	N, W, C
Nile Valley and Red Sea	4,581	42.4	16.2	17.5	14.2	2.4	A, O, D	N, W, C
Arabian Peninsula	2,374	24.0	34.3	20.2	32.4	6.0	A, O, D	N, W, C
South and West Asia	4,551	8.2	4.3	31.6	39.6	17.3	A, O, D, V	N, W, C, P
Central Asia and Caucasus	3,998	57.9	11.4	15.8	13.3	2.1	A, O, D	N, W, C, P

Cause: A = agriculture; D = deforestation; O = overgrazing; V = over exploitation of vegetation

Type: C = chemical deterioration; N = water erosion; P = physical deterioration; W = wind erosion

Source: Terrastat, FAO database, 2006.

has resulted in rising groundwater and soil salinization. For instance, nearly half the irrigated lands in Central Asia are affected by salinity (Kijne, 2005). The largest part of salt-affected soils and saline waters exists in the lower reaches of Amu-Darya and Syr-Darya basins, where salinity threatens food production.

2.1.5.2 Water-quality deterioration

Wastewater from household, municipal and industrial activities; agricultural drainage containing residues of pesticides, fertilizers and soil and reaction products of amendments; and overexploitation of brackish groundwater are the major contributors to deteriorating water quality in CWANA (Qadir, Sharma et al., 2007). In addition, drought has a negative influence on aquatic and land ecosystems and on the quantity and salinization of surface water and groundwater. More than half the major rivers in CWANA are seriously depleted and polluted, degrading surrounding ecosystems and threatening the health and livelihoods of the people depending on these rivers.

Wastewater and sludge. Water diverted for household, municipal and industrial activities generates wastewater containing undesirable constituents, depending on the source of the wastewater and its treatment. In most countries of CWANA, domestic wastewater is not segregated from industrial wastewater and other activities (Qadir et al., 2007). The wastewater is often discharged, untreated, into open drains, sometimes getting mixed with storm or fresh water. It is then channeled into natural water bodies or used in agriculture. In some cases, wastewater undergoes treatment.

Most wastewater treatment plants are of simple design or not adequately functioning in the region. Wastewater is diverted to farmers' fields—treated, partly treated, diluted or untreated. It is used by urban and periurban agriculture to grow crops, with vegetables being the most common, across CWANA (Lazarova and Bahri, 2005). Grain, fodders and industrial crops are cultivated as a secondary preference. In addition, parks, sports grounds and road plantings are irrigated with wastewater. Examples of wastewater aquaculture are also found in countries such as Kazakhstan.

Table 2-3. **Major soil constraints in CWANA.**

	Total area (1000 km²)	Sodicity	Shallowness	Erosion risk
			(%)	
North Africa	5,552	0.2	22.0	8.0
Nile Valley and Red Sea	4,581	1.5	21.8	8.6
Arabian peninsula	2,374	0.0	22.6	7.4
South and West Asia	4,551	0.9	25.8	19.6
Central Asia and Caucasus	3,997	28.4	16.3	5.0

Source: Terrastat, FAO database, 2006.

The use and disposal of partly treated or untreated wastewater has environmental and health implications, but despite some governmental restrictions in CWANA countries and potential health risks, farmers use it (Keraita and Drechsel, 2004). In Central Asia, wastewater is discharged into streams, rivers, lakes and natural depressions, significantly polluting the ecosystem and threatening human health through waterborne diseases such as typhoid fever and bacterial dysentery (Fayzieva et al., 2004). Except for a few national assessments, only scattered information exists on the volume of raw or diluted wastewater currently produced and used in agriculture in CWANA (Qadir, Wichelns et al., 2007). Although water-quality management is reported to be a high priority and a major concern of governments in the CWANA region, most countries do not have sufficient resources to avoid polluting water bodies. However, with increased investment and awareness, the safe use of recycled wastewater can be increased (Karajeh et al., 2004).

Much of the operating cost of wastewater treatment plants involves handling and disposing of the sludge. This sludge is rich in organic materials but may also contain heavy metals and pathogens. Sludge needs to be treated and its environmental effect reduced. Sludge may be treated using pasteurization, aerobic thermophilic treatment, thickening with lime and composting. In Morocco, sludge from drying beds was tested on Italian ray grass crops. Heavy metals in the soil or in the vegetation did not accumulate (Jamali and Kefati, 2002). In Jordan, anaerobic pond sludge of stabilized waste could be used as fertilizer in agriculture after drying for three months, when pathogens numbers were reduced enough to meet safety standards (Hindiyeh, 1995).

Saline and sodic water. Under irrigated agriculture, increases in cropping intensity, excessive use of agrochemicals, inappropriate irrigation methods and salt-affected soils for crops contribute to increased salt in drainage water (Skaggs and Van Schilfgaarde, 1999). This water is collected in artificial or natural drainage systems or penetrates through the soil to become groundwater. Water scarcity in several countries of CWANA has led to reusing drainage water and overexploiting groundwater to produce food, fodder and wood (Qadir, Sharma et al., 2007).

Changes in river runoff directly affect river water quality. In Central Asia, the rivers Amu-Darya and Syr-Darya are the major source of irrigation. Long-term monitoring of these rivers shows that, in the 1950s, the salinity varied around the year within the range of 0.33 to 0.72 g L^{-1}. Other river water-quality parameters, such as major cations and anions, organic compounds, pH and pesticide levels, were also within safe limits during 1950s. Since the 1970s, salts in river water have increased steadily as a result of a decrease in the flow of Amu-Darya and Syr-Darya and increased discharge of return water, particularly drainage water from irrigated schemes. Consequently, there has been a significant increase in river water salinity since the 1980s.

Although return flow of water to the rivers is an additional reserve for use, it has become a source of environmental pollution in Central Asia (Altiyev, 2005). About 95% of return water is drainage water, containing elevated salts and pesticides, herbicides, fertilizer and other agricultural chemical residues. It is estimated that annually about 140 10^6 tonnes of salt are discharged into the drainage water, 75% brought in by irrigation water. About one-quarter of the salt in drainage water is from the subsoil by mineral dissolution while some estimates reveal the average of mobilized salts at 40% of the total salt discharged (Kijne, 2005). About 51% of the total return flow goes into rivers, about 33% into depressions and 16% is reused in irrigation. As a result of returning water to natural depressions, hundreds of water bodies have been formed. Since these water bodies do not have an outflow, their water quality has deteriorated every year because of massive evaporation.

The Aral Sea, which is fed by two main rivers, Amu-Darya and Syr-Darya, is in the center of the Central Asian deserts and functions as a gigantic evaporator. This sea, which was the fourth largest inland lake in the world before 1960, is now the largest inland salty reservoir. It has become synonymous with environmental catastrophe, representing one of the world's worst ecological disasters. In the Soviet era, massive quantities of water from Amu-Darya and Syr-Darya were diverted for irrigating cotton, which decreased river water inflow to the Aral Sea. This led the sea to shrink dramatically. This seemingly irreversible process has continued, as irrigated agriculture has expanded and hydropower generation increased.

The disposal of agricultural drainage water containing many salts and agrochemicals into freshwater canals and rivers disperses salts and potentially toxic substances on a large scale. For instance, in the Euphrates Basin within Syria, about 1 10^9 m^3 of saline drainage water are put back into the Euphrates River, doubling the salinity in the river water (from ~ 0.5 to 1.0 dS m^{-1}) when it enters Iraq downstream. Inappropriate water management in the lower Euphrates Basin affects land and water quality in Iraq. In Jordan, water quality in the Amman-Zarqa Basin and Jordan Valley has been affected over the past few decades, with consequences for the downstream irrigated agriculture (McCormick et al., 2003). Anticipated increases in the basin's population and economic growth are expected to further affect the situation. Understanding the past dynamics and developing scenarios for the future that affect water quality for downstream users have been facilitated. National agencies have gathered extensive datasets, including water-quality data, for several years from strategic locations.

2.1.6 Agriculture and carbon sequestration

2.1.6.1 Soil organic stock and the potential of carbon sequestration

In CWANA, soil organic content is low, ranging from 0.2 to 0.8% in relation to aridity (Lal, 2002) for soils of Pakistan (Rashid, 2000), Iraq (Aziz et al., 1988) and elsewhere in the region (Ryan and Matar, 1988; ICARDA, 1991). However, when rainfall is high, soil humidification allows better nutrient status, and some soils may have a higher organic content, around 1.5 to 2.0% (Yurtserver and Gedikoglu, 1988).

The organic pool of most soils has been and is being depleted by soil degradation, erosion and subsistence and exploitative farming (Albergel et al., 2005). The historic loss of a soil organic content pool for CWANA may be 6 to 12 Pg. Assuming that 60% of the historic loss can be resequestered, the total soil carbon sink capacity of the region may be 3 to 7 Pg over 50 years (Cole, 1996; Lal, 2002). This may be realized by adopting measures to control desertification, restore degraded soils and ecosystems and improve soil and crop management techniques to enhance the soil organic content pool and improve soil quality. The strategies of soil carbon sequestration include integrated nutrient management and recycling, controlled grazing and growing of improved fodder on rangeland (Lal, 2001, 2002, 2006). In Morocco, Bessam and Mrabet (2001) show that switching from normal tillage to no tillage practices could increase carbon sequestration by 13.6% after 11 years. In Central Asia, the carbon studies component of the Livestock Management and Rangeland Conservation Tools project is providing data on rangeland carbon flux. First estimates in northern Kazakhstan (Wylie et al., 2004) show rangelands had an average of 1.27 tonnes C ha^{-1} sequestration of CO_2 during the 2000 growing season. The potential of soil carbon sequestration in different WANA ecosystems through desertification control and restoration of degraded ecosystems is 2.0 to 5.1 Pg C over 50 years (Table 2-4) (Lal, 2002).

Soils have the potential to reach an annual carbon sequestration rate of 0.2 to 0.4 Pg C, accounting for 24 to 30% of the potential of global dryland ecosystems (Table 2-5) This potential rate of carbon sequestration can be maintained over 25 to 50 years, provided coordinated efforts are made to adopt appropriate land use and recommended soil, water and crop management technology. However, agricultural intensification involves carbon-based inputs including tilling, pesticides, fertilizers and irrigation. Emission of carbon from all these inputs needs to be considered in evaluating the net soil organic content sequestration.

2.1.6.2 Incentives for land-use change

Promoting changes in land use that would increase the carbon sequestration rate will benefit the international community and the governments, international NGOs and private companies that now pay for these services. Opportunities for funding CO_2 sequestration through land-use change are limited to reforestation and afforestation under the Land Use, Land-Use Change and Forestry of the Clean Development Mechanisms. This might explain why it is not very popular in the dry areas of the world. Ninety percent of the Clean Development Mechanism projects are in Latin America and Asia and the Pacific. Morocco has registered 21 clean development projects and Egypt 28. However, these focus on clean energy, transport and waste management and only a fifth concern forestation and reforestation.

As the carbon market grows, new opportunities might rise through emerging carbon management programs or voluntary carbon markets (Taiyab, 2006). In addition, biodiversity, desertification control and improved water quality produced through sustainable land use are better valuated and should also be considered.

Biofuel crops and other nonfood crops grown for use in industry, chemicals (plastic, paint), industrial fibers (paper and textiles), pharmaceuticals, personal care products and biofuels can be an alternative to traditional food production. They cut across all development, with significant effects on the economy, society and the environment. They offer an opportunity for farmers to develop exports and industry with more diversified horticulture. Meanwhile, the world's ecosystem gains from a rich source of renewable materials that do not further deplete the earth's natural resources.

In CWANA, water scarcity is the main constraint for developing nonfood crops. Production of renewable biomass-suitable biofuel to be used as a substitute for fossil fuel need not compete with food production. The best opportunities are biomass resulting from agroindustry, biomass from wastelands and agroforestry based on oil trees not dedicated to food production: jatropha for example. It can reduce soil degradation and can be used for bioenergy (http://www.jatrophaworld.org/).

2.1.7 Agrobiodiversity

2.1.7.1 Changes in agrobiodiversity and agroecosystems

CWANA countries have about 10% of the world's endangered species, mainly among animals and about one-third of birds, mammals, reptiles and fishes (Table 2-6) for number of endangered species by subregion). Some of the world hotspots where unique biodiversity is threatened are in the Caucasus, Iran-Anatolia, Mediterranean Basin, mountains of Central Asia and part of the Horn of Africa. The development of key biodiversity areas, representing the most important sites for biodiversity conservation worldwide, is a new concept being tested in the area (http://www.iucn.org/themes/s sc/redlist2006/redlist2006.htm; http://www.biodiversityhotspots.org). With the exception of some countries in the region, like Turkey and Iran, most of the biodiversity areas are not yet well protected. National con-

Table 2-4. **Potential of soil carbon sequestration through desertification and restoration of degraded soils in the CWANA region.**

Ecosystem	Land area (10^6 ha)	Rate of soils C sequestration with improved management ($Mg\ C\ ha^{-1}\ yr^{-1}$)			Total potential over 50 years (Tg C)
		Soil organic C	Soil inorganic C	Total	
Irrigated Cropland	11.3	0.1-0.2	0.2-0.3	0.3-0.5	170-283
Rainfed cropland	50.1	0.05-0.1	0.01-0.1	0.06-0.2	150-500
Rangeland*	583.3	0.05-0.1	0.05-0.1	0.06-0.15	1,735-4,337

*Rangeland area has been reduced by 250 10^6 ha allocated for biofuel production

Source: Adapted from Lal, 2002.

Table 2-5. **Potential of the WANA region and global dryland ecosystems to sequester carbon.**

Strategy	WANA region[a]	Global dryland ecosystem[b]
	(Tg C y^{-1})	
Desertification control	40-100	200-300
Reclamation of salt-affected soils	9-18	200-400
Agricultural intensification on undergraded soils	6-12	10-20
Fuel C offset	88-175	300-500
Soil C sequestration under biofuel planting	25-75	n/a
Total	168-380	710-1,220

Source: [a] Lal 2002; [b] Lal 2001.

servation programs have not been initiated properly and biodiversity legislation has not been implemented to promote protection (http://www.cbd.int/reports).

The expansion of agricultural production and the intensive use of inputs over recent decades in CWANA countries are considered to be major contributors to the loss of biodiversity (FAO, 1996a). At the same time, certain agricultural ecosystems can serve to maintain biodiversity, which may create conditions favorable for many species that might be endangered by fallowing or changing to a different land use, such as forestry. Agricultural food and fiber production is also dependent on many biological services. This can include, for example, providing genes for developing improved crop varieties and livestock breeds, crop pollination and soil fertility provided by microorganisms.

The preservation and enhancement of biodiversity pose a major challenge for agricultural policy makers, as population and demand for food increase. Policy makers will need to find ways of minimizing the conflicts between expanding production and maintaining biodiversity, enhancing the many complements between agriculture and biodiversity and finding ways to prevent the loss of biodiversity on agricultural land (Pagiola and Kellenberg, 1997). For a growing number of CWANA countries, protecting and enhancing biodiversity are becoming important in their domestic and international agroenvironmental policy objectives and actions, in response to growing public concern. In practice, government policies towards biodiversity involve balancing the tradeoffs between socioeconomic values and biodiversity conservation. Typically, policies with low ambition can avoid short-term costs but may lead to more costs over the long term, such as risks to agricultural production from genetic erosion. More ambitious policies and targets towards biodiversity conservation will require scientific research, including developing biodiversity indicators. Indicators can help support decisions by providing information about the risks and degrees of sustainability associated with different options.

2.1.7.2 Introduction of modern varieties, case studies on wheat

For thousands of years, small-scale farmers have grown food for their own consumption—planting diverse crops, recycling organic matter and following nature's rainfall patterns. The trend of switching from traditional agriculture to cash crop agriculture and monoculture is leading to a decline in local crops and varieties and the loss of traditional knowledge, farming and old varieties and landraces. At present, some minor crops are maintained by farming households on a small scale to supply their traditional food cultures. Harlan (1951) noted that "crop germplasm in Vavilovian Centers are vulnerable to loss due to technological and economic changes."

Until the 1950s, farming in CWANA relied upon farmers' accumulated knowledge of the local physical and social environment. At the end of the 1960s, introduced improved wheat started to replace local varieties, causing the loss of old and traditional wheat cultivars, especially in areas suitable for extensive agriculture. This replacement is the major cause of genetic erosion, which frequently occurs because the genes and gene complexes found in the diverse local varieties are not all contained in the modern seed (FAO, 1996a). The lack of extension, no national planning and policies and no local training for maintaining unique, local varieties and a wide range of ecological problems associated with agricultural practices have caused environmental pollution and biodiversity lost. Having the first national program and storage facilities, Turkey carried out an intensive survey and collection program, in 1968 and 1969, at the coastal regions to maintain the local wheat cultivars, which were being replaced rapidly by improved Mexican wheat (Sencer, 1975).

2.1.7.3 Expansion of agriculture and crop and plant diversity change

CWANA has mainly dryland and mountain ecosystems. Both are fragile and open to a rapid decline of biodiversity. The expansion of agricultural production into formerly uncultivated mountain lands or forest reduces the habitat for other species and leads to a decline or deterioration of ecosystems, particularly where the lands are only marginally suitable for agriculture. The threat to traditional crops will increase as cropland available for each household is reduced (Tan, 2002).

Degradation of habitat and loss of related biodiversity are already leading to irreversible situations. They are re-

Table 2-6. **Red List category summary subregion totals of CWANA plants and animals.**

		North Africa	Nile Valley and Red Sea	Arabian Peninsula	Southwest Asia	Central Asia and Caucasus	Total
Plants	Extinct and extinct in the wild	0	6	0	0	0	6
	Threatened species	6	197	7	7	5	221
	Lower risk	27	222	9	97	43	398
	Total	33	425	16	104	48	625
Animals	Extinct and extinct in the wild	7	4	3	4	0	18
	Threatened species	267	233	197	512	201	1410
	Lower risk	2,831	3,833	2,067	4,673	2,176	15,580
	Total	3,105	4,070	2,267	5,189	2,377	17,008

Source: IUCN, 2008.

sponsible for migration of local communities, desertification and increasing mass poverty. It is difficult to separate those social factors causing habitat degradation from economic ones, since they are interrelated and have similar consequences. Important ones for Turkey are its location, agricultural activities, overexploitation of natural resources, population growth, large populations living close to natural resources, unregulated and overgrazing of pastures and high meadows, forest and stubble fires and incomplete cadastral works for determining ownership of land (Kaya, 2003). Overgrazing and extensive woodcutting, in addition to intensive agricultural practices, have caused a major threat to wildlife in Jordan by destroying natural habitat. Despite the economic importance of mining in Jordan, unplanned mining and quarrying can also destroy habitat (http://www.biodiv.org/reports/).

2.1.7.4 Market prospects and consumer preferences

The change in consumer and market demand and the loss of interest in some by-products of local cultivars is contributory to agrobiodiversity loss. When farmers become integrated into the market economy, they change from landraces or local cultivars to crops and fruits with higher production. The market demand for uniform varieties suitable for industrial processing is another cause for decrease in farming local varieties. However, landraces are often better suited for organic farming. Therefore, there is an increased market for some landraces (Tan, 2002).

The market for medicinal, aromatic and ornamental, species and traditional edible wild plants for food is high, with attractive prospects for the national market and for export. This creates options for additional income for the low-income rural population. In Turkey, there is a long tradition of eating edible wild plants. A recent study on wild medicinal plants of Turkey identified 346 taxa of commercially traded wild native plants. For households, many medicinal, aromatic and ornamental species are underpriced and overexploited. Villagers generally sell products without processing; the added value is captured elsewhere. Where resources are undervalued, prices or policy corrections could have an immediate beneficial effect (Ozhatay et al., 1997). To avoid overexploitation, some countries have legislation especially related to CITES (Convention on International Trade in Endangered Species of Wild Fauna and Flora) or a related project. For example, Turkey has its Regulation on the Collection, Production and Export of Wild Flower Bulbs (http://rega.basbakanlik.gov.tr).

Enhancing the multiple uses of underutilized species: case study 1. Various projects in the region are ongoing. For example, a pilot project in Syria aims to assess socioeconomic aspects related to producing and marketing selected neglected and underutilized species and shed light on challenges and opportunities in the economic exploitation of these species. The study looks at the market channel and product development of neglected species to identify the causes limiting their full deployment, along with the needs for their sustainable use. This investigation concentrates attention on actions to promote production, processing and marketing. It focuses on the needs of rural communities, where these species can become valuable in enhancing income generation (www.bioversityinternational.org).

Azraq Oasis, wetland reserve, Jordan: case study 2. The Azraq Oasis is located in Jordan's eastern desert near the border with Iraq. It once supported a rich biodiversity and was a stopover for hundreds of thousands of migrating birds. But increasing demand for water in the greater Amman area led to large-scale pumping of the Azraq Basin in recent decades. By 1993, after more than 20 years of intensive extraction, the underground springs giving life to the oasis had dried completely. Today, the Azraq ("blue" in Arabic) wetlands are experiencing a remarkable recovery thanks to a multifaceted project cofinanced and managed by UNDP, in its capacity as an implementing agency for the Global Environment Facility (GEF). Other partners include various Jordanian government agencies and the Royal Society for the Conservation of Nature / Jordan (www.gefweb.org/Outreach/outreach-PUblications/Project_factsheet/Jordan-cons-1-bd-undp-eng-ld.pdf).

Constructing dams and replacing dry farming with irrigated farming changes the cropping pattern and decreases the local varieties and weedy forms adapted to dry farming, as was the case for Turkey. But through planned and intensive collection missions the wild relatives and weedy forms of many species are collected and conserved as elsewhere (Tan, 1998). The on-site conservation of wild progenitors of legumes and cereals is also managed (Karagoz, 1998).

2.1.7.5 Change in rangeland composition

Because livestock use most of the primary production in agrobiodiversity systems of arid and semiarid regions, degradation has always been attributed to it (Sidahmed and Yazman, 1994; Squires and Sidahmed, 1997). Grazing helps maintain the composition and the diversity of rangelands. Overgrazing is the main cause for the change in rangeland composition. Countries like Turkey and Syria set up policies for managing rangelands (http://www.ifad.org/photo/region/pn/tr.htm). The absence of grazing livestock has some negative effects on vegetation. Presently, government supported programs maintain the original forest landscape with the help of goats. Goat keepers are paid, per day and per head, for grazing their goats in the forests of some countries.

2.1.7.6 Effects on animal breeds

Large numbers of indigenous livestock breeds are also threatened. Wildlife and livestock often symbiotically coexist. Plant biodiversity may decrease with the absence of grazing livestock. The replacement of the local breeds with exotic breeds is the main reason for breed extinction. Additionally, two main factors lead to the extinction of breeds: the expansion of crops and irrigation into marginal zones and the conversion of former pastures into protected areas. As a result, livestock keepers often lose first their traditional pastures, then their grazing livestock. Absence of market demand and inability to compete with improved breeds in production are other factors. When communities become integrated into the market economy, animal keepers switch to breeds that produce more milk, meat or eggs. If there is no demand for a local breed, related knowledge can vanish within a generation. Habitat loss also affects wild animals. For example, wetland decline has resulted in the eviction of buffaloes and the disappearance of nesting habitats for some migratory birds. Unbalanced water use and unplanned surface and underground water extraction are affecting the habitats and microecosystems of both animals and plants. In Jordan, pollution of surface and underground water and aquifers from agrochemicals, sewage discharge and solid waste disposal has caused an increased threat to the reproduction of many animals.

Conflicts and disasters also affect both bred and wild animals. Wars and natural disasters can cause massive loss of livestock. Aid agencies often try to help by restocking and importing animals from industrialized countries. Possibly there will be some effects from the wars in Afghanistan and Iraq and the earthquake in Pakistan.

2.1.7.7 Maintenance and conservation of agrobiodiversity in CWANA

The main focus of policy in biodiversity has been to protect and conserve endangered species and habitats. Some CWANA countries, such as Turkey, Pakistan, Morocco and Jordan, have introduced legislation for the protection of specific endangered species and habitats. They have also designated certain areas as biosphere reserves, nature parks and other protected sites. Most countries do not have legislation to protect agrobiodiversity. In some countries, the legislation has been prepared but not adopted or has not been enforced (http://www.biodiv.org/reports/). Turkey, for example, has various pieces of legislation to protect biodiversity, related to preserving agrobiodiversity and maintaining plant genetics. The most important direct legislation is the Regulation on the Collection, Conservation and Use of Plant Genetic Resources (http://rega.basbakanlik.gov.tr).

Given the importance of the region in both the richness and the uniqueness of its agrobiodiversity, in the 1950s and early 1960s FAO took the initiative to promote collecting and conserving genetic resources worldwide. The 1961 Technical Meeting on Plant Exploration and Introduction was the first multilateral initiative to recommend establishing exploration centers in the regions with the greatest genetic diversity. A pilot exploration center was established in 1964 at Izmir, Turkey, with an agreement between FAO and the Turkish government within a joint project of UNDP. The Crop Research and Introduction Centre with the inclusion of agricultural research, the Agricultural Research and Introduction Centre (ARIC) was the first regional center of Southwest Asia for collecting and conserving Southwest Asian plant genetics (FAO/UNDP, 1970; Kjellqvist, 1975, Frankel, 1985; Tan and Inal, 2003). This initiative was a good opportunity to preserve the unique agrobiodiversity in the first regional gene bank. ARIC (now known as the Aegean Agricultural Research Institute, AARI) did not remain a regional center but successfully began to work nationally in the mid-1970s and is still the Coordination Centre of National Plant Genetic Resources/Plant Diversity National Program of Turkey (Tan, 2000, 2001). In the other countries of this region plant genetic activities are not yet fully organized into a national program and strategy. Plant genetic conservation is mainly done through breeding and selection programs in research institutes and universities and also in the departments of forestry and livestock within the ministries of agriculture. Because of the lack of national policy and special budgets, plant resources are not receiving enough support. Moreover, there is often no coordination among different national institutions (www.bioversityinternational.org). In some countries, in addition to the national programs, many organizations are involved in conserving plant and animal resources. NGOs are active mainly in biodiversity conservation and do not take a key role in plant genetic resources in most countries.

Some CWANA countries have begun to develop national biodiversity plans, which usually incorporate agriculture in biodiversity conservation. These strategy plans set out the policy objectives and targets for managing and sustaining biodiversity. Most countries are addressing the threats to species loss and the need to address this issue through integrated local and distant conservation. They are highlighted in their National Biodiversity Strategy and Action Plans and National Reports prepared under the terms of Article 6 of the CBD (www.biodiv.org/reports). Almost all countries prepared reports for the preparation of the State of the World's Plant Genetic Resources for Food and Agriculture (FAO, 1996a) as an element of FAO. Turkey, as a member of the Organisation for Economic Cooperation and Development (OECD), has activity reports for both animal and plant diversity indicators, which are policy biodiversity indicators for measuring the performance of national policies and helping monitor progress in fulfilling international obligations (Tan, 2001).

Agricultural research and development institutions of the Arab League—ACSAD (Arab Center for Agricultural Research in the Dry Lands and Arid Zones) and AOAD (Arab Organization for Agricultural Development), COMSTECH-OIC (Scientific and Technical Committee of the Organization of Islamic Conference), the CIS (Commonwealth of Independent States among Central Asia and Caucasus countries) and the AU (African Union) are the principal intergovernmental bodies for international and regional collaboration in the region.

However, the way these conservation efforts are organized varies across countries, ranging from involvement of governmental and nongovernmental organizations to amateur collections and commercial companies. Some countries have national gene banks, others have several specialized agricultural research institutes responsible for maintaining agricultural genetic resources, while some countries work together in regional gene bank networks. The crop and regional networks are also related to agrobiodiversity protection. The Central Asia and Trans-Caucasus Network on Plant Genetic Resources (CA-TCN/PGR), established in 1996), The West Asia and North Africa Network on Plant Genetic Resources (WANANET, established in 1992-1998) have been active for the collection, conservation and sustainable use of the unique agrobiodiversity of the region. The other regional network, for rangeland seed information, was established with two subregion nodes: the Mashreq countries (Iraq, Jordan and Syria) and the Maghreb countries (Algeria, Morocco and Tunisia). Turkey is also a member of the European Cooperative Program for Plant Genetic Resources (EC/PGR).

The International Technical Conference on Plant Genetic Resources placed emphasis on the need for coordination between local and distant approaches. The Global Plan of Action agreed upon by 150 governments at the technical conference identifies the promotion of in situ conservation of wild crop relatives and wild plants for food production as one of its 20 priority areas (FAO, 1996b). Those activities are an excellent guide to the national programs of the region.

International agreements and conventions are also important in agriculture and biodiversity. Most notable was the international Convention on Biological Diversity (CBD), which was agreed upon at the UN Conference on Environment and Development at Rio in 1992. The CBD recognized the significance of biodiversity for agriculture. This has led the FAO to request member countries to negotiate, through the FAO Commission on Genetic Resources for Food and Agriculture, the revision of the international undertaking on plant genetic resources in agriculture. The International Treaty for Plant Genetic Resources for Food and Agriculture came into force on 29 June 2004; a few countries of the region participated, in harmony with the CBD. In addition, within the overall context of the CBD, the Biosafety Protocol was agreed upon in over 130 countries. This was the first major international agreement to control trade in genetically modified organisms (GMOs), covering food, animal feed and seeds. Other related international conventions include, for example, the Convention on International Trade in Endangered Species of Wild Fauna and Flora (CITES, 1973, www.cites.org), the Convention on Wetlands (Ramsar Convention, 1971, www.ramsar.org). Some countries of the region are signatory to those of the agreements and completed the ratification. The awareness of the countries of the region to the participation of those agreements is not enough. The Convention on the Conservation of European Wildlife and Natural Habitats (Bern Convention, 1971, www.coe.int/T/E//Cultural_Co-operation/Environment) is signed only by Turkey and enforced with various projects. Additionally, some countries are members of the World Trade Organization (WTO, www.wto.org) and bound to its rules, the World Intellectual Property Organization (WIPO, www.wipo.int) and the Patent Cooperation Treaty. The Uruguay Round Agreement of negotiations under the General Agreement on Tariffs and Trade (GATT, www.wto.org) produced the Trade-Related Intellectual Property Rights (TRIPS) Agreement, which commits all members of the WTO to adopt and enforce minimum protection for intellectual property rights (IPR). Article 27.3(b) of the agreement on TRIPS calls for members to develop plant variety protection legislation (www.wto.org).

Bioversity International's (former IPGRI) work in CWANA started as early as 1977 with an FAO project jointly conducted with the Aegean Agricultural Research Institute (AARI) in Izmir, Turkey. The Bioversity International regional office for the WANA region was located in Aleppo, Syria, in 1993, at the International Center for Research in the Dry Areas (ICARDA). In 1997, the office broadened its mandate to include Central Asia and changed its name accordingly into CWANA. A subregional office was opened in Tashkent, Uzbekistan, in 1998 at the ICARDA Facilitation Unit for Central Asia. The ultimate goal of IPGRI CWANA is to strengthen the national and regional capacities to achieve the effective conservation and sustainable use of plant genetic resources. Five other centers of the Consultative Group on International Agricultural Research (CGIAR) (CIAT, CIMMYT, CIP, ICRISAT and IRRI) have subregional offices and carry out activities on plant genetic resources on mandated crops and agrobiodiversity conservation.

Ex situ conservation has been predominant for conserving plant genetics for food and agriculture in the region. In recent years, the need for integrated conservation strategies for conserving plant genetics coordinating in situ and ex situ approaches has become clear. The first attempt at local conservation was the project implemented by Turkey in a multiple site and multispecies approach. Now there are various projects in the region for both in situ conservation of wild relatives of crops and farm conservation of the traditional crops in agrobiodiversity and some in ecosystems. The documentation of biodiversity has become important. National programs and formal and informal institutions of

CWANA countries have started to create databases on various agrobiodiversity topics. In Turkey, a comprehensive and complementary database management system linked with the geographic information system (GIS) exists for all related activities and for a better understanding of agrobiodiversity (Tan and Tan, 1998a, 1998b).

UNEP and the World Bank have a global program on biodiversity and genetic resources and have mobilized funds across CWANA, through the Global Environment Facility's support of biodiversity country strategies along with major in situ conservation projects, such as In Situ Conservation of Genetic Diversity in Turkey (Tan and Tan, 2002) and Design, Testing and Evaluation of Best Practices for In Situ Conservation of Economically Important Wild Species, with Demonstration in Egypt, Lebanon, Morocco, Turkey (UNEP, 2003). The UNDP regional office for the Near East approved an important regional collaborative project in 1998, Agrobiodiversity In Situ Conservation, involving Jordan, Lebanon, Palestine and Syria. It was coordinated by ICARDA, with technical backup by IPGRI and ACSAD and through GEF support, a regional date palm project for the Maghreb oases.

In situ conservation of genetic biodiversity of Turkey: case study 1. This five-year project, associated with the Bank's Eastern Anatolia Watershed Rehabilitation Project, started in 1993 and worked to address Turkey's natural resource degradation through in situ and ex situ conservation as supported by the World Bank GEF (Zencirci et al., 1998; Tan and Tan, 2002).

Project goals: Permit genetic evolution through in situ conservation (Kaya et al., 1998).

- Protect in situ the genetic resources and wild relatives of important crop and forest tree species. The southeastern, southern central Anatolian and Aegean regions have been identified as important biodiversity centers for wild relatives of cultured crops
- Establish the complementary conservation ex situ and in situ wild relatives of selected crops, fruit and forest species
- Establish ex situ conservation of associated species of target species in selected sites
- Establish natural reserves or gene management zones (GMZs) for target species
- Plan management for selected GMZs
- Prepare national plan for in situ conservation
- Create a comprehensive database management system linked with GIS

Strengthening the scientific basis of in situ conservation of agricultural biodiversity on farm: case study 2. In 1995, IGPRI and its nine national partners (Burkina Faso, Ethiopia, Hungary, Mexico, Morocco, Nepal, Peru, Turkey and Vietnam) formed the project Strengthen the Scientific Basis of In Situ Conservation of Agricultural Biodiversity On Farm (Jarvis and Hodghin, 1998; Jarvis, Myer et al., 2000). The nine countries are all within regions of primary diversity for crop genetic resources with world importance. The countries all have national programs organized to conserve crop genetic resources, including ex situ conservation facilities; additionally, all have a strong interest in developing the national capacity to support in situ conservation. The project serves to strengthen the relationships between formal institutions and farmers and local institutions. Project objectives are to develop global and national management frameworks for implementing in situ conservation, collecting and analyzing information to define the genetic diversity in farmers' fields and to develop the criteria for the successful maintenance of existing diversity. It intends to broaden and conserve agricultural biodiversity by farming communities and other groups.The IPGRI CWANA group is closely involved in this project and has two participating countries, Morocco (Birouk et al., 1998) and Turkey (Tan, 2002).

This project has created comprehensive database management systems for each national project. It provided a model to other countries for future projects. The Training Guide for In Situ Conservation On-Farm was produced (Jarvis, Sthapit et al., 2000) and tested through the GEF/UNDP Project on Conservation and Sustainable Use of Dry Land Agrobiodiversity in Jordan, Lebanon, the Palestinian Authority and Syria.

Conservation and sustainable use of dryland agrobiodiversity: case study 3. The five-year project was launched in 1999 to promote in situ conservation and sustainable dryland agrobiodiversity in Jordan, Lebanon, the Palestinian Authority and Syria, with financial support from the GEF/UNDP (www.icarda.org/Gef/P1.html). In addition to the country institutions, ICARDA, IPGRI and ACSAD are involved in the project.

Genetic erosion of some animal breeds has been occurring at an unabated rate from lack of incentives for conservation and population pressure, ecological changes, natural catastrophes and adverse economics. This depletion is an immense threat to the livelihood of the local pastoral communities. Conservation of these animal genetic resources by governments and other stakeholders would ensure the well-being of pastoralists and prevent genetic losses. Conservation of animal genetics is essential to enable farmers to adapt to different environments and consumer demand and to fully realize the investment made over generations in developing these breeds. Also, conservation of wild species will provide opportunity to develop the livestock sector. FAO has led efforts to sustainably use, develop and conserve animal genetic resources and since 1993 has engaged in preparing the Global Strategy for the Management of Farm Animal Genetic Resources. The global strategy is to serve as a strategic framework for international efforts in animal genetic resources.

The First Report on the State of the World's Animal Genetic Resources is an essential element of the global strategy. Countries were invited by FAO in March 2001 to produce and submit country reports (www.fao.org, www.dad.fao.org). The first report will provide a comprehensive review of current global livestock diversity and direction for better management of it. Country reports and the First Report of the State of the World's Animal Genetic Resources will provide a strategic planning framework for animal genetic resources, supporting the development and implementation of national, regional and global policies and programs. It will highlight opportunities, challenges, biological characteristics, institutional infrastructure and operational considerations influencing management of plant and animal genetic resources. It will also include main threats to livestock genetic resources and outline areas of greatest opportunity to better manage these resources. Some countries in the region have already prepared their country reports and set up projects to conserve farm animal genetics.

2.1.7.8 Exploitation and use of agrobiodiversity

Exploitation and use of genetic resources postulates knowledge and evaluates the characters expressed by the genome and identifies desirable characteristics for breeding. As indicated in the Global Plan of Agriculture for the Conservation and the Sustainable Utilization of Plant Genetic Resources for Food and Agriculture, "the broadening of the genetic base of crops will contribute to increasing crop stability and performance" (FAO, 1996b). In most crop species, populations have been reduced dramatically as a result of breeding and selection. For years, plant breeders have limited their programs to a small part of the diversity in the region. However, wild relatives and ancestors, old varieties, landraces and weedy forms were collected from different institutions and maintained in the national, regional and CGIAR gene banks. Those collections are a valuable but relatively unexploited source of genetic variability. To broaden the genetic base and enhance the ability to respond to abiotic and biotic stress, a systematic evaluation is needed of genetic diversity in the region. In well-organized CWANA countries national programs collaborate in breeding, evaluating and using genetic diversity (FAO, 1996a). In Turkey, many varieties are released from national collections (Tan and Inal, 2003). Collaboration with CGIAR has led countries to exploit and use their genetic diversity to adapt even to extreme conditions.

Participatory variety selection and plant breeding are effective at identifying new material for farmer conditions, preferences and needs. Participatory methods allow farmers to select new materials, enhancing diversity where traditional cultivars have been lost. Participation of farmers in the initial stages of breeding, when the genetic variability is untapped, will fully exploit the potential gains by adding farmers' perception of their needs and knowledge of the crop. Farmer participation has been established in some countries in the region by several agencies and ICARDA. Opportunities for interaction and cooperation between formal breeding station work and farmer expertise need to be fully explored. Research is also needed on transferring appropriate technology among farming systems to manage great diversity. Research support is needed for traditional seed production, emphasizing farmers and natural selection pressures, insect pests, diseases, storage conditions and soil fertility. Participatory plant breeding allows the farmers' diversity to be maintained on farms. This approach is a main component of in situ conservation programs.

Decentralized participatory barley breeding, ICARDA: case study. At ICARDA, the gradual change from centralized and nonparticipatory to decentralized participatory barley breeding was implemented in Syria between 1997 and 2003 in three steps. The model and concepts developed during this period were gradually applied in Egypt, Eritrea, Tunisia, Jordan, Morocco and Yemen.

The first was an exploratory step with the main objectives of building human relationships, understanding farmers' preferences, measuring farmers' selection efficiency, developing scoring methods and enhancing farmers' skills. The exploratory work included selecting farmers and sites and establishing one common experiment for all participants (Ceccarelli et al., 2000, 2003). The second step was primarily about methods and implementing the breeding plan, choosing and testing designs and analysis, refining farmer methods and planning village seed production.

A common finding of participatory breeding programs is that different farmers in different communities select different varieties. Data collected on barley suggested that farmer selection may narrow biodiversity in the breeding material. However, because different farmers select different material, the biodiversity is maintained or increased (Ceccarelli and Grando, 2002). Through participatory breeding several farmers became aware of the value of landraces and were interested in conserving them.

2.2 Policies, Institutions and Regulations

2.2.1 Development strategies and agricultural policies

In CWANA, as in other developing countries, agricultural development strategies have had successive shifts since the late 1950s. In the early post-independence era, the late 1950s to the early 1970s, development was strongly influenced by the "import substitution model," which was dominant and aimed to promote rapid industrialization. This meant heavy taxation of the agricultural sector, including taxes on commodity exports, overvalued currency exchange rates and high import tariffs.

Many governments attempted to correct the bias against agriculture by intervening in agricultural markets through price measures, setting up compulsory state monopolies, providing basic services, credit, essential inputs, technical and market information and by marketing and distributing (FAO, 2002). In the early 1980s, growing account deficits, external debt problems and foreign exchange crises imposed a shift in development strategies. Most developing countries implemented structural adjustment policies.

From the 1980s to the 1990s, structural adjustment involved import tariff reduction, market deregulation, privatization, fiscal stabilization through currency realignments and significant budget cuts. For agriculture, the primary objective was to make it more market oriented. Budget cuts were often made in subsidized credit, inputs, extension systems and in investment in research and infrastructure. Agricultural reforms often reduced or eliminated the state in trading, eliminated domestic price controls and gradually removed state procurement programs (FAO, 2002).

These policies had mixed results on the agricultural sector. While allowing for an increase of agricultural exports and the intensification of farm production in some countries of the region (Egypt, Morocco, Tunisia, Turkey), they disrupted markets and often resulted in deteriorating food security and increased poverty in many other countries.

While still focusing on improving the competitiveness through farm productivity growth, integrating more farmers in supply and developing standards and labels, current development policies increasingly integrate poverty reduction. Development policies that take into account the diversity of farmers are increasingly recognized. Still, the farming sector in CWANA is facing major national and international constraints, in particular unfavorable markets, persistent subsidies in the North, a growing technical divide between the North and the South and increasing environmental risks.

Agricultural development strategies in CWANA (Table 2-7) are faced with major challenges. How can agriculture ensure economic development objectives of food self-sufficiency and better position on the international market while reducing rural poverty and protecting the environment? What should be the connection between agricultural policies and rural development policies, especially where market orientation and intensifying farm production reduce the agricultural labor force? How can a multifunctional approach to agriculture take into account the social and environmental functions of agriculture and diversify economic activities in rural areas?

2.2.2 Land tenure including agrarian reform

Existing land-tenure systems in the region descend from Islamic and customary law, colonial legacy and national land policies. In the post-independence era, land reform extended state control over land through nationalization and granted rural communities land through privatization and reform of customary property rights. In most countries, institutional reforms were promoted to enhance the performance of farm households and communities.

Land policies in Morocco and Tunisia give priority to privatization to promote rural development by granting private rights to both farmers and tribal communities. The partial privatization approach was mainly applied in Jordan and Lebanon; the state retained landownership, the beneficiaries were granted use rights. To achieve better distribution of land and promote agricultural development, agrarian reforms were implemented in Algeria, Iraq, Libya and Syria. Many small farmers and herders received lands and were organized into cooperatives to facilitate access to credit and inputs.

In central Asia, under Soviet agricultural production, 95% of agricultural lands were controlled by large-scale collective and state farms. After independence in 1991, agrarian reforms favored the emergence of smallholder farming, and a large part of the agricultural output is from household plots. Agrarian policies have tended to focus on "decollectivization" of large state enterprises and privatization of holdings. Land reforms have not been adequately accompanied by other reforms, resulting in poor management of land and water.

In countries in the Middle East and North Africa, structural adjustment policies and liberalization movements since the late 1980s affected land-tenure systems. State withdrawal from direct involvement in agricultural production has been important in most countries. Morocco promoted the privatization of state, collective and religious-endowed land. Algeria sold "inefficient" state farms and reformed tenure arrangements in favor of family farms (Bush and Abdel Aal, 2004). Tunisia has achieved privatization of state and cooperative farms. Still, land-tenure systems and farmland distribution are major constraints for agricultural production and do not provide enough incentives to farmers to invest in enhancing productivity. In several countries, many farmers lack property titles and have limited access to bank loans. Demographic growth and no appropriate regulations have often fragmented land and contributed to a decrease in small farm viability. On the other hand, privatization and liberalization have favored larger plots and concentrated production, mainly to the detriment of family farming. Most countries are experiencing a consolidation of large farms, well integrated into national and international marketing and smallholders limited to household survival.

2.2.3 Trade policy, international and regional agreements and the WTO

2.2.3.1 Trade arrangements in the region

Although global trade liberalization is the goal of multilateral trade negotiations under the WTO, all WTO members have entered into regional or bilateral agreements. This subtle shift from WTO objectives is mainly from WTO failure to achieve consensus about trade agreements and the ease of forming regional blocs. The CWANA region is

Table 2-7. **Agricultural development strategies.**

	1950s-1970s	1980s-1990s	2000-
Dominant strategy	Import substitution Industrial development	Structural adjustment, liberalization of markets (Washington consensus)	Market orientation, governance, pro-poor growth (post–Washington consensus)
Agricultural policies	Urban bias Taxation of agriculture Subsidized inputs	Getting prices right, privatization abolishment of marketing boards, export promotion, reduced investments in extension, credit systems	Policies adapted to the diversity of situations, growth in productivity, integration in supply chains, standards and labels
Constraints	Loss of markets, expensive and inefficient government apparatus, budget deficits, debt crisis	Important functions of states not provided, private sector not ready, commodity markets depressed	Unfavorable market for products, persistence subsidies in the North, technological divide, environmental issues

Source: Giger, 2006.

not an exception. It has seen many regional and bilateral trade agreements emerge among neighboring countries. For instance, Egypt concluded 30 to 40 agreements (ESCWA, 1998b). Turkey entered into a customs union with the EU in 1996 and the Maghreb Union was established in February 1989. With the establishment of the WTO in 1994, it was expected these countries and regional blocs would review their trade policies to make them compatible with multilateral trading principles. According to article 24 of the General Agreement on Tariffs and Trade (GATT), regional blocs should facilitate trade among the members without restricting trade with other WTO members. The main concerns of these countries with the WTO are the compatibility of common tariff rates to which they are committed by joining WTO; specific binding restrictions in market access; domestic supports and exports subsidies; a common market with GATT provisions for regional blocs to foster their position in multilateral trade negotiations; and tariff structures under the most-favored-nation status in previous bilateral or regional trade agreements (Zaibet et al., 2003).

Since the 1990s, many countries of the Mediterranean region (namely North African and Middle Eastern countries) have signed partnership agreements with the European Union (EU). These countries are liberalizing their economies under the euro partnership conditions, a process which is strongly influenced by the EU's common agricultural policy (CAP). Since negotiations have started, trade policies have been revised, local and regional structural programs have been undertaken and a lot of changes have taken place at the international level. As a result of the above transformations the CWANA countries have expressed concern with regard to open market policies and access to industrialized country markets.

2.2.3.2 Trade negotiations and expected benefits

By joining WTO, developing countries in particular have sought more access to industrialized country markets and to gain advantages in international markets. The Doha Declaration has set milestones on a number of trade and nontrade issues known as the Doha Development Agenda. However, given the achievements of past negotiations, observers remain skeptical that a new comprehensive round can be completed in the coming years (Miner, 2001). The big players are expected to make additional policy reforms (e.g., trade legislation in the USA and CAP reforms in the EU) before undertaking strong concessions and commitments in the upcoming negotiations.

Benefits from agricultural trade liberalization have not materialized for two reasons. First, negotiations on agriculture alone do not consider comparative advantages. As a result, the Doha Declaration provided for broad trade negotiations to further trade liberalization for industrial products and services in which nations may take advantage (Merlinda, 2002). Second, national policies and legislations are creating additional transaction costs and limiting liberalization. Gerber (2000) pointed out that trade relations remain far denser within nations than between nations and indicate significant transaction costs across national boundaries.

Besides recurrent issues, new concerns are presumed to be on the table during coming negotiations (Zaibet et al., 2003). The main issues already identified in GATT on agriculture involved market access, export competition and domestic support. However, new trade and nontrade concerns are emerging (Tables 2-8 and 2-9). The agreement on agriculture included food security issues, food safety and quality, environment concerns, resource conservation and rural development (Miner, 2001). Additional issues raised in the last meetings included animal welfare, biotechnology, species preservation, safeguarding the landscape, poverty reduction and preservation of rural culture (Miner, 2001).

Newer border and trade topics included rules of origin, standards and technical barriers, intellectual property rights, SPS standards, dispute settlement and the role of small countries (Gerber, 2000). Among the nontrade, domestic policy issues are foreign investment, competition policies and labor and environmental standards.

Export markets for many developing countries are in a few countries in the North because of proximity and historic links (Diao et al., 2002). As a result, trade negotiations will be shaped by regional blocs. North African and Middle Eastern countries are increasingly interested in the EU agricultural markets and the EU agricultural reforms under the CAP agenda 2002 (Table 2-10).

The work program annexed to the Barcelona Declaration has the following objectives for the countries that signed the declaration:

- Integrate rural development
- Support policies implemented by the Mediterranean countries to diversify production
- Reduce food dependency
- Promote environmentally friendly agriculture

2.2.3.3 Challenges and relevance to AKST

Effects of the European Union enlargement. Enlargement of the EU to the Central and Eastern European countries is integral to the EU Agenda 2000. The process started following the decisions of the European Council in 1993 in Copenhagen and 1994 in Essen to be achieved by 2004. The enlargement of the EU could open new frontiers to more exports. It may, however, divert foreign investment to the eastern EU countries and prevent CWANA access to new technology.

Food safety and environmental quality standards. With the decline in traditional trade barriers, such as tariffs and quotas, there is evidence that technical and regulatory barriers are increasingly used instead (Wilson, 2001). In developed countries, many firms are moving toward adopting environmental standards. This move is relatively slow in CWANA countries and might be an obstacle to international trade.

Environmentally friendly agricultural practices. Current trends to protect the environment are illustrated by the EU provision of direct payment to farmers complying with environmental regulations and support of agricultural methods that protect the environment. These trends could spread low tillage or no tillage techniques along the region and could replace current practices.

2.2.4 Professional and community organizations

2.2.4.1 Turkey—case study

Although agricultural education and research were started

Table 2-8. **Traditional versus newer issues in trade and nontrade negotiations.**

Traditional issues	New trade issues	New non-trade issues
Access issues • Tariffs • Tariff rate quota • State trading enterprises • Safeguards and special treatment Export competition • Export subsidies • Export credits and assistance • Export state trading enterprises Domestic support • Aggregate • Measurement of Support AMS (direct and indirect) • Green box measures • Amber distorting measures • Blue box	• Dumping • Food safety and quality • Guaranteed Minimum Prices GMPs • Geographic indications and labeling • Sanitary, phytosanitary and related financial and technical assistance • Impact of two-price system and price pooling	• Food security • Biotechnology • Species preservation • Resource conservation • Animal welfare • Safeguarding landscape • Biodiversity • Poverty reduction • Preservation of rural culture • Environmental concerns • Rural development

Source: Zaibet et al., 2003.

Table 2-9. **Shallow versus deep integration measures.**

Trade issues (shallow integration)	Nontrade domestic policies (deep integration)	Relevance of nontrade concerns to the region
• Market access • Customs procedures • Rules of origin • Standards and technical barriers • Intellectual property rights • Subsidies and antidumping • Sanitary and phytosanitary • Services trade • Investment measures • Government procurement • Dispute settlement • Role of small economies	• Foreign investment • Competition policies • Limitation of horizontal restraints: price-fixing agreements • Formation of domestic cartels • Prohibition of vertical constraints • Exclusive agreements between producers and distributors • Labor and environmental standards • Harmonization of standards vs. mutual recognition • Institutional changes (long-term objective)	• Raise national investment levels • Increase the rate of technology transfer • Increase economic growth • Viewed as a form of protectionism • Need financial and technical aid

Source: Zaibet et al., 2003.

Table 2-10. **Relevance to the European Union and to the region.**

EU Agenda 2000	Relevance to the region
Trade and competitiveness • Reduction in market support prices for cereals, milk and beef • Direct aid payments to offset lower prices	Higher import prices New access concessions are sought
Rural development • Multifunctional nature of agriculture • Regeneration of rural areas and promotion of diversification	Included in the Barcelona Declaration
Protecting and enhancing the environment • Support agricultural methods which protect the environment • Direct payment conditional on compliance with environment targets	Included in the Barcelona Declaration
Food safety and consumers' protection • Quality assurance for hygiene, the environment and animal welfare • Improving traceability • Labeling • Organic farming	Impact on agricultural exports Needs technical and financial aid
Enlargement of the Union	Potential gains? Or losses?

Source: Zaibet et al., 2003.

80 years ago, agricultural organizations were set up in the last 50 years. Farmer chambers were first established in 1881; they became active mainly after the 1960s. Similarly, agricultural credit cooperatives were set up in the 1930s, but only started to act effectively in the 1950s. Extension services function under the Ministry of Agriculture and Rural Affairs in each province and county. Additionally, some agricultural cooperatives focus on one crop or crop group. Agricultural producer organizations in Turkey can be classified into cooperatives, producer unions and agricultural chambers (Table 2-11).

Turkey has over 700 agricultural chambers, with about four million producer members. They mostly provide vocational services and represent farmers. These organizations specialize in certain products or product groups and in provinces or districts. As the legal framework for these organizations is recent, the number of unions and members is rather low but shows a strong increase.

The Milk Producer Central Union has seven milk producer associations. Agricultural credit cooperatives are organized with a central association and 16 subassociations with about 1.5 million members. Agricultural cooperatives are composed of agricultural development cooperatives, irrigation cooperatives, fisheries cooperatives and sugar beet cooperatives. The agricultural development cooperatives promote activities related to producing and marketing crops, livestock and husbandry. These organizations are often multipurpose and usually not specialized.

Agricultural sales cooperatives and associations are generally specialized in crop products, processing and sales. The agricultural sales cooperatives purchase the products of their members. The unions take all the necessary measures for these products to be utilized in the best circumstances. They handle storage, standardization, first processing, transporting, packaging, export and domestic sales of finished and semi-finished products, provide all the inputs for agricultural production and support shareholders with credits and insurance for the producers.

In addition to these organizations, there are some small local farmer unions and cooperatives. The government has recently started to encourage farmers to become organized by setting up new regulations. Some professional associations have also been established.

2.2.5 Agricultural risk management policies, including drought risk

Agriculture is regarded as one of the most risky activities because of price, inelastic demand, short-run supply and exposure to natural shocks. In CWANA, agricultural risks also have extremely variable climate and recent economic changes that came with liberalization, which profoundly affected farm operations. Worldwide, economic changes and degradation of natural resources, diminishing water resources, pollution and climate change have prompted additional attention to risk. Interest in strategies and tools for managing market risk has increased in recent years.

2.2.5.1 Main risks affecting agricultural activity in CWANA

The diverse risks affecting agricultural activity in CWANA are (1) production risks, related to weather, including drought and pests and diseases, (2) ecological risks from managing natural resources, such as water, (3) market risks, mainly from variable output and input prices and from particular markets, such as quality and safety requirements for exports and (4) institutional risks linked to state intervention.

Although it has great diversity in climate and natural environment, CWANA in general has low, highly variable annual rainfall and a high degree of aridity. In the largest part of CWANA, especially in North Africa and the Near East, drought is recurrent, resulting from physical determinants and social factors. Increased cultivation of marginal and fragile arid lands, soil erosion and runoff exacerbate the region's vulnerability to drought and often lead to irreversible desertification.

In recent years, most countries of the region have had severe drought and consequently, growing water shortages. In North Africa, the Near East, Middle East, Afghanistan and Pakistan recent reports of the Intergovernmental Panel

Table 2-11. **Agricultural producer organizations by main types in 2006.**

Producer organizations	Number	Number of members	Number of subassociations	Number of central associations
Agricultural Development Cooperative	6,796	743,547	78	4
Irrigation Cooperative	2,349	276,246	11	1
Fisheries Cooperative	481	24,681	12	1
Sugar beet producers cooperative	31	1,587,324	1	1
Agricultural credit cooperative	1,948	1,500,000	16	1
Agricultural sales cooperative	350	671,928	17	—
Producers unions	133	8,566	—	—

Source: Ministry of Agriculture and Rural Affairs; www.tarim.gov.tr.

on Climate Change (IPCC) confirm some global warming and indicate that water scarcity, which was already a major constraint, may worsen substantially. Because of continued drought between 1999 and 2001, Algeria, Morocco, Tunisia, Turkey, Jordan, Iran and the Gulf countries saw an important decline in their agricultural output, especially in cereals and livestock. Drought adversely affected the livelihoods of much of the rural population, especially dryland farmers and nomadic livestock owners, particularly in Iran. The incidence of poverty in the region went up significantly toward the end of the decade because of drought with a proportion of the population living on less than $2 per day increasing from 25 to 30% (FAO, 2002).

The unpredictable and variable climate prevailing in the region and the different farming practices aggravate the risk of disease and pest epidemics across CWANA. Pests, including sunn pest, Hessian fly and cyst and root lesion nematodes, significantly damage cereal production. For this reason, the development of disease- and pest-resistant wheat varieties has been a key component of breeding programs to improve food security across CWANA. Agricultural production in CWANA is increasingly threatened by exotic pests, such as the peach fruit fly (*Bactrocera zonata*), red palm weevil (*Rhynchophorus ferrugineus*) and Bayoud disease of palm (*Fusarium oxysporum* fsp. *albedinis*), among others, indicating a lack of adequate phytosanitary controls.

Animal diseases are a major threat to livestock production in CWANA countries. In addition to screwworm infestation, at least three animal diseases have major economic impacts, especially in North Africa, the Middle East and the Arab Peninsula: foot-and-mouth disease, rinderpest in the Middle East, including Egypt and Sudan; and brucellosis, endemic in the whole region. These diseases seriously affect the potential in the region for livestock production. Their elimination would require well-focused pest and disease control, still lacking in most countries.

In CWANA, agricultural risks are related to the management of natural resources, such as water. The region's irrigation systems are under considerable environmental strain, with almost all countries experiencing problems with salinity and waterlogging. A major concern is the overexploitation of groundwater, particularly in the Persian Gulf region (FAO, 2002). The current water crisis calls into question the sustainability of most irrigation systems.

Drought and water scarcity place substantial strains on the environment, causing significant damage to biological diversity. As the FAO report points out, "wildlife has been severely affected as a result of the shortage of drinking water, lack of feed, dried wetlands and degradation of wildlife habitats. . . in the Hamoun wetlands of Iran, which are of international importance, aquatic life has disappeared. Herbivores are among the first animal species to be affected by a lack of feed. Drying of wetlands and natural lakes has also occurred in Morocco, as well as other countries of the region, causing similar and probably irreversible environmental damage. In Jordan, the continued drought during 1999 and 2000 caused visible damage to the natural and artificial forest" (FAO, 2002).

Risks affecting agriculture in CWANA increasingly result from rapid changes in input and output markets. Many farmers still practice subsistence agriculture. When a farm household's production is barely sufficient for its own consumption, market risks are clearly not important. But the increasing integration of farm producers in national and international markets place them at risk if there is price instability. Market risks faced by farm producers in CWANA are related to poorly organized national marketing circuits, significant increases in input prices and production costs, state intervention in pricing basic food products, difficult access to export markets from growing competition for fruit and vegetables and increasingly severe safety and quality requirements.

2.2.5.2 Risk management to reduce farm household income variation

The extremely variable climate prevailing in the region and economic liberalization affecting agricultural policies make farm producers particularly vulnerable. In this highly risky environment, farm households have developed strategies to mitigate risks and reduce income variation. Two strategies are ex ante risk management and ex post coping (Dercon, 2000). Ex ante involves crop management, technological choice, diversification of income by spreading risk among activities and market strategies. Risk coping deals with the consequences (ex post) and involves self-insurance through precautionary savings; informal insurance, such as kin sharing risk; and informal credit. Coping strategies may involve attempting to earn extra income to compensate for losses, selling livestock and making use of government aid.

In CWANA, drought is by far the greatest risk. Recent droughts have seriously affected dryland farmers and herders, resulting in severe loss of income through loss of harvests and partial loss of flocks. Sales of animals and off-farm activities are among the most common strategies adopted to cope with drought. Loss of harvests pushes farmers to rely on purchased animal feed to avoid further livestock losses. Harvest losses make it necessary for farms to rely more on short-term bank credit or informal credit to meet farm costs the following year, resulting in increased indebtedness.

Fall in income caused by drought leads small-scale farmers to give out their land to sharecroppers, a way to cope with little money and secure part of the expected farm production. Small-scale farmers also rely on state aid programs, such as seed and animal feed distribution, to reduce the hardship caused by drought. In hardship, family networks are used by farm households. They allow transfer of money from family members working in urban areas or abroad. These coping strategies, however, are not available to all farm households. The same holds for off-farm work, which has decreased in most countries of the region because of economic restructuring.

Among ex ante strategies in CWANA, farmers are diversifying farm production to reduce climate and economic shocks. Most farmers in the Maghreb countries combine livestock, mainly small ruminants, with cereal crops and olive trees. In this system, livestock allows better management of the farm treasury through the sale of animals to finance farm inputs and household expenses. In addition to combining livestock and crops to minimize risks, farmers in semiarid areas usually diversify animal production. Mixed species herds represent a way to spread risk, make better use of resources and reduce farm expenses by integrating low-cost production.

Diversifying income through wage labor and small trade is also a major risk-management strategy farmers of the region use. Small-scale and medium-scale farmers in the Maghreb and the Middle East have a high rate of engaging in a number of activities; almost 45% of Tunisian farmers have an off-farm activity. In the Maghreb countries, off-farm activities have been important for funding and developing agriculture. Now they have become rather "scarce, in the new national and international context due to emigration controls, decrease of national demand in nonskilled labour and high unemployment . . ." (Alary, 2005).

Ex ante risk management used by some groups of farmers in CWANA include crop management and improved farming techniques. These techniques include using drought-resistant crop varieties, fertilization and pest management to increase yields or minimize production failure. However, improved varieties can be more vulnerable to moisture stress and pests. They do better in assured rainfed or irrigated agriculture. Using new technology can generate environmental risk, such as pollution, and carries some long-term risks in soil depletion and genetic uniformity (Ramaswami et al., 2003). These are increasingly affecting farming in intensive production areas of the region, but farmer awareness and management strategies are still lacking.

In several areas of the region, minimizing farm risks includes developing irrigated farming. It stabilizes yields and allows for more intensive and more profitable production. This has led to development of surface and underground irrigation, which in several CWANA countries has led to overexploitation of water resources and increased soil degradation.

The shift from dryland to irrigated farming can generate new risks for farmers. They include environmental problems, the necessity to rely on the credit system, new farming techniques, integrating into collective water management, insufficiently organized marketing circuits and price instability.

Markets for horticulture products are liberalized in most CWANA countries, but they remain poorly organized. Farmers, especially small-scale and medium-scale ones, are usually vulnerable to market risks. These risks can be aggravated by state intervention geared to maintaining low food prices through importing fruits and vegetables. In several irrigated areas of Morocco and Tunisia many farmers have ceased irrigating because of the difficulty in selling their products profitably (Gana and El Amrani, 2006) and returned to dryland farming. Some have shifted the water from horticulture to cereals and forage.

Cereal crops benefit from more stable producer prices, as they are usually state controlled. The high variability of prices for fruits and vegetables, which is also due to the weakness of farmer organizations, is a major hindrance to developing high-value crops in the region. This explains the risk management favored by many farmers is cultivating cereal crops, even if they could diversify or develop other crops. However, a growing proportion of farmers in intensive irrigated farming in Morocco, Tunisia and Egypt have found forward contracting a way to reduce market risks.

Among livestock producers, risk management varies according to the size of the farm. Ex ante strategies can use integrating cropping and livestock by cultivating forage crops. These strategies are mostly available to medium and large farms in favored climates. In dry areas, herders often have to resort to the market for forage supply, where prices are unstable and vulnerable to market shock. Finally, ex ante risk management includes crop insurance, which is mostly available to farmers integrated into the bank credit system.

2.2.5.3 Agricultural risk management policies in CWANA

Despite the strategies farm households put in place to mitigate risks, they remain vulnerable to fluctuations in production, consumption and poverty. Therefore the state should intervene. Governmental intervention can include price supports, credit policy, natural resource management policy, promotion of technical change and development of insurance schemes and safety net programs.

In the CWANA region, recent droughts have pushed most countries to implement measures and policies to limit social and economic damage (see next section for drought management policies in CWANA). Policy in Iran, Jordan and Morocco established a national drought program monitored by an intergovernmental committee (National Drought Task Force), usually headed by the Ministry of Agriculture. This political body proposes a set of emergency measures and funds to ease the adverse effects of the drought and assist the affected rural populations. Emergency measures include emergency purchase and distribution of concentrate feed to livestock owners, seed distribution, veterinary prophylaxis, water development and wells for people and livestock, special access to credit, debt relief or agricultural tax relaxation and creation of job opportunities.

However, while these measures helped to mitigate the loss of animals from drought, they have been financially costly. Where they had untargeted distribution of subsidized livestock feed they primarily benefited the larger flock owners. The FAO report stresses that, "Moreover, they have created dependencies on feed supplements and have encouraged the maintenance of larger numbers of animals on the rangelands for longer periods each year, thus accelerating resource degradation. Consequently, the contribution of the natural grazings to total feed supply has fallen dramatically in nearly all Mashreq and Maghreb countries while concentrate feed use has escalated" (FAO, 2002).

Drought management and mitigation in the region consist mostly of short-term drought relief. Drought early warning systems are virtually nonexistent, and national integrated drought-monitoring programs are not operational. Mostly they have limited coordination of information on water supply from irrigation authorities, agricultural extension services and meteorological departments about the extent and impact of drought (De Pauw, 2001). Yet coordinating this information is essential for drought monitoring systems. Hence, there is an urgent need to establish national plans to manage drought more comprehensively and consistently and move from reacting to drought to managing it. Drought could be treated as an integral component of production and a structural feature of the climate.

Besides drought management, agricultural risk management policies in CWANA include programs to improve crop management and animal production techniques, crop and animal protection programs, irrigation facility and water management programs, price support programs (in particular guaranteed purchase prices for grains) and to improve input subsidy programs that have decreased substantially in recent years and to develop credit and insurance systems and safety net programs. These measures and development programs are unequally implemented in the region and substantial progress is needed in risk management policies.

In animal protection, risk management differs among North Africa, the Middle East and Central Asia, due to their different epidemiological situations. In Central Asia, diagnostic capacities for epidemics remain limited. Cattle are the main target of preventive vaccination, which is used more to prevent economic loss from disease than to prevent the spread of the infection. Turkey and Iran also vaccinate small ruminants in specific areas to prevent diseases being introduced from neighboring countries. In North Africa, risk management is more focused on emergency preparedness and limiting the diffusion of the disease when it is diagnosed. Effective control from quarantine and mass vaccination are used. Vaccination campaigns target cattle in Algeria and Morocco and both cattle and sheep in Tunisia. Still, animal diseases seriously affect the livestock production potential. Eliminating the diseases would require implementing focused pest and disease control operations, lacking in several countries.

Another important policy area to reduce agricultural market risks is food safety and quality. In several countries of the region, initiatives have been taken to reform and improve food control systems: development of a national strategy for food control (Morocco, Tunisia), implementation and development of food legislation complying with international requirements (Cyprus, Egypt, Jordan, Lebanon, Morocco, Oman, Pakistan, Sudan and United Arab Emirates) and review and update of food standards and regulations (Iran, Sudan and Syria). Some countries have harmonized their food standards with Codex, and Tunisia has introduced quality assurance systems. Despite the effort made by several countries to improve food control systems and to harmonize national food regulations with international standards, often with FAO support, further progress needs to be made to increase the efficiency of food safety systems, first in order to meet national public health requirements (for locally produced and imported products) and second to meet the food quality and safety requirements of export markets.

What is thus at stake for most countries of the CWANA region is the design and implementation of a comprehensive and proactive risk policy, which would include and coordinate the following elements:

- Establishment of drought early warning systems
- Development of crop insurance schemes, now available in only a few countries
- Increased public investment in public works, water management and agricultural research and extension
- Implementation of policies that protect the environment and discourage cultivation of marginal land
- Reinforcement of marketing systems and promotion of farmer organizations
- Development of new and improved food safety systems to comply with food safety standards in export markets

2.2.5.4 Drought risk management in CWANA

Drought is a recurrent phenomenon across CWANA countries. It has a severe effect on the populations and weighs heavily on all economic activities, particularly agriculture. Drought has a negative influence on aquatic and land ecosystems and on the quantity and quality of underground and surface water because of salinization. Regional and international organizations are putting in place various strategies to combat drought. The strategies may be divided into two groups:

- Improvement of countries' hydraulic equipment to collect and store water, rural water and soil water conservation development programs and range improvement schemes
- Reinforcement of institutions to integrate the risks of "drought" into economic planning, give rural zones the means to resist drought and start emergency programs as soon as a drought is detected

Hydraulic equipment and the fight against drought. During the 1970s and 1980s, significant efforts were made in the entire region to construct large dams (see section 2.1.4). In some countries, such as Tunisia, recent preference has been for small and medium hydraulic works (Albergel and Regeb, 1997), creating an agricultural revolution. The three-year drought in Tunisia, from 1993 to 1995, was overcome without rationing water to agriculture, towns, tourism or industry because the hydraulic infrastructure was well proportioned for the country's needs.

On the contrary, Afghanistan faces a food crisis each time there is inadequate rainfall during the winter or during the months of April and May. FAO assesses that 6.5 million people are seasonally or chronically food insecure in this country because of the lack of adequate hydraulic infrastructure. In 1999 to 2000, Afghanistan was hit by a serious drought as a consequence of low rainfall and snow melt over the winter. Central and southeastern Afghanistan was the worst affected, with the drought reaching crisis in some places when the population resorted to eating wild grasses and roots and drought deaths were reported (Marsden, 2000).

The Achilles heel of reservoirs in drought management is the high evaporation rate and, in particular, the sediments that come in the dams each year. It is estimated that in Morocco, 9 10^9 m^3 of water evaporate every year, or 33% of the 30 10^9 m^3 of rainfall. In Tunisia, large dams have an average volume loss of 25 million m^3 per year, or about 2%. For small dams, the volume lost is 5% (Boufaroua et al., 2000). By 2020, many countries in CWANA will have to manage an after period, already a problem in Algeria. To reduce the silting up of dams and the loss of agricultural land, countries have launched water and soil conservation policies.

Today, storing water in aquifers seems the best method for combating dry intervals. It protects water collected during excess rainfall years from evaporation. The reservoirs do not shrink in size. The only risk is pollution by compounds that are not stopped while the water is traversing the porous environment during infiltration. Many countries have converted some dams to refill groundwater. The El Aouareb Dam on the Merguelill, in Tunisia, now is managed to release water to refill the water table in the Kairouan Valley downstream. This experiment, which interests all countries in the region, is monitored within the research network, "Wadi Hydrology," of the International Hydrological Program of UNESCO.

The karstic systems of limestone rocks also show potential for storing water. The Figeh source, which supplies some potable water to Damascus, has a flow of 20 to 30 $m^3 s^{-1}$ in winter and only 3 $m^3 s^{-1}$ in summer. The plan is to stock the winter surplus, when demand is only 15 $m^3 s^{-1}$, in the subsoil (Miski and Shawaf, 2003). A technical study on this is now under way and more scientific research is proposed with the European Union's programs.

Institutional reinforcements: From past experience, emergency programs in case of drought should revolve around the following points:

- Provide potable water to cities and countryside and water to livestock
- Safeguard livestock using knowledge of the forage deficit
- Provide financial support for farmers most affected by the drought
- Provide seeds to farmers, keeping in mind that employment in the countryside prevents rural and agricultural exodus

To plan and implement these programs, governments should set up structures to forecast and identify droughts. They must have access to diverse, reliable data sufficiently processed to be easily interpreted by decision makers. Such information is either a forecast or an observation. Forecasts deal with climate trends, precipitation, evaporation, water that is available and collectible, grazing ranges and harvests. Observations are made at the first sign of the drought and deal with the crisis in each region and on the efficiency of the measures implemented; they must be made during the entire drought to better prepare for future droughts.

In an FAO study on planning antidrought strategies in Morocco, M. Bernardi (1996) recommends a four-level structure where the roles of each entity are well defined:

- A base includes the information providers who regularly monitor key indicators and forecasts (Agrometeorological Committee on Drought Monitoring)
- A second level determines the impact of the drought on different sectors in country (Drought Impact Evaluation Committee)
- A third level proposes measures based on the information received (Drought Monitoring Cell)
- A top level, the prime minister's cabinet, in coordination with the planning, finance and agricultural ministers, authorizes emergency actions and proposes medium- and long-term intervention plans to the government to mitigate the effects of the drought

The strategy depends greatly upon the first level, where tools remain the least effective and to which AKST could contribute greatly:

- *Long-term forecasting:* Reliable information on future seasons would facilitate preparing and executing the most effective policies to combat drought. The investments and international support needed to mitigate its effects should be foreseen and mobilized. Long-term forecasting always is difficult and remains at the continental and regional scale. On the northern shores of the Mediterranean, many programs have been started to research the consequences of global warming on water flows and on their new distribution amid the water cycle (European Environmental Research Programs).
- *Medium-term forecasting:* Medium-term forecasting is the area with the greatest expected benefits. These benefits are rapid alert systems, the rationalization of planning for strategic cereal reserves and improved exchanges of foodstuffs among countries. This gives governments the possibility of integrating climate variability into economic management (Bernardi, 1996). This forecasting relies particularly on hydraulic infrastructure, meteorological and hydrological networks and observation of agricultural production and range. WMO, UNESCO and Sahel and Sahara Observatory (OSS) programs encourage sharing information and forecasting: Med Hycos program, Alpine and Mediterranean Hydrology (AMHY) program and the environmental observatories of the Long Term Ecological Monitoring Observatories Network (ROSELT).
- *Short-term forecasting:* Forecasting during a single season is fundamental to improving forecasts for filling dams, the level of underground water tables and crop yields. Better performing models, with high spatial and temporal resolution, could furnish more reliable information during a season. This information, integrated with other data such as zones and land use, is the base of an early alert system. Progress in satellite imagery and in geographic information has contributed greatly to the development of these models.

2.2.5.5 Environmental policies and regulations

Environmental problems in CWANA are desertification, deforestation, rarefaction of water resources, pollution and disease and pest epidemics. They result from human activity, technical change and climate change. Global warming could drastically change the world's agroecological zones and destabilize weather patterns, leading to an increase in incidence of severe disasters, such as drought. The environmental problems of intensive and high-input agriculture are recognized globally. In the region, the main environmental problems related to agriculture are linked to farm mechanization (soil erosion), irrational use of chemical inputs and pesticides (water pollution) and irrigated farming (overexploitation of groundwater and salinization)

The increased awareness of the challenges to environmental sustainability has led to environmental regulations and policies, which, however, are unequally implemented by the CWANA countries. In several countries, measures are being taken to diversify agricultural practices and improve efficient resource use. Crop diversity will supply useful traits. Diversity of species can provide alternative crops for agricultural diversification.

Crop diversity provides the raw material for breeding new crop varieties that can adapt to climate change. It can also provide more flexible production, better adapted to stresses like drought or salinity and can reduce soil erosion. Crop improvement can meet the challenges posed by pests and diseases and can also help reduce chemical use. A more environmentally friendly agriculture requires both crop varieties and species that can grow with fewer fertilizers, pesticides and other agrochemicals. This is a shift in breeding programs away from yield alone and may require a rethinking of crop breeding. Among many things, it will require that farmers and breeders have access to a wider range of crop diversity—including traditional varieties—as sources of useful genes and genotypes.

Organic farming is often more environmentally friendly than conventional agriculture, but may require organic farming information, standards, certification and labeling, purchase of fertilizers, pesticides and animal health care products. Organic farming has developed in Egypt, Lebanon (http://www.earthfuture.com/economy/sekemegypt.asp), Morocco, Palestine, Tunisia and Turkey (Aksoy, 1999; Kenanoğlu and Karahan, 2002).

2.3 History of Public and Private Sector Investment in AKST

2.3.1 Investments in agricultural research and development

The investments and institutions of agricultural research and development (R&D) are undergoing rapid changes. Growth in public spending on agricultural research and development has not been consistent in CWANA. In some countries it has slowed, in others it has stalled and for some it has declined. In addition to the changes in public research, private sector investment in agricultural research has grown only in Jordan, Pakistan and Sudan.

In Jordan, only 6% of agricultural research and development is private and mainly involves high-value crops and fruit trees. In Sudan, private investment accounts for 8%, mainly in sugar cane. The most private research (>15%) is in Pakistan (Ahmad and Nagy, 2001).

Despite these rapid changes, information and policy analysis to inform and guide the changes under way in many CWANA countries is scant. Research is particularly lacking concerning public policies that can improve agricultural science and technology institutions, including their productivity, environmental and poverty consequences.

Among public agencies, most agricultural research is conducted by research institutions. The remaining public investment is done by higher education institutions (Table 2-12). In Syria nearly 83% of public investment is done by research institutions, the remaining 17% by agencies of higher education. In Morocco the contribution of higher education is as high as 36%, while major research, 64%, is done by research institutions.

In poorer countries, such as Mauritania and Somalia, public agricultural research is mostly done by research institutions. Higher education contributes little to public research in these countries. Similar patterns of research staff allocation are evident between public agencies and private enterprises and between research institutions and higher education. Most researchers are in public research institutions.

2.3.2 History of public agricultural research

Detailed historical information on agricultural R&D for all CWANA countries is not readily available. The agricultural research in Tunisia, for example, began over a century ago (Aubry et al., 1986) (Table 2-13). Formal research began later in other countries.

2.3.3 Human resources in public agricultural research and development

During the last three decades the number of agricultural research staff in many CWANA countries grew steadily. Also,

Table 2-12. **Agricultural research expenditures in selected CWANA countries, 2002.**

Country	Public agencies			Private enterprises	
	Research institutions	Higher education	Total		
	(%)				
Jordan	45.8 (58.5)	47.7 (38.1)	93.6 (96.6)	6.4 (3.4)	Mainly high value crops & fruit trees
Mauritania[a]	91.9 (91.9)	8.1 (8.1)	100 (100)	0.0 (0.0)	
Morocco[a]	63.6 (63.6)	36.4 (36.4)	100 (100)	0.0 (0.0)	
Sudan	65.3 (70.2)	26.4 (28.4)	91.7 (98.6)	8.1 (1.4)	Mainly sugar cane
Syria	83.4 (83.6)	15.8 (15.9)	99.3 (99.6)	0.7 (0.4)	
Tunisia[a]	73.6 (73.1)	26.4 (26.9)	100 (100)	0.0 (0.0)	

aPrivate-sector involvement in agricultural research is nonexistent

Note: Numbers in parentheses are the percentages of researchers

Sources: ASTI, 2003abc, 2004, 2005, 2006ab.

the quality of research staff has improved considerably over the last years.

In Jordan, nearly 61% of the 245 full-time researchers had postgraduate training and more than a third held a doctorate degree while in Morocco (ASTI, 2005) and in Tunisia (ASTI, 2006a) the number was over 90%. In contrast, only 25% of the agricultural researchers in Syria held master of science and doctorate degrees (ASTI, 2006b).

Despite a rise in the number of women pursuing scientific careers worldwide, female researchers are still underrepresented in senior scientific positions. In 2003, less than 13% of the researchers in Jordan were female. This is low compared with other countries, such as Morocco (18%), Syria (23%) and Tunisia (28%). In Jordan, women represented 5% of researchers with a doctorate degree, 17% with a master's and 19% with a bachelor's. In Syria, 23% of all researchers employed in public institutions in 2003 were female, including 5% holding a doctorate degree, 36% with a master's degree and 26% with a bachelor's (Table 2-14).

In Sudan, 79% of the 591 researchers had postgraduate training and one-third held a doctorate degree. In 2000, nearly 28% of full-time researchers were female, including 17% holding a doctorate degree and 26% with a master's degree.

In 2002, approximately 91% of the 362 researchers in Tunisia had done postgraduate work and 70% held doctorates. By comparison, 34% of agricultural researchers in Morocco held doctorates in 2002. Tunisia's particularly high PhD share is partly because the minimum qualification required for researchers in Tunisia's higher-education institutions is a master's in science (ASTI, 2006a). On average, 28% of all agricultural researchers were female. This is considerably higher than the 18% for Morocco in 2002. Both the share of female researchers overall and of those holding doctorate degrees are expected to rise in the near future. Over 50% of currently enrolled students of agriculture are female and many are finishing PhD degrees (ASTI, 2005).

2.3.4 Research intensity in public agricultural research and development

Total agricultural R&D spending as a percentage of agricultural output (Ag GDP), defined as research intensity, is commonly used to compare research investments across countries (Table 2-15). Jordan, for example, invested US$2.83 for every US$100 of agricultural output in 2003, which was a substantial increase over the 1996 ratio of US$1.61 (ASTI, 2006c). The 2003 ratio was also considerably higher than the average for CWANA, 0.66%, and for the industrialized world as a whole, 2.36%. The high ratio of research intensity in Jordan does not reflect high research investment in agriculture; rather it indicates agriculture's small share of the country's gross domestic product (GDP).

Syria invested US$0.53 in agricultural research for every US$100 of agricultural output in 2003. This was similar to the reported 2000 average for the developing world, but was lower than the average for CWANA, US$0.66. In 2000, Sudan invested US$0.17 for every US$100 of agricultural output. Sudan's research intensity declined, considerably lowering its ranking among other countries in the region. The 2000 intensity ratio was less than half of that in 1981

Table 2-13. **A short history of government-based agricultural research for selected CWANA countries.**

Country	History of agricultural R&D
Iran	The Razi Institute in 1925 was the first in the agricultural research system in Iran, conducting research and producing vaccines for contagious animal plague disease. In 1926, the first college of agriculture was founded in Karaj. In 1933, the first college of veterinary medicine was founded in Tehran.
Jordan	Formal agricultural R&D began in 1951 with the creation of the first agricultural research station in the Jordan Valley.
Mauritania	Agricultural research activities commenced in 1949 with the exploratory research by the French colonial government focusing on date palms and the production systems of the Senegal River and the country's oases.
Morocco	The first agricultural research activities in Morocco were carried out by the Agricultural Experimentation Service, established in 1919 by the French colonial government.
Pakistan	After the partition in 1947, only one agricultural college and one research station in three or four provinces of Pakistan remained. The research system was built progressively.
Syria	Formal agricultural R&D began in the early 1940s within the establishment of experiment farms at Deir Elhajar and Kharabo, close to Damascus.
Sudan	Agricultural research began under British rule in attempt to launch cotton production for the international market. Experimental research on irrigated cotton began in the northern part of the country in 1902.
Tajikistan	The Tajik Agrarian University was opened in 1931 on the base of the Central Asian State University as the Faculty of Agriculture.
Tunisia	Agricultural research began over a century ago with the creation of the Livestock Laboratory in 1897, the Colonial School of Agriculture in 1898, and the Botanic Service of Tunisia in 1913.
Turkey	Veterinary school in 1842 and agricultural school in 1881; veterinary research centers in Istanbul in 1914 and in Ankara in 1921. After 1930, several specialized research centers were opened.

Sources: Ahmad and Nagy, 2001; ASTI, 2003abc, 2004, 2005, 2006abc.

Table 2-14. **Educational attainment of researchers and share of female researchers for selected CWANA countries.**

Country	Year of data	Educational attainment			Share of females		
		(%)					
		BSc	MSc	PhD	BSc	MSc	PhD
Jordan	2003	39	28	33	19	17	5
Syria	2003	75	5	20	26	36	5
Sudan	2000	21	46	33	6	26	17
Tunisia	2002	9	21	70	3	6	20
Morocco	2002	11	55	34	28	18	14
Mauritania	2000	36	47	17	3	3	1

Sources: ASTI, 2003abc, 2004, 2005, 2006ab.

and 1995. Even though 1995 had 0.33% intensity, it was low compared with averages for Africa, 0.84% and for the developing world 0.62% (ASTI, 2003c).

In 2002, Tunisia invested US$1.04 for every US$100 of agricultural output. This was an increase over the 1996 intensity ratio of 0.78% and was slightly higher than the 0.95% in 2002 for Morocco (ASTI, 2005). Tunisia's and Morocco's 2002 intensity ratios were higher than the 2000 ratios for the CWANA region, 0.66%, and the developing world as a whole, 0.53%. The low world investment in agricultural research requires greater investment in CWANA countries.

2.3.5 Returns to investments

Investments in agricultural research have contributed greatly to the well-being of farmers, processors and consumers through new knowledge and technology. However, there remain more than 800 million undernourished people, mostly in developing countries, including CWANA, who need significant increases in local production to improve their food security (CGIAR, 2005). For CWANA countries and other developing countries increases in agricultural production and technology that improve disease resistance and drought tolerance and sustain natural resources are needed to lessen the widening food security gap.

The benefits of investing in agricultural research greatly outweigh the costs. To sustain research that will alleviate poverty and reduce food insecurity, governments must invest more in agricultural research. The effect of research achievements goes far beyond the outputs by research organizations. It involves all players between R&D, including research organizations, communities, extension systems, development agencies and policy makers.

Previous studies provided overwhelming evidence that investment in agricultural research has delivered real benefits to poor farmers and consumers through new crop, livestock, fish, forest and farming technology. These improve both productivity and farmer income and help protect the environment, thereby contributing to poverty reduction (Evenson and Gollin, 2003).

The Science Council of the CGIAR commissioned an independent study to compare the benefits from its research against the cost of operating the whole CGIAR system up to 2001. The most conservative assessment yielded a ben-

Table 2-15. **Research intensity in public agricultural R&D in selected CWANA countries.**

Country	Year of data	Research intensity (%)
Jordan	2003	2.83
Mauritania	2001	0.92
Morocco	2002	0.95
Sudan	2000	0.17
Syria	2003	0.53
Tunisia	2002	1.04
CWANA region	2000	0.66
Developing world	2000	0.53
Developed world	2000	2.36
Global	2000	0.80

Sources: ASTI, 2003abc, 2004, 2005, 2006ab.

efit to cost ratio of 1.9:1, meaning the CGIAR generated an indisputable return of nearly two US dollars for every US dollar invested (Raitzer, 2003; CGIAR, 2005). The most generous scenario yielded a benefit to cost ratio of 17.2:1 (CGIAR, 2005). This means the total investment in CGIAR from 1960 to 2001 of US$7 thousand million will generate US$123 thousand million in benefits by 2011 (in 1990 US dollars). Yet even this highly favorable result probably understates the total return on investment because it does not include the following points (Gregersen, 2003):

- Benefits from CGIAR's many research areas that are inadequately documented or inherently difficult to assign a value, such as influence on policy and natural resource management
- The multiplier effect, by which every US dollar of farm income contributes an additional US$0.5 to US$1 to the local nonfarm economy through higher demand for other products and services
- Land savings and their invaluable contribution to protecting biodiversity and watersheds, gained from intensified cropping of existing farmland

While the CGIAR system has demonstrated great international influence through scientific achievements and its pivotal role in the Green Revolution, it accounts for only a small fraction of the global agricultural R&D expenditures. In 2002, CGIAR accounted for 1.5% of the US$23 thousand million global, public investment in agricultural research and just 0.9% of all public and private agricultural research (CGIAR, 2005).

In line with food production trends in other developing countries, food production increases in CWANA the last four decades are attributed to many factors. These include crop genetic improvements and other research contributions, expansion in fertilizer use and pesticides, expansion in irrigation with improved efficiency, mechanization, better farmer education, improvement in transportation and marketing infrastructure and policy reform.

Evenson and Gollin (2003) assessed the effect of crop variety improvement on productivity:

- For all crops combined, the rate of improved varieties production increased each decade the last 40 years.
- Technological advances occurred in all crops, on all continents and in all agroecological zones, although these advances have been uneven.
- The progress achieved is related to the effort expended on research and the existing "stock" of research done on similar crops and growing environments. The internal rates of return on research suggest that public expenditures in agricultural research achieve high dividends. Studies of international and national investments in barley germplasm improvement, for example, indicate the return rate was up to 51% for Morocco. Iraq and Tunisia attained a return rate of 38% for their research investment, while Egypt had 32% and Jordan had 31%. Algeria, Ethiopia and Syria estimated return rates lower than 30% (Aw-Hassan and Shideed, 2003).

The rate of adopting improved barley varieties is growing in several CWANA countries. High adoption was reported in 1997 for Egypt (50%), Jordan (50%) and Tunisia (40%). Relatively low adoption was reported in Morocco (19%), Iraq (14%) and Ethiopia (11%). Algeria and Syria, two large producers, had the least adoption, 5% or less of the total barley (Aw-Hassan and Shideed, 2003).

Similar adoption patterns of improved lentils are reported for some CWANA countries. The national research program of Pakistan reported about 32% of the lentil area in the targeted region is planted with improved lentil varieties. About 25% of lentil area in Iraq and Syria is planted with improved varieties.

2.4 Market Trends and Socioeconomic Evolution

2.4.1 Agriculture market shares of CWANA in global and regional markets

The comparative advantages and natural endowments in some countries form the basis of their competitiveness in international markets. South Asia and West Asia stand out as important exporters of most agricultural products, while the Nile Valley and Red Sea and the North African countries are importers of agricultural products.

2.4.1.1 CWANA and international trade of cereals

South Asian and West Asian countries are leading players in grain exports. Their grain exports increased tremendously, climbing from less than 100,000 tonnes from 1961 to 1965 to more than 5 10^6 tonnes in 2001 to 2004. Grain exports also increased considerably between the mid-1980s and mid-1990s. Kazakhstan is the most important exporter in the region, followed by Pakistan and Turkey. The CWANA region's share in total world exports is only 4%, despite these three countries. North America had 37% of the total grain exports for 2001 to 2004, while Western Europe had about 20%. Together, these two blocs had half of the world's grain exports. However, the export grains differ from one country to another; Kazakhstan and Turkey mostly export durum wheat, Pakistan exports rice.

Despite these significant cereal exporters, CWANA stands out as a net cereals-importing region. Aside from Central Asia and the Caucasus, all CWANA subregions have negative trade balances, and the gap between the exports and imports of the CWANA region has widened annually at an average rate of 5.1% between 1961 and 2004. With a deficit of approximately 43 10^6 tonnes of cereals from 2001 to 2004, the region is just behind Asia. The North Africa, Nile Valley and Red Sea subregions have the highest demand for cereal imports (Figure 2-3). However, the biggest increase in cereal imports is in the North African countries. This dependence on international markets for food supply is a great economic constraint for the North African, Nile Valley and Red Sea countries, resulting in significant public budget deficits. Poor natural endowments for grain production coupled with poor rural livelihood and increasing rural to urban migration create considerable social and economic instability.

The processing industry is developing all over the world. Developing countries improved the technology and productivity of their food processing industries. Wheat flour milling is a primary industry that has flourished in CWANA from

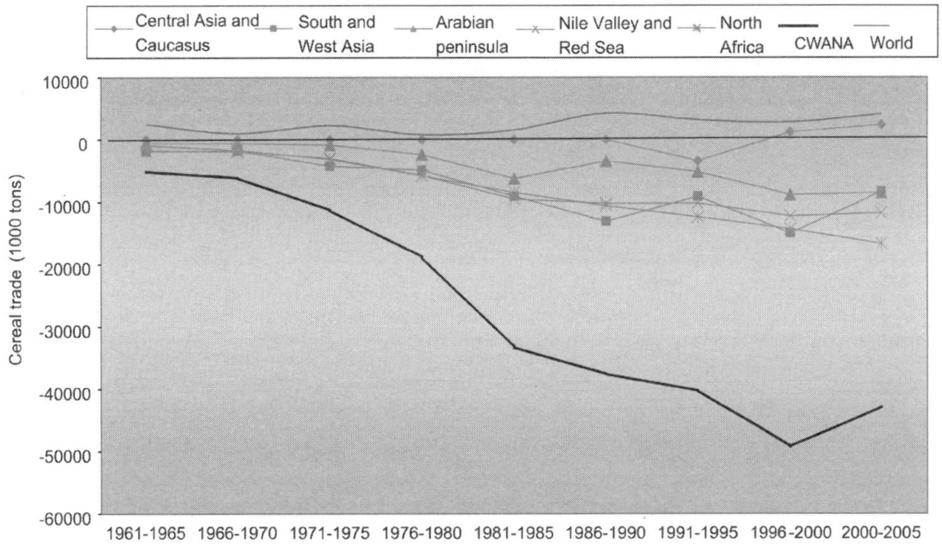

Figure 2-3. *Cereal trade in CWANA subregions, 1961–2005.* Source: FAO database

the 1960s. As a consequence, wheat flour exports increased considerably, to almost a fifth of the world total wheat flour exports in the 2000s. More than half (56%) of the CWANA exports are generated by South Asia and West Asia. Among the different regions worldwide, CWANA stands out with the highest annual average growth rate, 8.2%. This comes from the large increases in Central Asia, Caucasus, South Asia and West Asia (Table 2-16). North Africa increased its total exports during 1996 to 2000, while exports of wheat flour from the Nile Valley and Red Sea subregions witnessed a significant drop during the 2000s. Meanwhile, the market shares of developed regions, North America or Western Europe, saw a tremendous decrease since the early 1990s, a consequence of the development of the milling industry in developing regions. This U-turn will probably continue for food processing in industrialized regions specializing in high-value products. The developing world will invest more in primary processing for the domestic market and exports.

2.4.1.2 CWANA in international trade of oil crops

Most CWANA subregions that exported oilseeds in the 1960s became dependent on the international markets for their domestic supply. The downward trend started during the mid-1980s, pulling the region's trade balance toward the deficit side since the early 1990s. The abandonment of public policies encouraging oilseed production, mostly in Turkey, had a negative effect on production. The established oil-processing industries started to import most of their raw material from international markets. This deliberately chosen strategy, designed in Turkey and agreed upon by both public authorities and oil processors, saw advantage in the low world prices. Because of this deliberate choice in most CWANA countries, the entire region became dependent on international commodity markets for oilseed. Other developing regions more or less followed the same trend; North America, Latin America, the Caribbean and Oceania saw their shares increase tremendously. Western Europe, already dependent on international oilseed markets for its domestic supply, witnessed its deficit deepen. The most significant oilseed deficit is in Asia. This region had a surplus oilseed trade balance of more than 1.5 million tonnes in the early 1960s, but it had a 7.7 million tonnes oilseed deficit in 2001 to 2004, primarily from the sourcing strategies of multinational enterprises. This international division of world commodity markets may be an important constraint for CWANA, if these countries cannot develop activities for creating wealth to replace products abandoned in the last four decades.

2.4.1.3 CWANA in international trade of processed food

A direct result of urbanization and the increase in urban purchasing power is the increased demand for processed food. As economic employment of women increases, family structures move toward nuclear families. In nuclear families the parents and children form the nucleus, so meals prepared by female family members other than the mother decrease. At the same time, families are starting to live farther and farther away from work and school, causing a decrease in time spent at home and an increase in the demand for ready-to-eat meals. In addition, the percentage of meals eaten out by families increases. In the CWANA countries, a rather hybrid sociocultural structure can be observed. When there are extended family structures including grandparents and younger siblings of the parents, typical fast-food consumption patterns are developing. Accordingly, industrial processing enterprises focus on generic products; fresh dairy products, cheese, biscuits and pasta, beer and soft drinks are being manufactured at a rapid pace. Also, export industries, based on traditional agricultural products, are developing an industrial structure. In this panorama, exports increase as a result of the gain in international competitiveness among the national food processing enterprises, while processed food imports also increase from increasing domestic demand for more sophisticated, high-value food.

Table 2-16. **Evolution of the cereals trade balance of CWANA region and other regions of the world between 1961 and 2004.**

Region	1961-1965	1966-1970	1971-1975	1976-1980	1981-1985	1986-1990	1991-1995	1996-2000	2001-2004
	(1,000 tonnes)								
Central Asia and Caucasus	—	—	—	—	—	—	-3,456	1,173	2,357
South and West Asia	-1,877	-1,781	-4,221	-4,853	-9,072	-13,095	-9,080	-14,993	-8,303
Arabian Peninsula	-536	-590	-826	-2,383	-6,215	-3,428	-5,171	-8,875	-8,502
Nile Valley and Red Sea	-1,778	-1,956	-2,982	-5,730	-9,484	-10,214	-10,182	-12,225	-11,765
North Africa	-878	-1,700	-3,138	-5,764	-8,468	-10,580	-12,383	-14,310	-16,677
CWANA	**-5,136**	**-6,135**	**-11,307**	**-18,789**	**-33,305**	**-37,560**	**-40,272**	**-49,230**	**-42,890**
North America (developed)	52,500	56,731	82,219	117,290	129,712	116,448	116,031	106,382	97,821
W Europe	-29,446	-27,034	-26,121	-24,889	-1,495	18,462	23,315	16,812	3,880
EU (15)*						*20,996*	*23,279*	*15,629*	*4,767*
Oceania	6,705	7,060	10,465	13,546	15,889	16,881	12,744	20,905	17,782
E Europe	-5,527	-2,284	-4,703	-9,505	-3,134	-2,421	213	993	1,793
Latin America and Caribbean	2,591	4,602	550	-1,803	-1,280	-10,259	-17,120	-18,442	-17,000
Africa	261	-351	-270	-2,549	-7,122	-4,910	-11,138	-11,717	-17,838
Asia	-21,611	-34,460	-41,486	-52,344	-56,363	-60,637	-61,487	-61,328	-51,570
World	2,449	982	2,222	815	1,548	4,240	3,142	2,810	4,121

North African countries, Djibouti, Egypt, Somalia and Sudan are excluded from the Africa totals. South and West Asia, Central Asia and Caucasus, Arabian Peninsula countries and Yemen are excluded from Asia totals.

*EU 15 exports do not include the intra-EU trade

Source: FAOSTAT.

In CWANA, the total processed food exports doubled between 1961 and 2004, the highest increase realized by Turkey. However, as the other regions developed their processed food exports more rapidly than CWANA, the CWANA share in processed food exports worldwide fell from 14.3% in 1961 to 1965 to 2.7% from 2001 to 2004. The main winner in this development is Asia. China's industrialization in food now challenges the western European food processing industry. Latin America and the Caribbean also are promising challengers.

In the evolution of processed food imports, the CWANA share increased from 5.7% of world imports from 1961 to 1965 to 6.6% from 2001 to 2004. Just behind South Asia and West Asia, the Arabian Peninsula is an important importer. North Africa, Nile Valley and the Red Sea also drastically increased their imports in processed foods.

2.4.1.4 CWANA in international trade of fresh fruits and vegetables

Most CWANA countries have comparative advantages in fresh fruits and vegetables. Increasing demand from developed countries for fresh produce has had a positive spillover on organizing and developing exports. Morocco and Egypt have long had an export tradition. During the last three decades they have gained important organizational skills in this area. Turkey, Iran, Syria and Pakistan are relative newcomers, having entered the export scene in the early 1980s; their export volume, however, is increasing at a growing pace. As other regions of the world also have gained important market shares in fruits and vegetables, the CWANA world share fell from 12.2% to 8.9% between 1961 and 2004, and its annual average growth rate stayed slightly lower, 2.6%, than the annual average growth rate worldwide, 3.4%. The winners were in Asia. High growth came to China, India, Latin America and the Caribbean. Because of the high value of fruits and vegetables, North America and Western Europe have remained strong in world competition and have increased their exports with annual growth rates between 3 and 4%.

CWANA exports are not yet very significant: fresh fruit and vegetable exports from Central Asia and the Caucasus, 7.1% per year and the Arabian Peninsula, 9.5% per year. But these grew faster than exports from other CWANA subregions. Total export volumes of South Asia, West Asia, the Nile Valley and the Red Sea constantly increased, while those from North Africa trended downward during the last four and a half decades. Among the traditional exporting countries, Turkey increased its exports volume by 16 times and had 2.3% of the fresh fruits and vegetables exports worldwide from 2000 to 2004.

2.4.1.5 CWANA in worldwide imports of meat

Demographic pressure in the entire CWANA region had negative effects on its international trade balance. Meat imports increased with an annual growth rate of 8.2% from 1961 to 2004, with the most significant increase in the Arabian Peninsula. Despite big increases in total meat production within the different countries, the total dependence ratio of most CWANA countries increased from high demographic pressure. As a result, CWANA is the most important meat importer in the world, just behind Asia and just before Latin America and the Caribbean. The annual average increase of total imports in the region is two times, 8.2%, the world annual average increase, 4.3%. Large increases in poultry meat production are not enough to fill this gap, which widens each year. Meat imports in the Arabian Peninsula multiplied tenfold between 1961 and 2004. Meat import volumes of the other CWANA subregions multiplied 10 to 20 times. Lebanon, Jordan and Iran in South Asia and West Asia, Algeria and Morocco in North Africa, Kazakhstan in Central Asia and Egypt and Yemen in the Nile Valley and Red Sea stand out as the most significant meat importers. All regions in the world, developing and industrialized countries, increased their meat imports during these four decades, although the annual average increase rates in industrialized regions were lower than the rates in the developing regions.

2.4.1.6 CWANA in world imports of feed

While the meat and milk imports of CWANA increase at a significant pace, government subsidies and other measures to encourage the development of livestock production in most of the countries continue. As a result, feed imports increased considerably in volume and value in most countries. While milk husbandry and poultry became dependent on feed imports, extensive transhumant animal raising still continues to prevail as the significant system in Algeria, Egypt, Iran, Morocco and Yemen as well as in the Caucasus, Central Asia and West Asia. As a consequence of importing feed in CWANA, its share in world imports increased from 0.9% from 1961 to 1965 to 7.5% from 2001 to 2004.The imports took off in the mid-1970s. Despite this significant growth, CWANA lags far behind western Europe's and Asia's shares in world imports. Western Europe had a 52% share and Asia 20% of animal feed imports from 2001 to 2004. In light of the growing need for animal feed in livestock production, the upward trend of world imports seems likely to continue, most particularly imports in CWANA.

2.4.2 Changing lifestyles, consumer preferences and demands

Urbanization is, by far, behind the changes in people's lifestyles. Economic, social, cultural and spatial factors are pushing urban families to live and to consume differently. Changes in family structure, in work, residential patterns and improvement in urban infrastructure drive many urban consumers towards standardized, industrialized and globalized consumption patterns, even if their habits and preferences are largely influenced by local tastes and traditions.

Despite these changes, enhanced by urbanization and elite urban groups, purchasing power is still the main determinant of consumption in developing countries. Food expenditures are more than 40% of household expenditures in most CWANA countries but in rural households, food expenditures can exceed 60%. In comparison, approximately 15% is spent for food in developed countries. The high ratios show the vulnerability of consumption patterns in CWANA and the importance that food has in a transition toward a market economy. If households cannot achieve satisfactory disposable income, they will rapidly be exposed to undernourishment.

Most of the countries in CWANA have human development index (HDI) numbers lagging drastically behind industrialized countries. Highly skewed income distribution, lack of rural infrastructure and poor urban districts result in unequal access to education, health facilities and healthy food. The oil-rich countries of the Arabian Peninsula have HDIs higher than those of other CWANA countries. Oman showed the most dynamic evolution, nearly doubling its HDI between 1975 and 2004. The countries of Central Asia and the Caucasus have HDIs stagnating or falling since the 1990s, illustrating how difficult it is for these countries to develop free market economies. Uzbekistan and Tajikistan seem to be the worst in terms of development. In South Asia and West Asia, all countries have HDIs around 0.71 and 0.77 except Pakistan, which has an HDI well below the CWANA average of 0.539 (UNDP, 2004). Trends from 1975 to 2004 are positive, showing a dynamic evolution for most of the countries of South Asia and West Asia. In North Africa, the lowest HDIs are in Mauritania (0.486) and Morocco (0.64), far below the world average. In the Nile Valley and Red Sea subregions, all countries have HDIs below the world average. The high disparities in the standards of living, coupled with poor rural livelihoods, reinforce the high risks concerning the food security, especially in countries with difficult living conditions—Djibouti, Mauritania, Pakistan, Sudan and Yemen.

In 1969/1970, food production within some CWANA countries was almost adequate to meet demand; the self-sufficiency ratio for all cereals was nearly 90%. However, the food gap continued to widen over the last three decades. Expanding agricultural production, 2.9% annually, failed to keep pace with the rapid growth in demand, and self-sufficiency ratios declined. According to FAO estimates, this trend is expected to continue (Alexandratos, 1995). However, throughout CWANA, undernourishment has been under control since the late 1960s. In the Arabian Peninsula, North Africa, the Near East and Middle East, less than 10% of the population is declared undernourished. The exceptions are Pakistan and the Occupied Territories of Palestine. In the Nile Valley and Red Sea subregion, Egypt stands out as the country with the least of its population undernourished, but in both Djibouti and Sudan one-quarter of the population is undernourished and in Yemen more than one-third. Difficult economic conditions in Central Asian countries have negatively affected food security since the early 1990s. In Armenia, Tajikistan and Uzbekistan, one-quarter to one-half of the populations are estimated to be undernourished, according to preliminary FAO data. Azerbaijan and Kyrgyzstan seem to have reduced undernourishment since the early 1990s.

The general nutritional status of the CWANA countries has improved. Significant progress was made in raising the per capita daily food consumption in kcals per person, the key variable measuring and evaluating the world food situation. The average national food consumption per person in CWANA has increased quite satisfactorily since the 1960s and will likely continue. This is projected to increase from 3006 kcals in 1997-1999 to 3090 kcals in 2015 and close to 3170 kcals by 2030 (Figure 2-4).

In addition, the per capita daily food intake changed, particularly in the oil-rich Arabian Peninsula countries. The increased consumption of meat, particularly poultry and eggs, milk and milk products, fats, oils and nuts has been spectacular. North African countries, thanks to imports, increased dramatically their consumption of cereals, starchy roots (mostly potato), sugar and other sweeteners, milk and milk products. All, particularly South Asian and West Asian countries, are great consumers of fruits and vegetables, although the difference between the world average and the CWANA average seems narrow. While the average consumption worldwide of fruits and vegetables has increased, the consumption in CWANA has decreased from their high consumption in the 1960s. The average daily food intake in all CWANA continues to be dominated by vegetables, with animal products, especially meat, eggs, fish and seafood lagging far behind world averages. When compared with consumption in industrialized Western countries, this gap is even wider. The relatively high consumption of pulses somewhat narrows this protein gap.

Finally, it must be remembered that the progress shown by the positive figures for daily food intake between the 1960s to the 2000s do not reflect the uneven food distribution among the socioeconomic classes and poor rural areas.

2.4.3 Local markets and marketing channels

Market accessibility is becoming key to rural development. The lack of links between rural and urban areas is the largest constraint to improving rural livelihood. In addition, most farmers in CWANA are small landholders and have limited money to invest in new technology to improve their yields.

Since the mid-1980s, governments have improved the facilities for agricultural produce and market access for small landholders. In Tunisia and Morocco, milk collection centers have considerably helped increase marketed milk. In Turkey, successive governments oriented public credits through agricultural cooperatives to install greenhouses in horticultural regions and improve the conditioning facilities for fresh produce. However, despite these measures, most agricultural and livestock production is marketed through long marketing channels where middlemen keep much of the value created in supply chains; landholders mostly receive less than 30% of the final market price.

Traditional marketing channels, from the farmer to the consumer, comprise many marketing agents: the village tradesman, commissioner, wholesaler, industrial processor, retailer for the domestic market and the village tradesman, commissioner, wholesaler and exporter for exported products. These long marketing channels are harmful for the quality of agricultural and food products and for the food safety. The poor quality and small volume of agricultural produce create bottlenecks for industrial processors and ex-

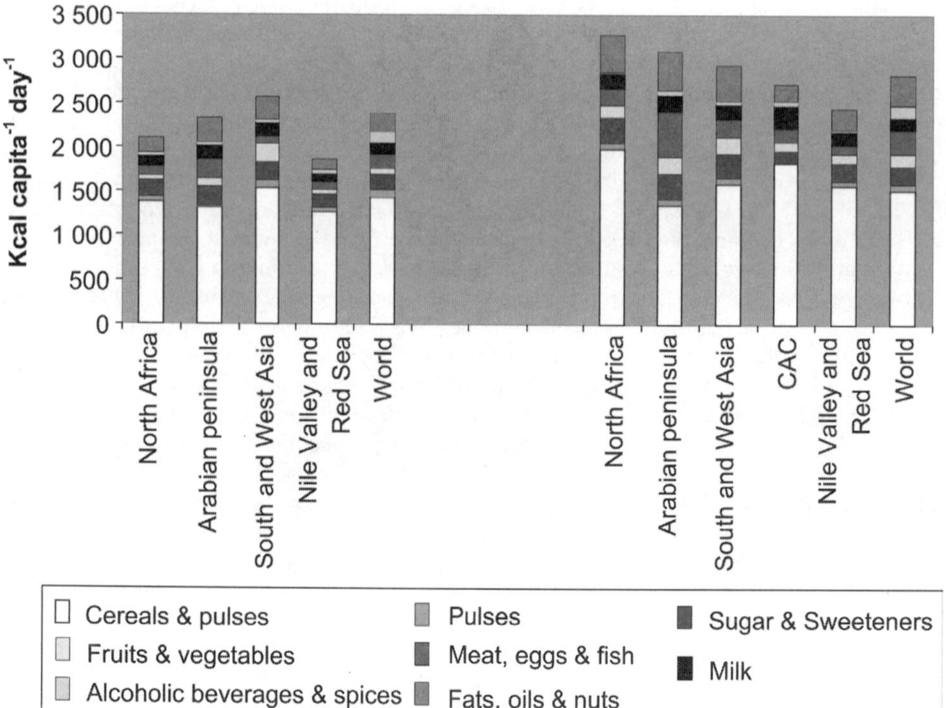

Figure 2-4. *Composition of per capita daily food intake.* Source: FAO database

porters. In addition, the many middlemen increase the price consumers pay for food, while farmers do not benefit.

The penetration of large Western retailers is changing the domestic markets of developing countries (Reardon and Berdegué, 2002). Countries of CWANA are no exception to this new trend. Large-scale retailers bring their logistic services, and the effects within domestic markets are rapid and spectacular. The high volume and standardized products demanded by these large retailers rapidly transform the structure and functioning of the supply chains, and much consolidation happens amid the domestic wholesalers and retailers. Suppliers diminish in number and the small, family-owned, traditional grocery shops quickly disappear. Agrifood market institutions are greatly affected by the rapid rise in private standards and a gradual rise in contract use (FAO, 2004).

2.4.4 Labor market

Rural to urban migration and urbanization in CWANA countries since the early 1970s changed the structure of labor markets considerably. The migration dropped significantly the percentage of the economically active population in agriculture (Figure 2-5). Even in the Nile Valley and Red Sea subregion, where the percentage of those active in agriculture is highest, the drop is considerable, an average of 72.6% from 1961 to 1965, 55.9% from 1986 to 1990 and 44.6% from 2001 to 2003. Central Asia and the Caucasus have 24% of the economically active population in agriculture and the Arabian Peninsula 9%. In North Africa, where rural-to-urban and cross-border migration is significant, the share of agriculture in the economically active population fell drastically, to less than 30%. In South Asia and West Asia this share is uneven from one country to another. In Afghanistan 66.3%, Pakistan 46.1% and Turkey 44.8% of the total economically active population was still occupied in agriculture in 2001 through 2003; in Lebanon only 3.2% for the same period.

Another prominent point is the increasing participation of women in economic activities in this region. The average for the economically active female population in the total economically active population for CWANA was around 33% for 2001 through 2003, while the world average was around 41%. However, there are important differences from one subregion to another and from one country to another. Central Asia and the Caucasus 46.9%, Egypt 45.7%, Mauritania 47.6% and Somalia 43.4% have ratios above the world average. Turkey has around 38%. In other CWANA countries the share of economically active women is between one-quarter and one-third of the total population. These two trends affect the overall labor supply growth figures. These figures continue to grow at high rates, reflecting high population growth, inflows from rural-to-urban migration and increasing female economic participation (Tzannatos, 2000). However, the urban formal sector is not sufficiently developed to absorb this labor excess because of the slow industrial growth rate. Informal activities are the largest source of income for the recently urbanized populations (Tzannatos, 2000). This large shift from high-paying formal jobs to low-paying informal jobs reduces the income of recently urbanized households considerably. A vicious cycle is in place; less educated people are employed only informally, diminishing household income and giving less chance for their children to become well educated and attain higher-paying formal jobs.

These negative trends are accentuated by macroeconomic changes in most of these countries. In some Arab countries relying on oil exports, the decline in oil prices during the 1980s negatively affected the investments in these countries. In Algeria, Egypt, Morocco, Turkey and Central Asian countries the retirement of the public sector from economic activities negatively affected public employees. Last but not least, after the Iran–Iraq and Gulf wars, countries with significant out-migration, Egypt, Jordan and Turkey, faced falling demand for their citizens abroad. Remittances and social compensations from workers in Western Europe and oil-rich Arab States declined in Algeria, Mauritania and Sudan and stagnated in Turkey. This result hurt the purchasing power of many urban and rural households in

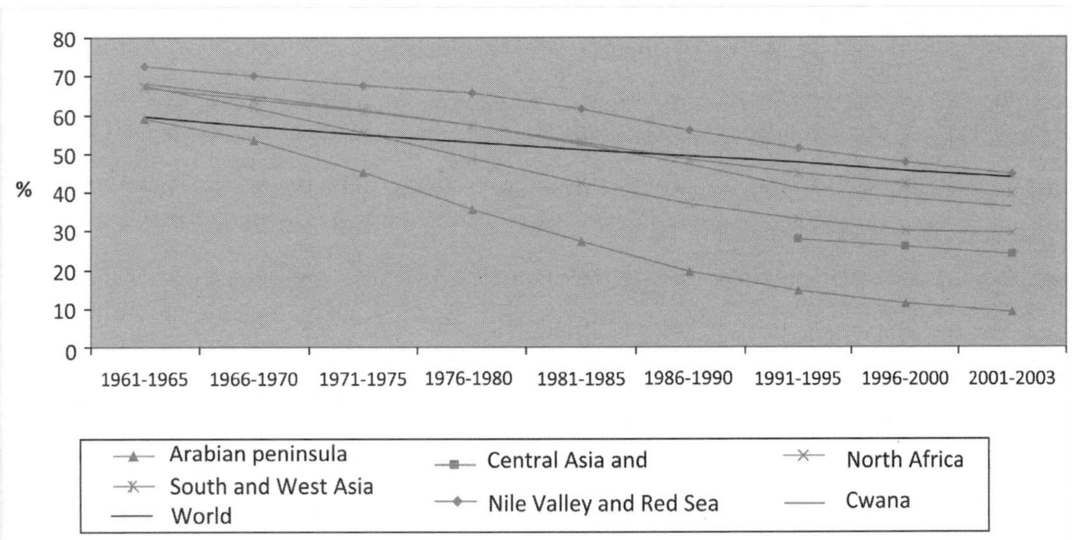

Figure 2-5. *Share of economically active population in total economically active population.* Source: FAO database

CWANA. Most CWANA countries are not ready to absorb the increasing labor supply and huge infrastructure investments needed to boost the nonagricultural labor demand. The quality of this labor is another huge problem for which simple solutions cannot be soon found; long-term investments are necessary.

2.4.5 Findings

The different development of CWANA countries must be reiterated. There is a great need to establish field studies using local surveys to analyze and assess the marketing conditions in the different regions of these countries. A global view does not take into account the important differences among countries and among the regions within countries. Problems in the Turkish agricultural sector cannot be considered similar to problems in Algeria, Djibouti or Sudan, nor can improvements in the agricultural sector of Pakistan be compared with improvements in Saudi Arabia.

This global assessment indicates that CWANA countries are losing their comparative advantage in most agricultural sectors. Recently developed sectors, such as livestock and fresh fruits and vegetables, create a certain dependence on modern inputs. No country can function with complete autonomy; all these countries need each other for products for which they do not have comparative advantage. Nonetheless, this need must create value and wealth for the developing CWANA countries and emphasize the urgent need to develop knowledge and technology to challenge the world status quo.

2.5 Technology Transfer and Adoption

2.5.1 Scaling-out approaches for technology adoption and transfer

The generation of improved technologies and assuring that farmers use them are key to crop productivity and improving farmer livelihoods. The lack of trained research and extension staff in most developing countries is largely responsible for these improved technologies being generated but not used. Technology transfer requires close cooperation among all concerned stakeholders, including farmers, research and extension staff and governmental and nongovernmental organizations. International agricultural research centers (IARCs) in the region, of which ICARDA is one, can work closely with national policy and decision makers and with various stakeholders.

NARS in the CWANA region differ in their human and physical resource capacities, needs and operational resources. In most countries of the region, priorities for agricultural research and human resource development depend, to a large extent, on the available financial resources and stem from a national desire to decrease dependence on food imports. Priorities, often set by policy makers belonging to different institutions, are transferred to research, extension and training specialists with limited room for interaction or feedback.

In most CWANA countries, agricultural research and extension are still handled by separate public institutions, with different mandates and operating systems. In the prevailing model for generating, transferring and adopting technologies, the new technologies—either superior genetic material or improved production packages—are developed by researchers then passed on to extension agents to demonstrate and disseminate to farmers. This model does not incorporate a feedback system from farmers to researchers or among research, extension and development agents. This drawback has deterred the development and transfer of technology appropriate for small-scale, resource-poor farmers, particularly those in low-potential, heterogeneous agroecological areas.

Adaptive research, such as on-farm verification and demonstration trials in farmer fields, must complement research to realistically evaluate any new technology. It provides an excellent opportunity for the farmers whom the new technology is meant to benefit to participate, learn about and gain confidence. Adaptive research also builds up and strengthens the research–extension–farmer–policy maker dialogue. Social and economic value and aspects of the tested technologies can also be evaluated and compared with existing practices. On-farm research ensures a feedback mechanism among farmers, scientists and policy makers. Such a mechanism, if effectively linked with extension services, helps research systems set priorities and adjust continuously and adequately to evolving farm circumstances (Swanson et al., 1988).

In collaboration with NARS in CWANA, ICARDA conducts adaptive trials and demonstrations in farmer fields that offer excellent opportunities for organizing field days and visits to promote new technologies. Scientists, extension specialists, farmers, seed specialists, government officials, representatives from international and regional organizations and representatives from universities and the donor community participate in the field days to discuss the new technologies applied under farmer conditions.

To achieve higher development and uptake of innovations, it has been recently proposed (Ceccarelli and Grando, 2007) that the traditional linear sequence of researcher to extension to farmer be replaced by a team approach, with scientists, extension staff and farmers all participating in all major steps of developing a new technology.

2.5.1.1 Agricultural extension and applied research project: case study

A pilot travel and visit (T&V) extension approach was introduced in Turkey in 16 towns through the Agricultural Extension and Applied Research Project (TYUAP) (Kumuk and van Crowder, 1996; www.aari.gov.tr). Reorganization of extension in towns and villages under the T&V system emphasized a group of specialists who do research and train in addition to extension work.

The major differences between conventional extension and the T&V approach are extension and research links, individual instead of group contacts and regular on-the-job training. The T&V approach tends not to involve extension workers in supplying inputs for farmers and focuses more on specific crops.

The T&V extension experience in Turkey suggests that the extension team approach is a more effective way to tackle the problems of mixed-crop farming systems. Multidisciplinary extension teams live and work in towns and villages, focusing on local problems using farming systems approaches and participatory methods. Extension workers and farmers supported by researchers who are subject matter specialists collaborate to make decisions and to analyze problems, plan solutions, implement activities and evaluate results. Farmers, collaborating with extension agents and researchers, participate in designing, evaluating and adapting proposed agricultural technologies. As this case study shows, the T&V approach has scope to contain a participatory team. The characteristics of this approach are described here (Kumuk and van Crowder, 1996):

- Emphasis in all activities on farmer participation to achieve extension relevance and sustainability; emphasis oriented toward issues and problems, on organizing farmers to participate in developing and disseminating technologies and on assessing farmer problems, needs and resources for proposing farming modifications
- Extension efforts oriented toward farming systems and household economy of groups of farmers as opposed to focusing on particular crops or commodities
- Use of both mass media and face-to-face communication, with farmers participating in designing and delivering the message, which is communicated to audiences with similar characteristics
- Strong links established with research, other development efforts in the area and farmers through a team approach that emphasizes consulting and collaborating with farmers
- Emphasis on providing advice, to educate rather than transfer technology, to provide regular in-service training for extension workers and team farmers and to assess the technology by using group activities

When basic elements of the T&V system are maintained, such as regular in-service training and an improved research-to-extension link, and when the extension team approach is introduced in towns and villages, the resulting extension system should be better suited to the needs of Turkish farmers, giving them an active role in generating, evaluating and diffusing technology. The design of such a system will require harmonizing T&V with the existing system and using a participatory team approach to extension.

During the project life, extension expenditures increased eightfold. In the project area, the extension cost per farmer was 27% higher than the national average and the expenditure per extension worker was about 28% higher. Project costs were associated with increases in extension staff, increases in coverage and intensity of extension activities and increases in operational and training costs. The impact was impressive: about 65% of the 85,300 farmers in the area were in direct contact with extension through the various field activities. In production increases, wheat increased 76%, barley 64%, rice 86%, cow milk 65% and goat milk 128%. Overall, agricultural productivity in the project area, measured by the value of the agricultural domestic product in the two provinces, increased 11-fold. While other factors undoubtedly contributed to this increase, improved extension was a major factor.

For the further improvement of TYUAP, a new program called TAYEK (Agricultural Research, Extension and Training Coordination) was organized and applied to better coordinate all sectors and stakeholders of agriculture, facilitating technology transfer for development. TAYEK also includes the farmer field school (FFS) approach. The collaboration with TYUAP and further with TAYEK was helpful for the public and for raising local awareness about in situ and on-farm conservation of genetic and plant diversity in Turkey.

2.5.1.2 Farmer field schools

Farmer field schools (FFS) have become an innovative, participatory and interactive model for educating farmers in Asia, many parts of Africa, Latin America and more recently in the Middle East, North Africa and Eastern and Central Europe. The approach was originally developed to help farmers tailor integrated pest management (IPM) practices in diverse and dynamic ecological conditions. The knowledge acquired as they learn enables farmers to adapt current technologies or to test and adopt new technologies to be more productive, profitable and responsive to changing conditions.

Farm field schools in IPM started in 1989 in Indonesia to reduce farmer reliance on pesticides in rice. Policy makers and donors were impressed with the results, and the program rapidly expanded. The experience generated in Asia was used to help initiate IPM FFS programs in other parts of the world. New commodities were added and these programs were encouraged to adapt them locally and institutionalize them. At present, various IPM FFS programs are being conducted in over 30 countries.

In the Near East and North Africa, FFS were first introduced in Egypt in 1996 with two Egyptian-German projects implementing IPM FFS on cucumber, tomato, citrus, mango and cotton. Several modifications were made to adapt to local conditions and the FFS were renamed farmer learning groups. Several other initiatives followed to organize pilot FFS based on the original concepts. In 2003 ICARDA started a regional FFS project in Syria, Iran and Turkey to extend IPM for pest management in wheat and barley. In Kyrgyzstan the FFS approach was introduced in 2003 for cotton. Uzbekistan introduced it through an FAO-supported project on managing irrigated lands that were salt affected and gypsiferous.

A two-year regional IPM project in the Near East started in 2004, funded by the Italian government, to develop a strategy adapted to local ecosystems that would achieve high-quality production in fruits and vegetables compatible with export requirements to target European markets. The project involved Egypt, Iran, Jordan, Lebanon, Palestinian Territory (Gaza and the West Bank) and Syria. It was expected to establish and strengthen the FFS extension approach to promote IPM technology among Near East farmers.

Another FAO-supported regional project began in 2004 in Algeria, Egypt, Ethiopia, Morocco, Sudan, Syria and Tunisia on training in management of a parasitic weed, orobanche, in leguminous crops (Braun et al., 2006).

2.5.2 Traditional knowledge in CWANA

Traditional knowledge (TK) is a cumulative body of knowledge, know-how and practices maintained and developed by people with extended histories of interaction with the natural environment (ICSU, 2002). It developed from experience gained and adaptations made to the local culture and environment. It is mainly practical in nature and provides the basis that enables communities to make decisions about many fundamental aspects of day-to-day life.

TK is the adaptive and decision-making skills of local people, learned and transmitted through family members over generations; strategies and techniques developed by local people to cope with sociocultural and environmental changes; time-tested natural resource management practices that farmers accumulate through experimentation and innovation (Warren and Rajasekaran, 1993). Traditional knowledge related to agriculture includes information on which farmers, consciously or unconsciously, base decisions related to their production systems (Brokensha et al., 1980; Warren et al., 1989, 1995).

Traditional knowledge is dynamic, resulting from continuous experimentation, innovation and adaptation. It is difficult to determine the historical depth of traditional practices when documentation on the past is lacking or insufficient. For example, since the birth of agriculture, farmers, fishers, pastoralists and forest dwellers have been managing genetic diversity by selecting plants and animals to meet environmental conditions and food needs in the Near East, North Africa and Central Asia. Farmers transfer this knowledge from one generation to the next. People originate TK; recognized and experienced people transmit it. Such knowledge supports diversity and enhances local resources.

Traditional and local knowledge is part of a complex system; it cannot be reduced to a list of technical solutions or restricted to a series of different applications for results to be attained. Its efficacy depends on the interaction among several factors that need to be carefully considered to understand the historical successes it has achieved and to use its internal logic to find modern solutions.

Each traditional practice is not an expedient to solve a specific problem but always a studied and often multifunctional method applied in an integrated approach including society, culture and economy. It is closely linked to a concept based on careful management of local resources.

Many experts and scientists have doubts about the basis of TK and do not give it enough credence in development planning (Howes and Chambers, 1980). Thus development projects may be designed without taking into consideration the effectiveness of traditional agricultural practices.

Ethnographic studies in CWANA on traditional farming systems indicate that local farmers have detailed knowledge of their local environment. Most practices are not arbitrary, even if some farmers may not be able to explain them. In traditional sustainable systems, the cumulative experience of generations of farmers shapes a wide range of practices that contribute to crop productivity and protection in different ways. It is important to examine the range of practices in traditional systems because they are key to sustainability. The system is often designed to prevent or minimize pest and disease problems through indirect methods.

Traditional knowledge represents the accumulated body of experience of people who are well aware of their situation, physical and biological environment and production systems. They are also aware of the possible effect a change in one factor will have on the other parts of the system. The quality and amount of TK vary among community members and also depend on age, gender, social status, intellectual capability and professional occupation (Warren and Rajasekaran, 1993).

Traditional knowledge is usually specific to locality. A good example is the agricultural calendar used by people of a region. Local farmers set the time for planting not by written schedules but by their observations of star risings and settings, the position of the sun's shadow and observable changes in the seasonal cycle, such as bird migration and the appearance of certain insects or plants. With this TK calendar the farmer determines a planting time that will provide a productive crop given the probability of rain and flooding and the menace of pests and diseases. In Yemen, for example, farmers use the local shadow scheme or local star calendars to define planting time. Those observations are rarely applied outside a specific context or local region (Serjeant, 1974; Varisco, 1985, 1993).

In recent years, working to recognize, validate and maintain traditional knowledge has been a substantial project component. Initiatives have been developed that strengthen traditional knowledge systems. The more extractive approaches of traditional ethnobotanists keep TK in context and not completely protected. In fact, the number of international forums considering how best to protect traditional technologies and knowledge has been rapidly increasing. The trend is growing toward recognizing or creating rights of control in farming communities over genetic resources and related knowledge.

Thus far TK has not been captured and stored systematically; the danger is that it may be lost altogether. Even now, TK about cultivated and wild species is rapidly being lost. Genetic information coded in wild species and traditional crop varieties could be lost as intensive monocultural production favors newer high-yielding crops. The collective knowledge of biodiversity and how to use and manage it are maintained in cultural diversity; conserving biodiversity often helps strengthen cultural integrity and values (WRI et al., 1992).

In an effort to conserve and promote a better understanding of indigenous knowledge systems, UNESCO launched the Local and Indigenous Knowledge Systems (LINKS) project in 2002. Since its inception, LINKS has supported several field documentation efforts. In addition to empowering communities in biodiversity governance by recognizing them as knowledge holders, the project seeks to maintain the vitality of local knowledge within communities. The key is to strengthen ties between elders and youth, to reinforce the transmission of indigenous knowledge and know-how from one generation to the next. The International Treaty on Plant Genetic Resources for Food and Agriculture (www.fao.org), already ratified by several countries in the region, recognizes the enormous contribution that farmers and their communities have made and continue to make to the conservation and development of plant genetic resources. This is the basis for farmers' rights, which include protection of TK and the right to participate equitably in sharing benefits and in national decision making about plant genetic resources. Farmers possess invaluable knowledge, including the ability to choose appropriate varieties or breed for particular agricultural ecosystems. Their contribution is increasingly being recognized, as is their right to receive more benefits, including monetary benefits.

In CWANA the number of publications on the relevance

of TK in several areas has grown exponentially. Helping local people use their own knowledge of indigenous foods and agriculture provides better prospects for long-term sustainability than imposing solutions from outside. To date, however, little has been documented about the foods grown and used in poorer parts of the region, particularly as to how these foods are preserved for later use in a hostile environment. Today in rural Sudan various foods are being considered from the perspective of nutrition and food microbiology and for their part in the social fabric and the struggle for survival (Dirar, 1993). Information was gathered from elderly rural women who traditionally hand down such knowledge from generation to generation. With increased urbanization and dislocation of family structures, there is danger that such knowledge will be lost unless it is documented.

2.5.2.1 Plants

North Africa has one of the oldest and richest traditions using medicinal plants, important especially in rural areas, because they are frequently the only medicine available. Even in many urban areas, the price of modern medicine is increasing and people are turning back to traditional plant remedies.

The demand for medicinal plants is currently increasing in both developed and developing countries because of their accessibility, affordable cost and the growing recognition that natural products have fewer side effects. Therefore, a number of important plant species have become scarce in areas where they were previously abundant and some species may become threatened with extinction if their collection is not regulated. The theme of medicinal plants, relevant in most countries in North Africa, is a good entry point for biodiversity conservation. Use depends on local knowledge, which is based on traditional techniques linked to local identity.

Local communities, such as the Bedouins in Egypt, possess invaluable knowledge of nature. This TK is being gathered, documented and fed into a regional compendium on medicinal plants. Most Egyptians rely on modern medicines, although herbalists and their shops still thrive. The Bedouin communities, with much stronger traditional culture, have a real interest in medicinal plants. The demand for medicinal plants in Egypt is big, but most are for export to the USA and Europe. Of the 2,000 species of plants in Egypt, 1,000 occur within 30 km of the Mediterranean coast. Many of Egypt's plants have become rare or extinct from habitat destruction, overgrazing and overharvesting.

The Center and Garden for Conservation of Threatened Plants was built near El Hammam to conserve medicinal plants under threat in North Africa and to serve as an education and awareness center for the entire region. The garden undertook trials to cultivate plants under different conditions and propagate them. Transplants and propagules were exchanged with Bedouin nurseries, so they could cultivate plants in micronurseries. Four micronurseries and about 20 smaller ones established with the Bedouin communities on Bedouin lands focus on sustainable use of medicinal plants.

The cultivation of these plants, a new concept for the Bedouins, has slowly caught on because plants in the wild are diminishing in number and they realize that a market can be found for medicinal and culinary plants. These nurseries have been decisive in significantly reducing the uncontrolled gathering of endangered plant species (http://iucn.org/places/medoffice/nabp/index.html).

Food barley importance, uses and local knowledge

Case study 1. This ICARDA study highlights food barley production in over 14 countries (Grando and Gomez Macpherson, 2005). It includes a review of food barley farming systems, bottlenecks in production, research efforts in improvement, major cultivated varieties, quality characteristics desired by consumers and constraints to production and research. Local crop development is based on farmer knowledge of local crop varieties, their skills in adapting them to their environmental and socio-economic conditions and contributions of local seed systems. Papers presented in the book focus on describing varietal characteristics important to farmers; how farmers observe, select and experiment with crop varieties; and the techniques they employ for storing and distributing seed.

Barley grain is used as feed, malt and food. Our ancestors depended on barley as a staple food more than we do now. Barley was important in the origin and development of the Neolithic culture. Early barley remnants from Mesopotamia and Egypt suggest that barley was more important than wheat in the human diet. Nowadays, barley is an important staple food in several developing countries; generally it is the most viable option in places with harsh living conditions and home to some of the poorest farmers in the world.

Barley is still a major staple food in several regions of the world: some areas of North Africa and the Near East, the highlands of Central Asia, the Horn of Africa, the Andean countries and the Baltic states. Food barley is often found in regions where other cereals grow poorly because of altitude, low rainfall or soil salinity. It remains the most viable option in dry areas (< 300 mm of rainfall) and in production systems where alternatives for food crops are limited or absent, such as in highlands and mountains.

Food barley consumption has decreased considerably in the last 40 years with the increase of urban populations and often with the introduction of national policies supporting wheat consumption. In Morocco, food barley consumption decreased from 87 kg per person per year in 1961 to 57 kg in 1999. In 1961, yearly consumption per person was 27 kg in Algeria, 35 kg in Libya and 15 kg in Tunisia.

Food barley use is associated with local knowledge on preparation, health and nutritious attributes. Food barley is used either to make bread, usually mixed with bread wheat, or in specific recipes. Its cultivars have particular characteristics consumers appreciate that make them irreplaceable by feed or malting barley. Now local knowledge and unique genetic material are under risk of being lost for future generations.

Archaeobotanical and archaeozoological analyses of archaeological sites, in addition to ethnobotanical and subsistence base studies in contemporary rural societies, have indicated the likelihood that most of the ancient ways of obtaining food and materials have remained in use (Anderson and Ertuğ-Yaras, 1998). In Anatolia, for instance, about ten to twelve thousand years after domestication was successfully accomplished, wild plant gathering is still an active tradition in several parts of the country (Ertuğ, 1998, 1999, 2000a, 2000b, 2000c).

Wild plant gathering in agricultural societies

Case studies 2 and 3. Two case studies, conducted in Aksaray in central Anatolia and Muğla in southwestern Anatolia, indicate that people have adopted various ways of using their environment for food, medicine, fuel, fodder, building materials and many other purposes (Ertuğ, 2006).

In a single village and its surroundings in Aksaray, 300 locally used and named plants have been recorded. The villagers consider over 100 plant species edible, while others are used for medicine, fuel, fodder, building material, dye, gum and glue. In the Bodrum Peninsula in the Muğla area, about 360 useful wild plants have been recorded during a two-year study; 140 were used for food, about 100 were medicinal and others were used for various purposes such as making baskets, brooms and mats.

It is almost impossible to find two identical patterns of managing faunal and floral resources; some variations are apparent even within the same unit of study. Different wild plants are gathered in two adjacent regions and even within the villages in both regions. In both regions, however, although they now have access to fresh vegetables all year round, wild plants available in winter and spring continued to supplement the villagers' diets, during periods when, historically, fresh vegetables were scarce. This continuity of gathering may well be explained by nutritional needs of people and their search for "traditional tastes". While several plant uses, such as plaiting mats and fuel, are decreasing, gathering wild edibles is still more or less consistent in rural areas.

Inventory of traditional knowledge to combat desertification

Case study 4. UNESCO launched a global program, the Traditional Knowledge World Bank, for an inventory assigned to the IPOGEA Research Centre on Traditional and Local Knowledge to Combat Desertification. The project gathers and protects historical knowledge and promotes and certifies innovative practices based on modern restoration of tradition (Laureano, 2005).

Traditional knowledge and techniques were identified by surveys and studies in the field as well as by collecting photographs and current project documents. An iconographic system has been designed to show and easily identify the techniques and their use. Each technique matches an icon.

Each technique is linked to photographs, charts and drawings, project reports, bibliographical documents, analysis of exact references and geographic and chronological dissemination maps. All this information is in clusters of competence and in several categories, including Agriculture, Water Management, Soil and Environment Protection, Breeding, Hunting and Harvesting.

A case study in Wadi Mzab in Algeria classified traditional techniques according to natural context, rural settlement and urban settlement.

2.5.2.2 Water management

Large-scale water management techniques were developed by the ancient empires that flourished on the alluvial sediments of silt, loess and sand along the Afro-Asian river basins in the five subregions of CWANA. Great civilizations known as hydraulic societies prospered near rivers like the Nile, Euphrates and Tiger and in arid areas and oases. They developed hydraulic infrastructures to elevate water from rivers, such as the noria system of lifting water in buckets on a current-driven wheel in the Orontes River in Southwest Asia, for stoking and transporting water and for rain harvesting. The knowledge concerning water management and irrigation was transmitted down generations.

The validity of traditional knowledge on water management and the practices derived from it have been studied and documented since the 1980s. For the last 20 years research on traditional water techniques has aimed at overcoming a top-down approach to transferring water management technologies and at achieving a participatory relationship to foster sustainability (Brokensha et al., 1980). Many international bodies, such as the International Labour Organisation (ILO) (Bhalla, 1977), the Organisation for Economic Cooperation and Development (OECD), the Food and Agriculture Organization of the United Nations (FAO), the United Nations Educational, Scientific and Cultural Organization (UNESCO), the United Nations Environment Programme (UNEP) and the World Bank, have declared TK validity in research and documents. The interest of the United Nations conventions is clearly highlighted in the report entitled "Building Linkage between Environmental Conventions and Initiatives" (UNCCD, 1999).

2.5.2.3 Water harvesting

Water harvesting can be traced through human history almost as far as the origin of agriculture. This ancient practice sustained populations when conditions would have otherwise totally prevented agriculture, and many peoples in the world have continued to rely on water harvesting. Harvested water is used for drinking (although less commonly now, because even rainwater is less safe), irrigation, livestock drinking and groundwater recharge.

Various forms of water harvesting have been used throughout the centuries. Some of the earliest Middle East agriculture diverted wadi flow (spate flow from normally dry watercourses) onto agricultural fields. Reviewing archaeological evidence, Prinz (1994) notes indications of water-harvesting structures in Jordan, believed to have been constructed over 9,000 years ago and in southern Mesopotamia from 4,500 BCE (Bruins et al., 1986). A number of distinctive historical examples that incorporate effective water-harvesting systems survive in many CWANA countries. These include the cut-stone reservoirs of the Nabatean city of Petra in Jordan and the underground cisterns found in the country's Umayyad desert palaces, Crusader period castles and traditional village houses.

A sequence of reviews and manuals produced over the last 20 years provides a good inventory of old and new water-harvesting techniques, as well as essential information for their implementation (Frasier, 1974; GDRC, 1983; Pacey and Cullis, 1986; Reij et al., 1988; Critchley and Siegert, 1991; FAO, 1994; Prinz, 1999). Farmer innovations, ancient and modern, have stimulated research and research has started to solve problems on the farm.

In the early 1990s, several studies on traditional water-harvesting infrastructures were published (Prinz, 1996; Prinz and Wolfer, 1999). These techniques, which deeply mark the landscapes of the arid and semiarid areas, are regarded as part of our world inheritance. In North Africa, Saharan tourism has vigorously helped to promote them. In South Tunisia for example, several water-harvesting techniques given up in the 1960s have been reestablished to produce fresh fruits and vegetables for hotels and ecotourism.

In poorer regions, the productivity of land and water in rainfed areas is greatly enhanced by harvesting water. Marginal lands with annual rainfall of less than 300 mm a year can be cultivated if limited but controlled additional water is made available (Oweis et al., 1999; Rodriguez, 1996; Rodriguez et al., 1996). In many instances, appropriate water-harvesting techniques can provide an incremental water supply.

Modernization and diffusion of these ancient technologies have to be sought to increase agricultural productivity and provide a sustained economic base. The choice of technique depends on the rainfall and its distribution, land topography, soil type and depth and local socioeconomic factors, so these systems tend to be very site specific. The water-harvesting methods strongly depend on local conditions and include widely differing practices—bunding, pitting, microcatchments and harvesting flood- and groundwater (Critchley et al., 1992; Prinz, 1996).

Rainwater harvesting areas are not well mapped and few statistics are available nationally or regionally. Several experiences are quoted in specialized literature but little information is given about their importance in concerned areas, benefits to people and economic return. AQUASTAT FAO databases for CWANA are available only for Egypt (133,000 ha), Iran (40,000 ha), Lebanon (500 ha) and Tunisia (898,000 ha).

Work in Tunisia may be divided broadly into two types: description and rehabilitation of indigenous systems and the large-scale technical development program of the Departement de la Conservation des Eaux et du Sol. This program is one of the few in CWANA that integrates soil and water conservation into hydrological priorities (Selmi, 1994). As well as constructing bunds and terraces for conservation, it includes building small dams on watercourses high in the catchments of major rivers. Purposes include flood control, recharge of shallow groundwater for irrigation and reduction of siltation of major dams supplying domestic and industrial needs. Among the spinoffs is that hill farmers will have water from small dams to use in supplemental irrigation. This program seeks to conserve soil and water by focusing primarily on engineering works. Unfortunately, the socioeconomic problems and different options for land users are essentially neglected.

Indigenous systems in Tunisia have recently been described in two monographs, by Ennabli (1993) and Alaya et al. (1993). The former provides detailed descriptions of nearly 30 traditional systems for capturing and using water in the dry areas of Tunisia. The water interception, concentration, conveyance and storage techniques reported (many still in use) illustrate the wealth of ingenuity in human adaptation to dry environments. Alaya et al. (1993) focus on *tabia*, the earthen bunds widely and variously used in Tunisia to intercept and redirect runoff water to crops and trees. Though primarily an implementation manual, this book is also rich in descriptions of traditional practices.

The *meskat* system, using tabias to support olive plantations, covers about 300,000 ha in central Tunisia (Prinz, 1994). It comprises catchments of about 500 m^2 surrounded by tabia and spillways to control runoff flow into bunded plots of trees. This is a successful system, but according to Reij et al. (1988) it suffers heavily from increasing land pressure, resulting in a decrease of catchment areas and leading to lower efficiency.

The *jessour* system is based upon cultivating sediments built up behind large tabia, often stone-reinforced and with stone spillways, constructed in water cascading down narrow mountain valleys in southern Tunisia. Akrimi et al. (1993), from the Institut des Regions Arides (IRA) near Medinine, reported a multidisciplinary study involving jessour cultivators in the Matmata Mountains. Maintaining the tabia and spillways is a major problem in some areas, due partly to the degree of outmigration. It is Tunisian government policy to assist with jessour rehabilitation. Proposals for further research by the same IRA team note the launching of major development schemes for soil conservation and rainwater harvesting, but comment that community participation has been weak because the schemes have failed to take account of local traditions and existing production systems.

Development schemes in Jordan involve building earth dams that divert runoff to improve pastures and bunds to conserve soil and moisture on steep land. Research was started in 1987 by the University of Jordan to explore the development potential, particularly the water-harvesting potential of a 70-km^2 catchment under low rainfall (100-250 mm per annum) east of Amman (Taimeh, 1988). Irrigation from wadi flows trapped by earthen dams and microplots supporting fruit trees are two techniques that have shown promise, both socioeconomically and technically. Currently data collected on this catchment are used to develop a coupled prediction-optimization model for harvesting, storing and using water in similar dry areas of Jordan and elsewhere (Sarraf and Taimeh, 1994).

Other ongoing regional activities include a relatively large development project, with an included research component, in a steppe in southern Syria using the integrated management of soil, water and vegetation (Rashed, 1993). The project uses water supplied by various harvesting techniques and a limited groundwater supply to enhance production, particularly of forage crops and shrubs.

In Yemen, a major research focus is to conserve the ancient terrace system, parts of which have fallen into disrepair following socioeconomic changes. The terraces are not just to conserve soil and water but also to control water, including harvesting water for human consumption, flowing from the high, often degraded pasture lands and to protect the low-lying intensively cultivated banks of the main wadis and the flood irrigation systems. A new multidisciplinary project with a participatory approach addresses the socioeconomic, institutional and policy issues that are involved (Mouhred, 1994).

The rainfed coastal areas of Egypt have received considerable R&D attention over recent decades. Initially the aim was to settle the Bedouin population. Projects were undertaken to rehabilitate degraded rangeland and increase use of runoff, through terracing wadis, similar to Tunisian jessours and enhancing indigenous runoff farming systems (Perrier, 1986). More recently, the coastal areas have come to be seen as another small but potentially productive national agricultural resource and emphasis has shifted toward more intensive development. However, natural resource issues —water quantity and quality, population growth, environmental deterioration—remain the same (Abdel-Kader et al., 1994).

In highland Balochistan, in Pakistan, an indigenous *khuskaba* system uses bunds to guide runoff water and promote infiltration. Rodriguez et al. (1996) found 1:1 treatments (catchment:production area ratio) in valley floors increased seven-year wheat yields over controls, higher ratios having a risk of waterlogging in wetter years. Farmers practicing the indigenous khuskaba system adjust the size of the catchment according to soil moisture at planting and rainfall expectations for the season.

Several water storage practices have been passed down from generation to generation. The individual cistern is an ancient method that has been in continuous use with some modifications. Cisterns have long been used by people without access to adequate and safe water or villages lacking a local water source or not connected to a water supply network. Harvested rainwater stored in cisterns during the short rainy months can adequately sustain the water supply in isolated habitations. Cisterns can also be a multiuse resource; besides water for drinking and cooking, households can use extra water for irrigating productive home gardens and for watering livestock. In 1982, studies were conducted on traditional cisterns and what was necessary to build, modernize and manage more of them (Bourges et al., 1979; Fujimura, 1982).

The region is also rich in traditional knowledge related to irrigation—*kharez* in Pakistan and Afghanistan, *qanat* in Iran, *foggarras* in Tunisia and Algeria and *khettaras* in Morocco. The survival of these ancient irrigation systems is testimony to brilliant local engineering. Presumed to be of Persian origin and introduced to the Maghreb during the Arab conquest, these systems were partly responsible for the wealth of the former *ksours* along the trans-Saharan trade routes of the past.

A kharez (qanat) is an unlined tunnel in the hillside, bringing water by free flow from underground aquifers to be used for surface irrigation. Dug by local craftsmen from shafts at close intervals, they are small in size but may be many kilometers long. In Afghanistan, data of the last inventory, conducted in 1967, estimated that 6,470 kharez still supply water to 167,750 ha. Kharez are often used for the domestic water supply.

In North Africa, the simplicity and ingenuity of these underground systems allow the capture and distribution of groundwater over thousands of kilometers. The system works through a complex network of underground channels and storage chambers set 10 to 15 meters deep, to avoid loss through evaporation. Hundreds of conduits (*seguias*) carry water, bringing it eventually to the surface and thanks to a slight slope, leading it to gardens at a flow of 3 to 12 liters per second.

2.5.2.4 Intellectual property rights

Several proposals have been made, within and outside the IPR system, to "protect" traditional knowledge (Correa, 2001). Such proposals often fail to set out clearly the rationale for its protection. Any system of protection, however, is an instrument for achieving certain objectives. Therefore, a fundamental question, before considering how traditional knowledge may be protected, is to define why it should be.

Some understand the concept of protecting IPR, where protection means to exclude unauthorized use. Others regard protection as a tool to preserve traditional knowledge from uses that may erode it or negatively affect the life or culture of the communities that have developed and applied it. Overall, the main arguments for protecting traditional knowledge include:

- Equity considerations
- Conservation concerns
- Preservation of traditional practices and culture
- Prevention of unauthorized parties appropriating traditional knowledge components
- Promotion of its use and its importance in development

2.5.3 Human capacity enhancement

Enhancing human capacity is important for agricultural development; therefore, capacity building is primary in development programs. Capacity is built so that country scientists and extension staff become more able to carry out integrated agricultural research, disseminate the information, demonstrate techniques and transfer technologies.

National programs in the region vary widely in their development, capability and needs. Countries benefit through collaborating with regional international institutions operating in CWANA and by networking to improve and strengthen the capacity for adopting and transferring technology. ICARDA and other CGIAR centers in the region play a catalytic role in helping various regional countries.

Since its establishment in 1977, ICARDA has considered training, capacity building and networking as essential for institutions and individuals to keep pace academically and professionally with the rapid development in agricultural sciences, especially in developing countries. ICARDA recognizes that a well-trained cadre of agricultural technicians, scientists and managers is essential to develop effective and sustainable national agricultural research systems (NARS). The center has responded by working closely with NARS to develop and implement training programs that address their changing needs.

Based on the needs of NARS, the center offers many training options, including long-term courses, specialized short-term courses, individual non-degree training and MSc and PhD degree-related studies. ICARDA organizes regional, subregional and country courses, which are usually conducted in close collaboration with NARS. International courses are also organized in collaboration with other international and regional organizations on subjects of mutual interest.

Training at ICARDA changes annually in response to NARS training priorities. These priorities are usually presented by national scientists and discussed during the annual national, subregional and regional meetings with NARS and during regular work visits.

ICARDA has improved its training program to better address human capacity development (www.icarda.org) by

- Refining selection procedures of training participants
- Decentralizing large parts of training activities from its headquarters to national programs
- Placing more emphasis on specialized training courses, including the degree-related training programs

- Conducting regular follow-up studies on training
- Improving training materials and creating and updating a computerized database of training participants

Using national, regional and global networks is an effective way to develop, transfer, adopt and use new technology. Research and training networks are effective for linking national scientists with each other and with regional and international organizations. These networks also insure a continuous flow of information among interested scientists; they provide opportunities for donor organizations to allocate financial support to networks that suit their priorities and interest. A number of donor organizations and cooperating countries support and coordinate these networks in CWANA.

Training at CGIAR centers and most regional institutions has been considered an integral component of the overall activities. It is recognized as an educational process that requires more than information giving or skill development; it also requires a thorough understanding of training and the value of continuous, vigorous evaluation. Success in reaching training objectives can only be judged when those who receive training apply what they have learned and when changes can be observed in practice.

Regional institutions include the International Center for Advanced Mediterranean Agronomic Studies (CIHEAM), an international organization dedicated to postgraduate and specialized education, applied research and the development of Mediterranean agriculture.

IARCs training programs are not based on a professor–student relationship but rather on a mature partnership and are regarded as a two-way learning process through which exchange of experience becomes a natural outcome. So the training participants, regardless of their positions or duties, become future collaborators. Participants are mostly the future leaders of their national projects or programs. They can certainly play an active role in technology transfer and therefore in improving food production in their own countries. Centers offer a wide variety of training activities to meet the evolving needs of client countries. These include long-term group courses, specialized short-term courses, individual non-degree and degree courses, regional and subregional courses and in-country training courses. The last three types are usually conducted in close collaboration with the concerned NARS. Each of these training programs is aimed toward improving the professional skills of the training participants and hence developing their national programs (Bunting and Araujo, 1987).

Workshops, traveling workshops, seminars, meetings and exchange of visits of national program representatives comprise important components for strengthening national programs and serve as a forum for exchanging ideas and deciding on future collaborative activities with NARS staff. The national programs in most countries in the region conduct these activities, some of which, at the request of national scientists, are organized either at ICARDA headquarters or outside.

Some crop-based expert systems have recently been established in the WANA region, like the Wheat Expert System (WANA Wheat). Its aims are to establish an expert system that all countries in the WANA region can use, to disseminate information about the system on the Web and to train extension workers on how to use the developed system (www.claes.sci.org).

Information and communications technology (ICT) has played an important role in disseminating information and knowledge in the last decade. Many institutions have investigated using this technology to transfer information and knowledge in the agriculture domain. Both formal and informal sectors in most of the involved countries have established a Web-enabled system for transferring agricultural information to scientists, extension services and growers to inform and train them in how to adopt these new technologies. Regional and international agricultural research centers strengthen cooperation with their partners involved in technology transfer by providing improved services in the areas of publications, translation, library search and training. The centers also contribute to and participate in most of the regional and international agricultural information networks. Through its Communication, Documentation and Information Services (CODIS), ICARDA places high priority on increasing and further improving the quality of agricultural information and its subsequent dissemination and adoption by national programs in the WANA region and beyond.

IARCs in the region do not conduct specific research on agricultural extension or offer training in it. However, they recognize that unless farmers adopt an improved technology, it is almost useless. Therefore, IARCs play an important if indirect role in developing and transferring technology by various means, including on-farm testing, organizing field days and visits for farmers and policy makers and organizing traveling workshops, training courses and roundtable discussions for farmers, researchers, extension workers and government officials. They also assist in producing field guides and extension publications related to using the new technology. Such joint activities help bridge the gap between researchers and extension specialists and improve technology transfer and use.

2.5.3.1 Technology transfer and adoption activities in Central Asia and the Caucasus

Under the ICARDA coordinated program for Central Asia and the Caucasus (CAC), NARS are being strengthened to become more efficient and responsive to the emerging challenges in the region. So far, over 2,500 scientists have either been trained or given the opportunity to participate in various meetings, workshops and conferences (see www.icarda.org).

Plant genetic resources and germplasm development. In cooperation with Bioversity International (formerly IPGRI), plant genetic resource units have been established in each of the eight CAC countries. Collection missions have been undertaken and different crops collected and added to their genebank collections. New varieties of winter wheat have been developed based on material from the joint CIMMYT/ICARDA/Turkey Program on Winter Wheat Improvement.

2.5.3.2 Natural resource management

Encouraging progress has been made with the introduction and adoption of improved soil and water management technologies. This has been achieved under the project on Soil and Water Management initiated in Kazakhstan, Kyrgyzstan, Tajikistan, Turkmenistan and Uzbekistan and financed by the Asian Development Bank.

- Improved irrigation technologies, developed and tested on-farm, increased the average yield of winter wheat by more than 40%, reduced soil erosion by almost 60% and increased water use efficiency by 50-100%. These technologies are ready for adoption on approximately 1.4 million ha in Uzbekistan.
- In southern Kazakhstan, improved irrigation technologies have led to about 30% less water being used than with traditional furrow irrigation. It has also reduced the pressure on the drainage system by 40%.
- Experiments using treated wastewater to irrigate fodder and industrial crops and tree plantations in Kazakhstan and Tajikistan have led to promising potentials for saving scarce water resources.
- Under the rainfed semiarid conditions of northern Kazakhstan, minimum and zero tillage techniques have resulted in grain yields 15% higher than those from deep plowing. Zero tillage has already been adopted by farmers on approximately 10,000 ha.
- Reduced tillage has led to promising results in Turkmenistan; water productivity increased by 25% compared with traditional deep-plowing practices.
- In Uzbekistan and Tajikistan, cotton planted as a double crop after winter wheat gave similar yield under no-tillage when compared with traditional deep plowing and monocropping. The no-tillage practice has now been introduced on about 4,000 ha in the two countries.
- Wheat–cotton rotation is becoming popular with the introduction of conservation tillage, varietal adjustments and alternate furrow irrigation technologies. It is expected that the area under this rotation in the CAC region will increase to about 200,000 ha in the next two to three years.

2.6 AKST and Its Impact on Agricultural Production and Development Goals

2.6.1 Impact on agricultural production and development goals

Increased agricultural productivity in the twentieth century has greatly contributed to the alleviation of poverty and hunger and enhanced economic growth. These results have mainly been attributed to increased investments in agricultural R&D. Globally, nearly US$731 thousand million (or 1.7% of the world's GDP) was invested in all the sciences in 2000, including research conducted by public and private institutions (Pardey et al., 2006a).

Among the developing countries, real research expenditures between 1995 and 2000 increased the most in the Asia-Pacific (11.9%) and the Middle East and North Africa (11.5%) regions. The overall average of the annual growth rate in research spending for developing countries was 8.6% between 1995-2000. The lowest annual growth rates were 1.9% for "other developing countries" (which includes several former Soviet states) and 3% for sub-Saharan Africa. China and India achieved the highest annual growth rates in research expenditures of 19.7% and 12.2%, respectively.

Similarly, trends of public spending in agricultural R&D reveal that investments increased by 51% worldwide over the last two decades, from US$15.2 thousand million in 1981 to US$23 thousand million in 2000 (Pardey et al., 2006b). During the 1990s, developing countries as a group undertook more of the world's public agricultural research than did industrialized countries. The Asia-Pacific region has accounted for the largest share of the developing-country total since 1981, accounting for 32.7% of the global total agricultural spending on research in 2000. China and India alone accounted for 39.1% of the developing world's agricultural R&D expenditure in 2000, a large increase from their combined share of 22.9% in 1981.

Five developing countries (Brazil, China, India, South Africa and Thailand) accounted for 53.3% of the developing world's public investments in agricultural research in 2000, up from their 40% share in 1981. Meanwhile, only 6.3% of the global investment in agricultural R&D was conducted in 80 countries (mainly low income and home to 625 million people).

Research intensities (that is, agricultural R&D spending expressed as a percentage of agricultural GDP) provide relative measures of R&D investments. Industrial countries as a group spent 2.36% of agricultural GDP in 2000 on R&D, a noticeable increase over the 1.41% in 1981. Developing countries, on the contrary, have not experienced a measurable growth in the intensity of agricultural research since 1981. These countries spent only 0.53% of their agricultural GDP on R&D in 2000. These figures indicate that the scientific or knowledge intensity of agricultural production grew at a much faster rate in rich countries than in poor ones, suggesting an increased intensity gap over the past three decades (Pardey et al., 2006b). The Asia-Pacific region has experienced the lowest research intensity, <0.5%, since 1981. The West Asia and North Africa region is the second lowest region in terms of research intensity. Although most sub-Saharan countries had lower 2000 intensity ratios than in 1981, the research intensity in this region is still higher than it is in the Asia-Pacific and WANA regions.

Per capita agricultural R&D spending is another research intensity ratio. It reveals that rich countries spent US$692 per agricultural worker in 2000, while poor countries spent just US$10.

Historically, agricultural innovations in the form of improved crop varieties, livestock breeds and farm management practices were typically the result of farmer experimentation through adapting and developing earlier ideas and then passing on inventions to younger generations. Publicly funded research is relatively new. It began in the early to mid-1700s as part of the efforts of the agrarian societies that formed throughout the United Kingdom and Europe at that time. Consequently, the publicly funded and operated agricultural experiment stations developed around the mid-1800s (Pardey et al., 2006b). Both public and private agricultural R&D continued to evolve, from trial-and-error efforts of many individuals to large-scale input supply firms investing in their own private R&D facilities.

In agriculture, however, it is difficult for individuals to gain full advantage from their research investments. Thus it is widely held that government needs to invest in R&D for the public good. Even so, private investments are evident in agricultural R&D. About 36% of global spending on it in

2000 was by private firms and the remaining 64% by public agencies. Most of the private R&D investment (about 93%) was in rich countries and is extremely limited in developing countries at 6.3%. In industrialized countries 54% of the agricultural R&D is private; in developing countries, it is predominantly public and there are large disparities in the private contribution figures among the different regions of the developing countries. In the Asia-Pacific region, nearly 8% of the agricultural R&D investments are private compared with 3.5% in the Middle East and North Africa region. Among developing countries, private investment in agricultural R&D is lowest (1.7%) in sub-Saharan Africa (Table 2-17).

This pattern of private contributions to agricultural R&D investments has important implications for the intensity of agricultural research in all countries. In 2000, developing countries as a group had an agricultural R&D intensity ratio of 0.53% compared with 5.16% for industrial countries. This results in an intensity ratio of 9.2:1 compared with a 4.5:1 ratio if only public research investment were considered (Pardey et al., 2006a). Previous information on agricultural R&D expenditures suggests the following conclusions:

- There has been a slowdown of support for publicly funded agricultural research among developed countries. This is partially attributed to a shifting emphasis from publicly to privately funded agricultural R&D and to a shift in government spending priorities. In developing countries, including CWANA countries, the public sector undertakes most of the investment in agricultural R&D. The contribution of private funding is and will continue to be, limited. Thus, the public sector needs to fund future expansion in agricultural R&D investments.
- There is a clear reorientation of agricultural R&D in industrialized countries away from intensifying productivity in food staples toward concerns over the environmental effects of agriculture, as well as food quality and medical, energy and industrial applications of agricultural commodities. Such research reorientation has important implications for the links between industrialized and developing countries in improving the productivity of staples, which is still a priority research area for developing countries. This is particularly in line with current trends in research expenditure of international agricultural research centers toward environmental sustainability and policy at the expense of research on increasing productivity.
- Although limited, most private agricultural R&D in developing countries is oriented toward research on crop improvement or on export crops such as cotton, corn and sugar cane. This implies that the private R&D contribution is expected to stay minimal in research to increase productivity of staple crops. Publicly funded agricultural R&D will continue as the main source of such research in CWANA countries.

2.6.2 Options and insights for making more effective use of agricultural science and technology

To enhance the effectiveness of public investments in agricultural science and technology in the CWANA region, we suggest the following:

- Enhance technology strategies and priority setting. CWANA countries are invited to develop their strategies and research priority settings in line with their comparative advantages, resource endowments and contribution to the developmental goals of poverty alleviation, food security enhancement and natural resource sustainability. Regional research priorities for CWANA have already been developed by ICARDA in 2002 (Belaid et al., 2003). New efforts to orient national research priorities in CWANA countries need to capitalize on the new research focus of international agricultural research centers (represented by the CGIAR centers), which is directed toward agricultural development in developing countries.
- Define options and opportunities for optimizing the contribution of agricultural R&D and determine the best application of resources to meet research priorities.
- Develop and maintain appropriate agricultural science and technology databases. These include quantitative and qualitative information on changing research and funding environments as well as national, regional and global institutional changes.
- Identify complementary roles of different research partners, including NARS, advanced research institutions and CGIAR centers. ICARDA in its R&D continuum clearly draws the roles of different partners in the whole

Table 2-17. **Estimated global public and private agricultural R&D investments circa 2000.**

	Expenditures (million international $, year 2000)			Share (%)	
Region	Public	Private	Total	Public	Private
Asia–Pacific	7,523	663	8,186	91.9	8.1
Latin America and the Caribbean	2,454	124	2,578	95.2	4.8
Sub-Saharan Africa	1,461	26	1,486	98.3	1.7
Middle East and North Africa	1,382	50	1,432	96.5	3.5
Developing-country subtotal	12,819	862	13,682	93.7	6.3
High-income country subtotal	10,191	12,086	22,277	45.7	54.3
Total	23,010	12,948	35,958	64.0	36.0

Source: Pardey, et al., 2006b.

R&D chain. It also monitors changes in the research environments at all levels for implications on strategies and priorities of different organizations.

- Carry out ex ante and ex post research evaluation for accountability and resource allocation. This evaluation should lead to developing appropriate processes and mechanisms for allocating research resources for maximum effectiveness.
- Improve incentives to generate, access and use new technology. Investments in agricultural R&D can contribute significantly to feeding poor people. The potential benefit can be greatly enhanced if successful partnerships are further developed.

2.6.3 Dynamics influencing the role of women in agriculture

2.6.3.1 Land and agrarian reforms

According to the first resolution of the United Nations Subcommission on the Prevention of Discrimination and the Protection of Minorities "continued discrimination faced by women in all matters [related] to land and property is the single most critical factor in the perpetuation of gender inequality and poverty" (United Nations, 1995). Laws and social norms in many CWANA countries restrict women's ability to buy or inherit land, particularly agricultural land and resources, negatively affecting women's participation in agriculture.

In Iraq, land and agrarian reforms assigned plots to men and women alike and the law guaranteed gender-equal inheritance rights. The state recognized and supported women's roles as landowners and farmers. (Customary law, however, often prevails over state law and ownership of land continues to be predominantly exclusive to men.) In Syria, on the contrary, land reform assigned plots only to the male heads of household. Women became "helpers" rather than farmers in their own right. Their access to agricultural basics was limited and thus they lost independent access to food production and their control over produce revenue.

Since women lack control of the means for production and entitlement to what they produce, their access to loans and social security is often restricted, their autonomy and decision-making power are limited and consequently their ability to achieve food security is curtailed. Women's limited access to markets also curtails their control of farm income. As shown in a study on Jordan, women working on land they own, rent or sharecrop, rather than on household land, are much more likely to engage in marketing activities, control the income earned on the land and allocate household expenditures. Agriculture, however, is mainly a male activity in Jordan and land is predominantly owned by men. The percentage of women farming their own land is low, approximately 1% of Jordanian population and 11% of the female agricultural labor force (Flynn and Oldham, 1999).

2.6.3.2 Migration

Many countries in the CWANA region have been characterized by male rural-to-urban migration and by out-migration, mainly to the Gulf states. As a consequence, the number of female-headed households has increased substantially over the years. This has often been paralleled by agricultural intensification trends that in Jordan as in Egypt (Taylor, 1984), Gaza (Esim and Kuttab, 2002), Lebanon and Syria have caused an increasing demand for women's labour in agriculture (World Bank, 2005, 2006). Women more and more work as unpaid family laborers, their agricultural duties added to their domestic ones. In some countries female farmers have started also working off-farm in agriculture since revenue sent by migrated relatives is often not sufficient for survival and plots are too small to sustain the family. These situations have led to growing feminization of agriculture with increasing rates of women working in unpaid, informal systems. These systems are characterized by gender-based wage differentials, precariousness and lack of social services, all of which contribute to women's economic vulnerability. The increase in household workload also involves children, affecting their school attendance, free time and health.

These changes in the management of rural households have not been followed by adjustments to legal rights—such as property ownership, assets entitlements or labor rights—or to the agrarian systems—such as distribution of agricultural basics, market arrangements, technology introduction—that generally assume farmers to be male, thus favoring their needs, preferences and rights. These inequalities negatively affect women's agricultural work and arguably their agricultural productivity.

Migration also influences intrahousehold dynamics. Women may gain independence because of men's absence. They participate in decision making by managing small household budgets and their mobility is increased as they sometimes go to the market to sell their products even if they still rely on male relatives for major decisions such as the sale of an animal (cow, calf, sheep) (CNEA, 1996). Or women may lose independence if a male relative manages the household during the absence of the migrant man. In Syria, women seem to perform most of the agricultural work but do not have management or decision-making control, which has remained in the hands of male relatives (Abdelali-Martini et al., 2003). A study on Egypt in the 1980s reported that only women in independent households gained more control of their own lives if their husband migrated. In extended families, their autonomy was reduced by the increased control of the mother-in-law (Taylor, 1984).

2.6.3.3 Conflict

Women's rights to property, access to land and entitlement to agricultural basics are not effectively protected by either legal structures or social norms. In conflict and postconflict situations, when the number of female-headed households increases, these rights are even more difficult to demand and women's means for a sustainable livelihood are undermined. Women thus often resort to working in the informal sector, despite the constraints with regard to assets, markets, services, regulatory frameworks and the larger gender-based wage differential (Esim and Kuttab, 2002).

According to a study on the Palestinian conflict, women face the repercussions of the occupation, the gender-based discrimination to property rights and the obstacles due to traditional, patriarchal practices (Esim and Kuttab, 2002). Agriculture is the second most important sector of employment for women, and feminization of agriculture is a growing phenomenon. Apart from problems in claiming their rights to land and resources, women have to deal with an old agricultural system and techniques, since not much investment was ever made in agriculture because of the continuous occupation. Moreover, in 2001, women's agricultural activities were shrinking because land was confiscated and donor support for agriculture decreased. "In this context informal employment has become a survival mechanism, especially for households maintained by women" (Esim and Kuttab, 2002).

In Iraq, women-headed households are numerous in the rural areas and women are increasingly becoming a vulnerable group because of the ongoing violence. Women farmers are particularly vulnerable because they have limited control over production resources such as land and technology and reduced access to support services (UN/World Bank, 2003).

2.6.3.4 Mechanization and technology

Mechanization and labor-saving technology have radically changed agricultural production and work organization in rural areas. These changes have been beneficial to women in some cases and detrimental in others. Home-based technology, such as piped water and electricity, has helped reduce female domestic drudgery by reducing the amount of work necessary to collect fuel and water. Agricultural machinery, however, is usually designed for male users, thus reinforcing the gender division of labor. Big handles and heavy levers can impede women's use of machines. Social biases that associate machinery with men further limit women's use of technological improvements (Brandth, 2006). This is confirmed by a research project on Lebanese agriculture, according to which the low involvement of women in technology is due to practical difficulties in access and cultural restrictions on use. In addition, women's crops and livestock are usually disregarded as research priorities (ESCWA, 2001).

In the 1960s, when Egypt started mechanizing agricultural production, men's work began to change radically while women's work remained labor intensive (Saunders and Mehenna, 1986). The introduction of new agricultural technology in the Syrian countryside brought many farmers who had migrated back to the fields with prospects of increased production. Men took over the use of the machinery for land preparation or harvesting while women and children were assigned tedious manual jobs such as weeding and thinning. In some cases, the new machines have freed women from performing time-consuming tasks (FAO, 1995).

2.6.3.5 Globalization trends

Many countries of the CWANA region, such as Egypt, Jordan, Syria and Turkey, are moving toward structural adjustment policies that reduce agricultural subsidies, increase the role of the private sector and free market, decrease government expenditures and increase efficiency. Evidence shows that liberalization measures have mainly disfavored small-scale farmers and unskilled and informal workers. Women constitute a large part of these categories and are increasingly suffering from job insecurity. The increasing precariousness of work has affected mainly women, who are the first to be discriminated against in employment patterns. At the same time, the potential benefits connected to globalizing the labor force do not benefit women, whose working choices are restricted, for social reasons, to the internal labor market and eventually to conditions of limited reward.

Policies of market liberalization suffer from gender biases, and market dynamics have marginalized petty trading, which primarily involves women. Gender discrimination in state and market institutions and intrahousehold inequalities all reduce women's control over the income from their work (Baden, 1998). Social policies to counteract the marginalization of disadvantaged sectors have not been put in place. On the contrary, the retreat of the state from providing social security has greatly affected women, who have suffered from the lack of support. For example, women and girls are forced to compensate for the weakened public health system by caring for the old and sick at home. Migration trends have continued to intensify the female labor load in rural areas. Environmental degradation is adding pressure by affecting the ecosystem many depend on for their livelihoods (Sindzingre, 2004).

References

Abdelali-Martini M., P. Goldey, G. Jones, and E. Bailey. 2003. Towards a feminization of agricultural labour in northwest Syria. J. Peasant Studies 30(2):71-94.

Abdel-Kader, F., H. Schrier, and J. Fitzsimmons. 1994. Water/land management: North-west coast of Egypt. Pp. 19-22. *In* G. Cliche (ed) Roundtable on land and water management: Proc., Cairo. 13-15 Dec 1993. IDRC Reg. Off. North Africa Middle East, Cairo.

Ahmad, M., and J.G. Nagy. 2001. Pakistan private investment in agricultural research. AER-805. ERS/USDA, Washington, DC.

Akrimi, N., M. Neffati, H. Taamallah, and M. Sghaier. 1993. Resource management in the Graguer Basin of southern Tunisia. *In* Proc. Workshop on Dryland resource management and the improvement of rainfed agriculture in the drier areas of West Asia and North Africa, 23-27 May 1993. ICARDA, Aleppo.

Aksoy, U. 1999. Organic agriculture in Turkey. Biomed 2(1):8-9.

Alary, V. 2005. Obstacles to technology adoption for small and medium farms in the arid and semiarid areas of the Maghreb. Femise Res. Proj. FEM 21-05. Available at http://www.femise.net/activites/pub_ao21.html. ICARDA, Tunis.

Alaya, K., W. Viertmann, and T. Waibel. 1993. Les tabias. Direction Generale des Forets, Ministere de l'Agric., Tunis.

Albergel, J., T. Mansouri, P. Zante, A. Ben Mamou, and S. Abdeljaoued. 2005. Organic carbon in the sediments of hill dams in semiarid Mediterranean area. Pp. 289-299. *In* E.J. Roose et al. (eds) Soil erosion and carbon dynamics. Adv. Soil Sci. Series. Taylor and Francis, Boca Raton.

Albergel, J., and S. Nasri. 2001. Rapport final du programme de recherche sur les lacs collinaires dans les zones semiarides du pourtour méditerranéen. IRD/INRGREF, Tunis.

Albergel, J., and N. Regeb. 1997. Les lacs collinaires en Tunisie: Enjeux, contraintes et perspectives. Comptes rendus à l'Académie d'Agriculture de France. Séance du 19 Mars 83(2):77-88.

Alexandratos, N. (ed) 1995. World agriculture towards 2010. FAO, Rome.

Al Hadidi, K. 2002. Valuation of irrigation water—the case of Jordan. Water demand management forum on water valuation. Beirut, 25-27 June 2002.

Altiyev, T. 2005. Water for development in Central Asia. Pap. presented at subregional workshop Implementation of integrated water resources management for well-being and future development in Central Asia. Almaty, Kazakhstan, 31 Oct-2 Nov 2005.

Anderson, S., and F. Ertuğ-Yaras. 1998. Fuel, fodder and faeces: An ethnographic and botanical study of dung fuel use in central Anatolia. Environ. Archaeol. 1:99-108.

Anonymous. 2001. Le Gharb, hier, aujourd'hui et demain. Terre et Vie n° 47, Maroc Agri, Février.

AQUASTAT. 2002. FAO's information database on water and agriculture. Available at http://www.fao.org/ag/AGL/aglw/aquastat/. FAO, Rome.

ASTI. 2003a. ASTI Country Brief No. 2. IFPRI, Washington, DC.

ASTI. 2003b. ASTI Country Brief No. 9. IFPRI, Washington, DC.

ASTI. 2003c. ASTI Country Brief No. 11. IFPRI, Washington, DC.

ASTI. 2004. ASTI Country Brief No. 15. IFPRI, Washington, DC.

ASTI. 2005. ASTI Country Brief No. 27. IFPRI, Washington, DC.

ASTI. 2006a. ASTI Country Brief No. 29. IFPRI, Washington, DC.

ASTI. 2006b. ASTI Country Brief No. 35. IFPRI, Washington, DC.

ASTI. 2006c. ASTI Country Brief No. 36. IFPRI, Washington, DC.

Aubry, C., M. Elloumi, M. Essamet, A. Gana, M. Gara, C. Gitton et al. 1986. Les systèmes de production dans le semiaride: première approche de la dynamique des exploitations dans la région de Zaghouan. Annales de l'INRAT 59.

Aw-Hassan, A., and K. Shideed. 2003. The impact of international and national investment in barley germplasm improvement in the developing countries. *In* R.E. Evenson and D. Gollin (eds) Crop variety improvement and its effect on productivity: the impact of international agricultural research, CABI Publ., Wallingford.

Aw-Hassan, A., S. Farouk, and L. Iñiguez. 2005. Helping small enterprises capture the livestock products market in West Asia and North Africa. *In* ICARDA Caravan, Issue No. 22. ICARDA, Aleppo.

Aziz, F.C., B.I. Mohammed, and H.M. Sali. 1988. Nitrogen forms in calcareous soils and their correlation to nitrogen indices of wheat. Pp. 175-187. *In* J. Ryan and A. Matar (eds) Soil test calibration in WANA. ICARDA, Aleppo.

Baden, S. 1998. Gender issues in agricultural market liberalisation. BRIDGE Rep. No. 41. Inst. Dev. Studies (IDS), Sussex.

Baroudy, E., A. Abid Lahlou, and B. Attia. 2005. Managing water demand: policies, practices and lessons from the Middle East and North Africa. IDRC and IWA Publ., London.

Belaid, A., M. Solh, and A. Mazid. 2003. Setting agricultural research for the Central and West Asia and North Africa Region (CWANA): Toward a new NARS/NARS and CGIAR/NARS Collaboration Spirit. ICARDA, Aleppo.

Bernardi, M. 1996. La planification des stratégies contre les effets de la sécheresse. SD Dimensions. FAO, Rome.

Bessam, F., and R. Mrabet. 2001. Time influence of no tillage on organic matter and its quality of vertic calcixeroll in a semiarid area of Morocco. *In* Proc. Int. Congr. Conserv. Agric., Madrid. 1-5 Oct 2001.

Bhalla, S. 1977. Agricultural growth: Role of institutional and infrastructural factors. Econ. Polit. Weekly 12(45-46):5-12.

Birouk, A., A. Amri, and M. Sadiki. 1998. Establishing multidiciplinary groups and project preparation in Morocco. *In* D.I. Jarvis and T. Hodghin (eds) Strengthening the scientific basis of in situ conservation of agricultural biodiversity on-farm. Options for data collecting and analysis. Proc. Workshop to develop tools and procedures for in situ conservation on-farm, 25-29 Aug 1997. IPGRI, Rome.

Boers, T. M., and J. Ben-Asher. 1982. A review of rainwater harvesting. Agric. Water Manage. 5:145-158.

Bossio, D., W. Critchley, K. Geheb, G. van Lynden, and B. Mati. 2007. Conserving land: protecting water. *In* D. Molden (ed) Water for food, water for life: a comprehensive assessment of water management in agriculture. Earthscan, London.

Boufaroua, M., A. Alouini, M. Slimani, and S. Selmi. 2000. Lacs collinaires et conflits de gestion des ressources en eau, 2000. Available at http://afeid.montpellier.cemagref.fr/mpl2003/Conf/Boufaroua.pdf.

Bourges, J., C. Floret, G. Girard, and R. Pontanier. 1979. Etude de la citerne Telman (1972-77). ORSTOM-DRES, Tunis.

Brandth, B. 2006. Agricultural body-building: Incorporations of gender, body and work. J. Rural Studies 22(1):17-27.

Braun, A., J. Jiggins, N. Röling, H. van den Berg, and P. Snijders. 2006. A global survey and review of farmer field school experiences. ILRI, Nairobi.

Brokensha, D.W., D.M. Warren, and O. Werner (eds) 1980. Indigenous knowledge systems and development. Univ. Press America, Lanham, MD.

Bruins, H. J., M. Evenari and U. Nessler. 1986. Rainwater-harvesting agriculture for food production in arid zones: The challenge of the African famine. Appl. Geograp. 6:13-32.

Bunting, A.H., and E. Araujo. 1987. Training in the CGIAR system. CGIAR Secretariat, Washington, DC.

Bush, R., and M. Abdel Aal. 2004. Land and conflict in the Middle East and North Africa. Workshop 11 of the 5th Mediterranean Soc. and Pol. Res. Meeting, Florence. 24-28 March 2004. Available at http://www.iue.it/RSCAS/Research/Mediterranean/mspr2004/WS11.pdf.

Ceccarelli, S., and S. Grando. 2002. Plant breeding with farmers requires testing the assumptions of conventional plant breeding: Lessons from the ICARDA Barley Program. Pp. 297-332. *In* D.A. Cleveland and D. Soleri (ed) Farmers, scientists and plant breeding: Integrating knowledge and practice. CABI Publ., Wallingford.

Ceccarelli, S., and S. Grando. 2007. Decentralized-participatory plant breeding: an example of demand driven research. Euphytica 155(3):349-360.

Ceccarelli, S., S. Grando, M. Singh, M. Michael, A. Shikho, M. Al Issa et al. 2003. A methodological study on participatory barley breeding. II. Response to selection. Euphytica 133:185-200.

Ceccarelli, S., S. Grando, R. Tutwiler, J. Baha, A.M. Martini, H. Salahieh et al. 2000. A methodological study on participatory barley breeding. I. Selection phase. Euphytica 111:91-104.

CGIAR (Consultative Group for International Agricultural Research). 2005. Science for agricultural development: Changing contexts, new opportunities. CGIAR Sci. Council Secretariat, Rome.

Clarke, C. 2000. Cross-check survey: Final report. A WCD survey prepared as an input to the World Commission on Dams, Cape Town.

CNEA (Centre National des Etudes Agricoles). 1996. Impact assessment of the northwestern integrated development project on women. World Bank, Washington, DC.

Cole, C.V. 1996. Agricultural options for mitigation of greenhouse gas emissions. Pp. 745-772. *In* R.T. Watson et al. (eds) Climate change 1995: Impacts, adaptations and mitigation of climate change—scientific-technical analyses. Contribution of Working Group II to the Second Assessment of the Intergovernmental Panel on Climate Change. Cambridge Univ. Press, Cambridge, UK.

Correa, C.M. 2001. Traditional knowledge and intellectual property: Issues and options surrounding the protection of traditional knowledge. Disc. Pap., Nov 2001. Available at http://www.iucn.org/themes/pbia/themes/trade/training/TK%20and%20Intellectual%20Property.pdf. Quaker UN Off., Geneva.

Critchley, W., C. Reij, and A. Seznec. 1992. Water harvesting for plant production. Vol 2: Case studies and conclusions for sub-Saharan Africa. Tech. Pap. 157. World Bank, Washington, DC.

Critchley, W., and K. Siegert. 1991. Water harvesting: a manual for the design and construction of water harvesting schemes for plant production. AGL Misc. Pap. 17. FAO, Rome.

De Pauw, E. 2001. Drought early warning systems in West Asia and North Africa. *In* D.A. Wilhite et al. (eds) Early warning systems for drought preparedness and drought management. Proc. expert group meeting. Lisbon. 5-7 Sept 2000. WMO Publ. TD-1037. Available at http://www.drought.unl.edu/monitor/EWS/EWS_WMO.html. WMO, Geneva.

Dercon, S. 2000. Income risk, coping strategies and safety nets. Working Pap. No. 136. Centre Study African Econ., Oxford Univ., Oxford.

Diao, X., T. Roe, and A. Somwaru. 2002. Developing country interests in agricultural reforms under the World Trade Organization. Am. J. Agr. Econ. 84(3):782-790.

Dirar, H.A. 1993. The indigenous fermented food of the Sudan: a study in African food and nutrition. CABI Publ., Wallingford.

Easter, K.W., and Y. Liu. 2005. Cost recovery and water pricing for irrigation and drainage projects. World Bank, Washington, DC.

El Amami, S. 1983. Les aménagements hydrauliques traditionnels en Tunisie. CRGR (Centre de Recherche du génie Rural), Tunis.

Ennabli, N. 1993. Les amenagements hydrauliques et hydro-agricoles en Tunisie. Institut National agronomique de Tunis, Tunisia.

Ertuğ, F. 1998. Plant gathering versus plant domestication: An ethnobotanical focus on leafy plants. Pp. 218-223. *In* A.B. Damania et al. (eds) The origins of agriculture and crop domestication. Proc. Harlan Symp, Aleppo, 10-14 May 1997. ICARDA, Aleppo.

Ertuğ, F. 1999. Plants used in domestic handicrafts in central Turkey. OT Sistematik Botanik Dergisi 6(2):57-68.

Ertuğ, F. 2000a. Baharın Müjdecisi: Harbinger of spring: Crocus (çiğdem) or AN.TAH.SUM. SAR: a small contribution to the Anatolian flora of the Hittite period. In Turkish. Turk. Acad. Sci. J. Archaeol. (TÜBA-AR) 3:129-136.

Ertuğ, F. 2000b. An ethnobotanical study in central Anatolia (Turkey). Econ. Bot. 54(2):155-182.

Ertuğ, F. 2000c. Linseed oil and oil mills in central Turkey: Flax/linum and eruca, important oil plants of Anatolia. Anatolian Studies 50:171-185.

Ertuğ, F. (ed) 2006. Ethnobotany: at the junction of the continents and the disciplines. Proc. Fourth Int. Congr. Ethnobotany (ICEB 2005). Ege Yayinlari, Istanbul.

ESCWA (Econ. Social Commission for Western Asia). 1998a. Transboundary water resources in the ESCWA region: Utilization, management and cooperation. United Nations, NY.

ESCWA (Econ. Social Commission for Western Asia). 1998b. Challenges and opportunities of the new international trade agreements for the ESCWA member countries in selected sectors. United Nations, NY.

ESCWA (Econ. Social Commission for Western Asia). 2001. Gender in agriculture and agro-processing in Lebanon. United Nations, NY.

Esim, S., and E. Kuttab. 2002. Women's informal employment in Palestine: Securing a livelihood against all odds. Background Pap. for the ERF 9th annual conference, Sharjah, United Arab Emirates. 26-28 Oct 2002. Available at http://www.erf.org.eg/9th%20annual%20conf/9th%20conference_main_background.php (verified 17 Sep. 2007). Econ. Res. Forum, Cairo.

Evanari, M., L. Shanan, and N. Tadmor. 1971. The Negev, the challenge of a desert. Harvard Univ. Press, Cambridge, MA.

Evenson, R.E., and D. Gollin (ed) 2003. Crop variety improvement and its effect on productivity: the impact of international agricultural research. CABI Publ., Wallingford.

Falkenmark, M. 1997. Meeting water requirements of an expending world population. Phil. Trans. R. Soc. B 352: 929-936.

FAO. 1994. Water harvesting for improved agricultural production. Proc. Expert Consultation. Cairo, 21-25 Nov 1993. Water Rep. 3. FAO, Rome.

FAO. 1995. Women, agriculture and rural development: A synthesis report of the Near East Region. FAO, Rome.

FAO. 1996a. The state of the world's plant genetic resources for food and agriculture. FAO, Rome.

FAO. 1996b. Global plan of action for the conservation and sustainable utilisation of plant genetic resources for food and agriculture and Leipzig declaration, adopted by the Int. Tech. Conf. Plant Genetic Resources, Leipzig. 17-23 June 1996. FAO, Rome.

FAO. 2002. The state of food and agriculture 2002. FAO, Rome.

FAO. 2004. Food safety and international trade in the Near East region, Doha, Qatar, 13-17 March 2004. 27th FAO Reg. Conf. Near East. FAO, Rome.

FAO. 2007. Country data for Turkey extracted from FAOSTAT. Available at: http://www.fao.org/es/ess/top/country.html. FAO, Rome.

FAO/UNDP. 1970. Turkey: Technical report prepared by the FAO Agric. Res. Intro. Centre, Izmir, for the government of Turkey. FAO, Rome.

Fayzieva, D.Kh., N.S. Atabekov, I.A. Usmanov, A.A. Azizov, S.N. Steblyanko, S. Aslov, and A.S. Khamzina. 2004. Hydrosphere and health of population in the Aral Sea Basin. p. 51-100. *In* D. Fayzieva (ed) Environmental health in Central Asia: The present and future. WIT Press, Southampton.

Flynn, D.K., and L. Oldham 1999. Women's economic activities in Jordan, Research findings on women's participation in microenterprise, agriculture, and the formal sector. WIDTECH, Washington, DC.

Frankel, O.H. 1985. Genetic resources: The founding years. Diversity 7:26-29.

Frasier, G.W. (ed) 1974. Proc. Water Harvesting Symp. Phoenix AZ. 26-28 Mar 1974. USDA, Washington, DC.

Fujimura, F.N. (ed) 1982. Rain water cistern systems. Proc. Int. Conf. 15-17 Jun 1982. Water Res. Center, Univ. Hawaii, Honolulu.

Gana, A., and M. El Amrani. 2006. Crise hydraulique au Maghreb: Raréfaction de la ressource ou problèmes de gestion ? Géocarrefour 81.

GDRC. 1983. Rain and stormwater harvesting in rural areas. Global Dev. Res. Center, Tycooly Int. Publ., Dublin.

Gerber, J. 2000. National policies and the limits of international integration. Estey Centre J. Int. Law Trade Policy. 1(1):11-22.

Giger, M. 2006. Sustainable development in relation to rural contexts and policies. Available at http://.www.cde.ch/university/pdf/sd_rural_context_policies(1).pdf.

Grando, S., and H. Gomez Macpherson (ed) 2005. Food barley: Importance, uses and local knowledge. ICARDA, Aleppo.

Gregersen, H. 2003. Impact of the CGIAR: Everybody's business. CGIAR Ann. Rep. Sect. 2. CGIAR, Washington, DC.

Hamdi, A., and C. Lacirignola. 1994. Water resources management in the Mediterranean basin. p. 1-28. *In* A. Hamdi (ed) Land and water resources management in the Mediterranean region. Int. Conf., Valencano, Bari. 4-8 Sept 1994. CIHEAM/IAM, Valencano, Italy.

Harlan, J.R. 1951. Anatomy of gene centres. Am. Naturalist 85:97-105.

Hindiyeh, M.Y. 1995. Enumeration and survival studies on helminth eggs in relation to treatment of anaerobic and aerobic sludges in Jordan. PhD. thesis. Univ. Newcastle upon Tyne, UK.

Howes, M., and R. Chambers. 1980. Indigenous technical knowledge: analysis, implications and issues. p. 329-340. *In* D.W. Brokensha et al. (eds) Indigenous knowledge systems and development. Univ. Press America, Lanham, MD.

ICARDA. 1991. Annual Rep., farm resource management program. ICARDA, Aleppo.

ICSU (International Council for Science). 2002. Science and traditional knowledge. Rep. from the ICSU Study Group on Sci. Trad. Knowl. ICSU, Paris.

IPCC (Intergovernmental Panel on Climate Change). 1998. The regional impacts of climate change: an assessment of vulnerability. Cambridge Univ. Press, Cambridge.

IPCC (Intergovernmental Panel on Climate Change). 2007. Climate change 2007: The physical science basis. Summary for policy makers. Contribution of Working Group I to the Fourth Assessment Report of the IPCC. IPCC, Geneva.

Jaber, B. 1997. La problématique de l'eau au Liban. Ministère des Ressources Hydrauliques, Beirut, Lebanon.

Jamali, A., and A. Kefati. 2002. Wastewater reuse in Morocco. Water demand management forum on wastewater reuse. Rabat, 26-27 Mar 2002.

Jarvis, D.I., and T. Hodghin (eds) 1998. Strengthening the scientific basis of in situ conservation of agricultural biodiversity on-farm: Options for data collecting and analysis. *In* Proc. Workshop to develop tools and procedures for in situ conservation on-farm. 25-29 Aug 1997. IPGRI, Rome.

Jarvis, D.I., L. Myer, H. Kelmick, L. Guarino, M. Smale, A.H.D. Brown, et al. 2000. A training guide for in situ conservation on-farm. Vers. 1. IPGRI, Rome.

Jarvis, D.I., B. Sthapit, and L. Sears (ed) 2000. Conserving agricultural biodiversity in situ: a scientific basis for sustainable agriculture. IPGRI, Rome.

Job, J.O. 1992. Les premiers aménagements hydroagricoles en Mésopotamie et les problèmes de sels. p. 474-480. *In* A. Cornet and M. Grouzis (ed) L'aridité, une contrainte au développement. Collection Didactiques, Orstom, Paris.

Karagoz, A. 1998. In situ conservation of plant genetic resources in Ceylanpinar State Farm. Pp. 87-93. *In* N. Zencirci et al. (eds) Proc. Int. Symp. In situ Conservation of Plant Genetic Diversity, Antalya, 4-8 Nov 1996.

Karajeh, F., R.A. Mills, F. BenJemaa, and M. Prillwitz. 2004. Issues confronting water recycling in California. Int. Water Demand Manage. Conf. Dead Sea, Jordan. Min. Water Irrigation, Jordan and USAID, Washington, DC.

Kaya, Z. 2003. The effects of social and environmental problems on conservation programs of Turkey. MEDRAP (Concerted Action to Support the Elaboration Northern Mediterr. Reg. Action Prog.) Workshop, Montpellier, 15-20 Jan 2003.

Kaya, Z., E. Kun, and A. Guner. 1998. National plan for in situ conservation of plant genetic diversity of Turkey. Çevre Bakanlığı, Ankara.

Kenanoğlu Z. and O. Karahan 0. 2002. Policy implementations for organic agriculture in Turkey. Brit. Food J. 104(3/4/5):300-318.

Keraita, B.N., and P. Drechsel. 2004. Agricultural use of untreated urban wastewater in Ghana. p 101-112. *In* C.A. Scott et al. (eds) Wastewater use in irrigated agriculture. CABI Publ., UK.

Khouri, J., A. Amer, and A. Salih. 1995. Rainfall water management in the Arab region. UNESCO/ROSTA working group. ROSTA, Cairo.

Kijne, J.W. 2005. Aral Sea Basin Initiative: towards a strategy for feasible investment in drainage for the Aral Sea Basin. IPRTID/FAO, Rome.

Kjellqvist, E. 1975. A regional plan for collection, conservation and evaluation of genetic resources. Pp. 455-456. *In* O.H. Frankel and J.G. Hawkes (ed) Crop genetic resources for today and tomorrow. Cambridge Univ. Press, Cambridge.

Kumuk, T., and L. van Crowder. 1996. Harmonizing training and visit extension: some experiences from Turkey. Available at http://www.fao.org/sd/EXdirect/EXan0011 .htm. SD Dimensions. FAO, Rome.

Lal, R. 2001. Potential of desertification control to sequester carbon and mitigate the greenhouse effect. Clim. Change 51:35-72.

Lal, R. 2002. Carbon sequestration in dryland ecosystems of West Asia and North Africa. Land Degrad. Dev. 13:45-59.

Lal, R. 2006. Enhancing crop yields in the developing countries through restoration of the soil organic carbon pool in agricultural lands. Land Degrad. Dev. 17(2):197-209.

Latiri, K. 2005. Les céréales en Tunisie. p 259-279. *In* Les défis de la terre : l'agriculture en Espagne et en Tunisie face aux défis de la libéralisation. CERES, IRESA.

Laureano, P. 2005. The water atlas: traditional knowledge to combat desertification. IPOGEA, UNESCO and Laia Libros, Barcelona.

Lazarova, V., and A. Bahri (ed) 2005. Water reuse for irrigation: Agriculture, landscapes, and turf grass. CRC Press, Baton Rouge, LA.

Marsden,T. 2000. Rural sustainability. TCPA Tomorrow Ser. Pap. 4. Town Country Planning Assoc., London.

McCormick, P.G., S.R. Grattan, and I. Abu-Eisheh. 2003. Water quality challenges to irrigated agriculture productivity in the Jordan Valley. *In* A.J. Clemmens and S.S. Anderson (eds) Water for a sustainable world: limited supplies and expanding demand. Proc. Second Int. Conf. Irrigation and Drainage. US Comm. Irrigation and Drainage, Phoenix, AZ.

Merlinda, I. 2002. Agricultural policy reforms in the new multilateral trade round. Am. J. Agr. Econ. 84:798-799.

Miner, W.L. 2001. An overview of the issues and positions of the major countries in the WTO negotiations. Estey Centre J. Int. Law Trade Policy 2(1):1-23.

Miski and Shawaf. 2003. Aspects of groundwater resources management, protection and monitoring in Syria. Rep. prepared for ACSAD, ACSAD-BGR Technical cooperation project Management, protection and sustainable use of groundwater and soil resources in the Arab Region between BGR and ACSAD. ACSAD, Damascus.

Mouhred, A.W. 1994. Dryland resource management in northern highlands of Yemen. Pp. 49-56. *In* G. Cliche (ed) Roundtable on land and water management. Cairo. 13-15 Dec 1993. IDRC Reg. Off. North Africa and the Middle East, Cairo.

Nasr, M. 1999. Assessing desertification and water harvesting in the Middle East and North Africa: Policy implications, ZEF Disc. Pap. Dev. Policy No. 10. Center for Dev. Res., Bonn.

Nasri, S., A. Albergel, C. Cudennec, and R. Berndtsson. 2004. Hydrological processes in macrocatchment water harvesting in the arid region of Tunisia: The traditional system of tabias. Hydrol. Sci. 49(2):261-272.

Oldeman, L.R. (ed) 1988. Guidelines for general assessment of the status of human-induced soil degradation: global assessment of soil degradation (GLASOD). Int. Soil Ref. Information Centre Working Pap. Reprint No. 88/4. ISRIC, Wageningen.

Oweis, T., A. Hachum, and J. Kijne. 1999. Water harvesting and supplemental irrigation for improved water-use efficiency in dry areas. System-wide Initiative on Water Management. Pap. No. 7. IWMI, Colombo.

Oweis, T., M. Pala, and J. Ryan. 1998. Stabilizing rainfed wheat yields with supplemental irrigation and nitrogen in a Mediterranean climate. Agron. J. 90:672-681.

Ozhatay, N., M. Koyuncu, S. Atay, and A. Byfield. 1997. The trade in natural medicinal plants in Turkey. Dogal Hayati Koruma Dernegi, Istanbul and Fauna Flora Int., Cambridge.

Pacey, A., and A. Cullis. 1986. Rainwater harvesting: the collection of rainfall and runoff in rural areas. IT Publ., London.

Pagiola, S., and J. Kellenberg. 1997. Mainstreaming biodiversity in agricultural development: toward good practice. Environ. Pap. No. 15. World Bank, Washington, DC.

Pardey, P.G., J.M. Alston, and R.R. Piggott (eds) 2006a. Agricultural R&D in the developing world: too little, too late? IFPRI, Washington, DC.

Pardey, P.G., N. Beintema, S. Dehmer, and S. Wood. 2006b. Agricultural research: A growing global divide? ASTI Initiative. IFPRI, Washington, DC.

Perrier, E.R. 1986. Notes prepared for a workshop: Investigation of alternative agricultural water management practices for the northwestern coastal region of Egypt. ICARDA, Aleppo.

Prinz, D. 1994. Water harvesting: past and future. *In* L.S. Pereira (ed) Sustainability of irrigated agriculture. Proc. NATO Advanced Res. Workshop, Vimiero, 21-26 Mar 1994. Balkeman, Rotterdam.

Prinz, D. 1996. Water harvesting: History, techniques, trends. Z.f. Bewaesserungswirtschaft 31(1):64-105.

Prinz, D. 1999. Water harvesting techniques in the Mediterranean. Proc. Sem. Water Harvesting and Water Storage in the Mediterranean. Lund Univ., Sweden, June 1998.

Prinz, D., T. Oweis, and A. Oberle. 1998. Rainwater harvesting for dry land agriculture: developing a methodology based on remote sensing and GIS. Proc. 13th Int. Congr. Agric. Engineering, 2-6 Feb 1998. ANAFID, Rabat.

Prinz, D., and S. Wolfer. 1999. Traditional techniques of water management to cover future irrigation water demand. Z. Bewässerungswirtschaft 34(1):41-60.

Qadir, M., and S. Schubert. 2002. Degradation processes and nutrient constraints in sodic soils. Land Degrad. Dev. 13:275-294.

Qadir, M., B.R. Sharma, A. Bruggeman, R. Choukr-Allah, and F. Karajeh. 2007. Non-conventional water resources and opportunities for water augmentation to achieve food security in water scarce countries. Agric. Water Manage. 87:2-22.

Qadir, M., D. Wichelns, L. Raschid-Sally, P.S. Minhas, P. Drechsel, A. Bahri, et al.

2007. Agricultural use of marginal-quality water: opportunities and challenges. *In* D. Molden (ed) Water for food, water for life: a comprehensive assessment of water management in agriculture. Earthscan, London.

Raitzer, D. 2003. Benefit-cost-meta analysis of investment in the international agricultural research centers of the CGIAR. CGIAR Sci. Council Sec. FAO, Rome.

Ramaswami, B., S. Ravi, and S.D. Chopra. 2003. Risk management in agriculture. Disc. Pap. in Econ. No. 03-08. June 2003. Indian Statist. Inst., Delhi.

Rashed, E. 1993. Integrated watershed development: Syria. IDRC Project Doc. IDRC Middle East Reg. Off., Cairo.

Rashid, A. 2000. Taliban—militant Islam: Oil and fundamentalism in Central Asia. Yale Univ. Press, New Haven.

Reardon, T., and J. Berdegué. 2002. The rapid rise of supermarkets in Latin America: Challenges and opportunities for development. Dev. Policy Rev. 20(4):371-388.

Reij, C., P. Mulder, and L. Begemann. 1988. Water harvesting for plant production. Tech. Pap. No. 91. World Bank, Washington, DC.

Rodriguez, A. 1996. Sustainability and economic viability of cereals grown under alternative treatments of water-harvesting in highland Balochistan. Pakistan J. Sustain. Agric. 8(1):47-59.

Rodriguez, A., N.A. Shah, M. Afzal, U. Muatafa, and I. Ali. 1996. Is water-harvesting in valley floors a viable option for increasing cereal production in highland Balochistan, Pakistan? Exp. Agric. 32:305-315.

Roose, E. 1996. Land husbandry: components and strategy. FAO Soils Bull. 70. FAO, Rome.

Ryan, J., and A. Matar. 1988. Soil test calibration in West Asia and North Africa. ICARDA, Aleppo.

Sarraf, S., and A. Taimeh. 1994. Development of optimal strategies for water harvesting and utilization in arid lands, Jordan. Pp. 23-29. *In* G. Cliche (ed) Roundtable on land and water management. Cairo, 13-15 Dec 1993. IDRC Reg. Off. North Africa and the Middle East, Cairo.

Sartre, M. 2001. D'Alexandre à Zenobie: histoire du Levant antique. Fayard, Paris.

Saunders, L.W. and S. Mehenna. 1986. Unseen hands: Women's farm work in an Egyptian village. Anthro. Q. 55(3):105-114.

Selmi, S. 1994. L'Operation las colliniare dans les gouvernorats de Siliana, Kassine et Kairouan. ORSTOM, Tunis.

Sencer, H.A. 1975. Recent and proposed exploration activities of the Izmir Centre, Turkey. p. 151-157. *In* O.H. Frankel and J.G. Hawkes (eds) Crop genetic resources for today and tomorrow. Cambridge Univ. Press, Cambridge.

Serjeant, R.B. 1974. Cultivation of cereals in medieval Yemen. Arabian Studies 1:25-74.

Sidahmed, A.E., and J. Yazman. 1994. Livestock production and the environment in lesser developed countries. Pp. 13-31. *In* J. Yazman and A.G. Light (ed) Proc. Int. Telecomputer Conf. Perspectives on Livestock Res. Dev. Lesser Developed Countries, Nov 1992-April 1993. IDRC, INFORUM, Winrock.

Sindzingre, A. 2004. Explaining threshold effects of globalisation on poverty: an institutional perspective. UNU-WIDER Conf. Impact of Globalization on the World's Poor. First Project Meeting on Conceptual Issues. Helsinki, 29-30 Oct 2004.

Skaggs, R.W., and J. Van Schilfgaarde (eds). 1999. Agricultural drainage. ASA-CSSA-SSSA, Madison.

Skouri, M., and K. Latiri. 2001. L'agriculture pluviale dans les pays du Maghreb. Comptes rendus de l'Académie d'Agriculture, Séance du 14 Mars. 87(2):147-160.

Squires, V.R., and A. Sidahmed. 1997. Livestock management in dryland pastoral systems: prospects and problems. Ann. Arid Zone 36(2):79-96.

Swanson, B.E., C.M. Sands, and W.E. Peterson. 1988. Analyzing agricultural technology systems: some methodological tools. INTERPAKS, Univ. Illinois, Urbana-Champaign.

Taimeh, A.Y. 1988. Agricultural production in the semiarid to arid land and areas suffering from desertification. Project Rep. Univ. Jordan, Amman.

Taiyab, N. 2006. Exploring the market for voluntary carbon offsets. IIED, London.

Tan, A. 1998. Current status of plant genetic resources conservation in Turkey. Pp. 5-16. *In* N. Zencirci et al. (eds) Proc. Int. Symp. In-situ Conservation of Plant Genetic Diversity. Antalya, Turkey. 4-8 Nov 1996.

Tan, A. 2000. Biodiversity conservation: ex situ and in situ conservation—a case in Turkey. p.151-157. *In* K. Watanabe and A. Komamine (eds) Challenge of plant and agricultural sciences to the crisis of biosphere on the Earth in the 21st Century. Landes Bioscience—Eurekah, Austin TX.

Tan, A. 2001. Agricultural plant diversity in Turkey. OECD Expert Meeting on Agri-Biodiversity Indicators. 5-8 Nov 2001. Swiss Fed. Res. Station Agroecol. Agric. (FAL), Zurich.

Tan, A. 2002. Türkiye (Geçit Bölgesi) Genetik çeşitliliğinin in situ (çiftçi şartlarinda) muhafaza olanaklarinin araştirilmasi (In-situ on-farm conservation of landraces grown in north-western transitional zone of Turkey. In Turkish.). Final Rep. Tubitak-togtag-2347. Tubitak, Ankara.

Tan, A., and A. Inal. 2003. Ege tarimsal araştirma enstitüsü bitki genetik kaynaklari çalişmalari Plant genetic resources activities of Aegean Agricultural Research Institute. In Turkish. ETAE Publ. 112. Ege Tarımsal Araştırma Enstitüsü. Aegean Agric. Res. Inst., Izmir.

Tan, A., and A.S. Tan. 1998a. Database management systems for conservation of genetic diversity in Turkey. *In* N. Zencirci et al. (eds) Proc. Int. Symp. In situ Conservation of Plant Genetic Diversity, Antalya, 4-8 Nov 1996.

Tan, A., and A.S. Tan.1998b. Data collecting and analysis: for in situ, on farm, conservation. *In* D.I. Jarvis and T. Hodghin (eds) Strengthening the scientific basis of in situ conservation of agricultural biodiversity on-farm: Options for data collecting and analysis. Proc. workshop to develop tools and procedures for in situ conservation on-farm. 25-29 Aug 1997. IPGRI, Rome.

Tan, A., and A.S. Tan. 2002. In situ conservation of wild species related to crop plants: the case of Turkey. Pp. 199-204. *In* J.M.M. Engels et al. (eds) Managing plant genetic resources. CABI Publ., Wallingford.

Taylor, E. 1984. Women and labor migration. Mid. East Res. Inform. Proj. (MERIP) Rep. 124:3-10.

Tropp, H., M. Falkenmark, and J. Lundqvist. 2006. Water governance challenges: Managing competition and scarcity for hunger and poverty reduction and environmental sustainability. Human Development Report 2006. Available at http://hdr.undp.org/hdr2006/pdfs/background-docs/Background_papers/SIWI.pdf. UNDP, NY.

Tzannatos, Z. 2000. Social protection in the Middle East and North Africa: A review. Pap. presented at the Mediterranean Development Forum. Cairo, Mar 2000.

UNCCD. 1999. Building linkages between environmental conventions and initiatives. UNCCD, Bonn.

UNDP. 2004. Human development report. UNDP, New York.

UNEP. 2002. Global environmental outlook 3. UNEP, Nairobi.

UNEP. 2003. Design, testing and evaluation of best practices for in situ conservation of economically important wild species, with demonstration in Egypt, Lebanon, Morocco, Turkey. PDF-B project. UNEP, Nairobi.

United Nations. 1995. The right to adequate housing. Final Rep. submitted by Mr. Rajindar Sachar, Special Rapporteur. UN Doc. E/CN.4/SUB.2/1995/12. United Nations, NY.

UN/World Bank. 2003. Joint Iraq needs assessment working paper: Agriculture, water resources and food security. Available at http://lnweb18.worldbank.org/mna/mena.nsf/Attachments/IQ-AGRICULTURE/$File/AGRICULTURE+final+sector+report+16+October.pdf.

Varisco, D.M. 1985. The production of sorghum (dhurah) in highland Yemen. Arab. Stud. 7:53-88.

Varisco, D.M. 1993.The agricultural marker stars in Yemeni folklore. Asian Folkl. Stud. 52:120-142.

Warren, D.M., D. Brokensha, and L.J. Slikkerveer (ed) 1995. The cultural dimension of development: indigenous knowledge systems. Intermediate Tech. Publ., London.

Warren, D.M., and B. Rajasekaran. 1993. Using indigenous knowledge for sustainable dry-land management: a global perspective. Pp. 89-100. *In* D. Stiles (ed) Listening to the people: social aspects of dry-land management. UNEP, Nairobi.

Warren, D.M., L.J. Slikkerveer, and S.O. Titilola (eds). 1989. Indigenous knowledge systems: implications for agriculture and international development. Studies Tech. Social Change No. 11. Iowa State Univ., Ames.

Wilson, J.S. 2001. Bridging the standards divide: Recommendations for reform from a development perspective. World Bank, Washington, DC.

World Bank. 2005. The economic advancement of women in Jordan: A country gender assessment. Available at http://siteresources.worldbank.org/INTMNAREGTOPGENDER/Resources/JordanCGA2005.pdf. World Bank, Washington, DC.

World Bank. 2006. World development indicators 05. CD-ROM. World Bank, Washington, DC.

WRI (World Resources Institute), World Conservation Union, UNEP. 1992. Global biodiversity strategy: Policy-makers' guide. WRI Publ., Baltimore, MD.

Wylie, B.K., T.G. Gilmanov, D.A. Johnson, N.Z. Saliendra, K. Akshalov, L.L. Tieszen et al. 2004. Intra-seasonal mapping of CO_2 flux in rangelands of northern Kazakhstan at one-kilometer resolution. Environ. Manage. 33:S482-S491.

Yurtserver, N., and I. Gedikoglu. 1988. A tentative calibration of the Olson bicarbonate soil test with wheat responses to P in rainfed conditions of Turkey. Pp. 111-124. *In* J. Ryan and A. Matar (eds) Soil test calibration in WANA. ICARDA, Aleppo.

Zaibet, L., M.S. Bachta and A. Chaffai. 2003. Concerns about WTO negotiations: a developing country perspective. Working Pap. Dep. Agric. Econ., ESA Mograne, Zaghouan.

Zencirci, N., Z. Kaya, Y. Anikster, and W.T. Adams (ed) 1998. Proc. Int. Symp. In situ Conservation of Plant Genetic Diversity. Antalya, Turkey. 4-8 Nov. 1996. Central Res. Inst. Field Crops Publ., Ankara.

3

Agricultural Change and Plausible Futures

Coordinating Lead Authors:
Abdin Zein El-Abdin (Sudan), Syed Wajid H. Pirzada (Pakistan)

Lead Authors:
Jamal M. Abo Omar (Palestine), Sajid Kazmi (Pakistan), Rafael G. Litvak (Kyrgyz Republic)

Contributing Authors:
Younis Al Akhzami (Oman), Abdallah Muhammad Omezzine (Oman)
Review Editors:
Mustapha Guellouz (Tunisia) and Hamid Siadat (Iran)

Key Messages

1. The trend in population growth will continue, leading to strong increase in demand for food and intensified pressure on natural resources, high rates of unemployment, and increased urbanization and migration. The anticipated increase in urbanization and labor migration will speed up the shift of labor from agriculture to other sectors.

2. Hunger and malnutrition, especially child malnutrition, will continue to worsen, leading to a weakened workforce. The incidence of malaria, HIV and AIDS, tuberculosis, maternal mortality, and water-borne diseases will be on the rise in the region because of a shortage of proper health care.

3. With the increased pace of trade liberalization, agricultural producers will face problems of access to the market and competition in terms of price and quality vis-à-vis industrial countries. New barriers in terms of quality, social standards and intellectual property rights will further limit market access. Countries in the CWANA region have been denied market access for their agricultural output; labor and environmental standards in the multilateral trading system are also an obstacle for the region.

4. Although investment and improved management of AKST will proceed at a slow pace, it will have a greater role in agricultural development because of expanded research programs and better relations of research with extension agents and farmers. CWANA will need to improve and develop proper educational systems in agriculture and related sciences and reverse the trend of downgrading agriculture as an engine of development.

5. Indigenous knowledge will continue to lose ground to new technologies. Transgenics will continue to be a source of controversy because of cultural and traditional beliefs.

6. Good governance, democratization, decentralization and other sociopolitical reforms can remove stumbling blocks and speed up development. Pressures from globalization can contribute to adoption of democratic thinking in some countries, consequently reducing corruption in developing economies.

7. There is room for improvement in coordination at all levels—public, private and civil society—to facilitate development. Better coordination and collaboration will enhance development objectives, especially poverty alleviation and amelioration of hunger.

8. Despite the substantial role of women in agriculture, they will continue to have limited authority and ownership of resources such as land, labor, credit or capital. Nevertheless, through higher education for women and change in attitude, society will develop institutions of governance, legal systems and policies that are socially aware and gender sensitive, which will decrease the disparities between women and men.

9. Natural resources will face severe degradation and unsustainable productivity as a result of intensified exploitation. Increasing land degradation will limit the ability of agricultural systems to provide food security for the region. However, alternative scenarios show that there will be other options for improving the positive role of agriculture by minimizing its negative environmental effects.

10. Agrobiodiversity will continue to be endangered and many plant and animal species will be lost for various reasons. This will adversely affect food production and food security.

11. Use of agrochemicals will increase because of an expansion of intensive agriculture, which will lead to increased risk of pollution and adverse effects on human health. Nonetheless, increased awareness and modern AKST options for pest control, i.e., IPM, will strengthen the regulatory control of pesticide application at the regional level.

12. Water will continue to be the most limiting factor for agriculture because of competition from other sectors and more frequent droughts in the region. Water quality will also be at risk because ground water has been overdrawn and surface sources have been polluted.

13. Agriculture will be severely and adversely affected by climate change, especially in low-income regions. Developments in AKST will determine the ability of agricultural systems to respond to the consequent results of climate change.

14. Higher energy prices could encourage the use of more energy-efficient technologies in agricultural production.

15. Indicators show that agriculture may become an important producer of bioenergy. However, in the CWANA region, the issue of bioenergy is controversial and unpredictable because of its link to food production and biodiversity.

3.1 Introduction and Scope

This chapter builds on the major issues and challenges identified in chapter 2 and projects the corresponding future status. In other words, it explores a plausible future of the CWANA region, perceived for the next 50 years, in the short, medium and long term, and in this context takes the *reference world* scenario (i.e., business-as-usual) outlining plausible futures as a benchmark and adds value while prospecting the situation 50 years ahead. With this backdrop, the present scenario has a logical plot and narrative storyline—the manner in which events unfold over the next five decades. To this end, the scenario draws heavily upon the existing body of knowledge and learns from the historical analysis of trends, in qualitative terms. Although the scenario is not prescriptive in nature, it suggests the most likely outcomes for CWANA and agricultural knowledge, science and technology (AKST) in the region (World Bank, 2006b).

Apart from the business-as-usual scenario, we may perceive other plausible scenarios, such as

- Sunset agriculture, interfacing authoritative governance and weaker AKST institutions
- Sunrise agriculture, interfacing democratic governance and strong AKST institutions
- People-centered and development-oriented agriculture, interfacing globalization

Since the role of agriculture is likely to further decline as key employer in the long term, its contribution to national output may concomitantly decrease, as has happened in industrial countries (Fresco, 2002; Inter Academy Council, 2004b). However, given the political will to change, agriculture can potentially reemerge as a promising sector with value-added contributions toward economic growth and development in the region through AKST. In this context, it has a future as a sunrise sector (Fresco, 2002; Lipton, 2005). The world in general, and CWANA in particular, has witnessed the ills brought on by authoritative regimes and the models of development that have been implemented. Hence the focus is likely to shift on the learning curve toward more people-centered development. Historical evidence suggests that development models drawn in consultation with the people for whom those interventions are made hold promise of yielding more sustainable solutions in terms of development objectives, since they place people at the center of the development paradigm (Hardi and Zdan, 1997).

This chapter is based on a pragmatic and rational understanding of direct and indirect drivers, and how these are likely to behave in future, learning from historical trends and analyzing them in qualitative terms. Thus the chapter sets a future scene with a logical plot describing how things are likely to unfold under different socioeconomic and political conditions and an AKST environment in the short, through the medium, to the long term.

3.2 Method

The approach uses the business-as-usual scenario as its benchmark and adds value while prospecting the situation 50 years ahead. Scenario development and analysis provide a logical way of thinking through a range of plausible futures. Scenarios are "plausible and often simplified descriptions of how the future may develop, based on a coherent and internally consistent set of assumptions about key driving forces and relationships" (MEA, 2005). Scenarios can be developed for various purposes: to explore a range of plausible futures, to analyze possible response strategies, or to provoke creative thinking (Alcamo, 2001).

A number of recent international environmental assessments have made use of scenario development and analysis, and the scenarios introduced by international environmental and agricultural assessments have influenced both scientific and public debate. Prominent examples of global scenario-based assessments include the Intergovernmental Panel on Climate Change (IPCC, 2007), the Global Environmental Outlook of the United Nations Environment Programme (UNEP, 2002) and the Millennium Ecosystem Assessment (MEA, 2005).

3.3 Key Drivers of Agricultural Change

The factors that drive the change—that regard the future of agriculture in the CWANA region—need to be looked at in more detail. We must understand these drivers to assess the plausibility of the scenarios.

A driver is any natural or human-induced factor that directly or indirectly causes a change in an ecosystem. A direct driver unequivocally influences ecosystem processes. An indirect driver operates more diffusely, by altering one or more direct drivers.

For the purpose of this assessment, a driver is a factor that can to a certain degree potentially change the development landscape in a given sector or subsector of economy with regard to agricultural research and development (R&D). A direct driver influences agricultural production and services and can therefore be identified and measured to a degree of accuracy. The influence of an indirect driver is established by understanding the nature of its effect on a direct driver. Drivers can be influenced by policy choices. However, the distinction between indirect and direct drivers may at times not be clear. Many implicit links exist between and across the different drivers, and the discussion in this chapter needs to be viewed in that context.

A host of direct and indirect drivers is relevant to agricultural systems and AKST (Conway, 1997; Dixon et al. 2001; FAO, 2004; Das and Laub, 2005; DFID, 2005). Following is a prioritized set of direct drivers.

Economic drivers. Economic growth and development, national and per capita income, macroeconomic policies, international trade, trade policies, trade liberalization and capital flow, marketing chains, market access opportunities, market distortions and support, food security, competition between different crops.

Food demand and consumption patterns. Population dynamics, consumption levels, dietary preferences, food quality, nutritional values and standards.

Agricultural and natural resource management. Land tenure, agricultural inputs, pest and disease management, use of agrochemicals, cropping patterns, role of livestock, agricultural biodiversity, transgenic crops (GMOs), impact of agriculture on natural resources, constraints of management, indigenous knowledge.

Land and water resource management. Land use, land cover change, land degradation, land availability and productivity, water allocation, water quality, transfer and transboundary water management, surface and groundwater management and protection.

Climate change. Effect of climate change (global warming, change in precipitation) on agriculture, drought, floods and famine, other climate-driven changes.

Energy and biofuels. Relationship between energy (cost, production, distribution, access) and agriculture; hydroelectric energy, bioenergy.

Human resources. Education, training, role of women, rural labor migration, social capital, cultural and religious factors.

Investment in AKST. Scientific and technological developments, private vs. public investments in R&D, rate of adoption of new technologies (biotechnology, information technology), training of agricultural scientists, interregional research cooperation.

Following are the main indirect drivers of change.

Demographics. Demographic dynamics of population size and growth, age and gender structure, spatial distribution.

Sociopolitical. Sociopolitical developments, governance and democratization, corruption, enforcement of legislation, traditional norms, civil society and the private sector, conflicts, international politics.

CWANA countries with inward-looking development policies will not be able to fulfill their international development commitments, like those reflected in the Millennium Development Goals (Global Monitoring Report, 2006; World Bank and IFPRI, 2006).

Where this is the case, AKST will not effectively play its role in attaining the cherished objectives of development goals. However, the national policies may be influenced by international economic and geopolitical conditions (Table 3-1).

Table 3-1. **Key drivers affecting food security in the CWANA region and assessment of their role in the future.**

Drivers	Future impact
Demographics and human health	–
Trade and markets	+
AKST investment	++
Indigenous knowledge	–
Transgenics	+
Governance and coordination	+
Women in agriculture	+
Agrobiodiversity	–
Agrochemicals	–
Water	–
Climate change	– –
Energy and biofuels	–

Legend

++ very strong positive impact

\+ strong positive impact

– strong negative impact

– – very strong negative impact

Source: Authors' elaboration.

3.3.1 Direct drivers

3.3.1.1 Economic drivers

Under the reference world scenario, the capacity and capability of countries may readjust to change in the global arena in a dynamic way. Democratic institutions will gradually evolve and national policies will get more focused in the long term. Economic growth and development in CWANA will help the region attain its development goals, albeit somewhat more slowly than the rest of the world. Present-day economic paradigms like trade investment and innovation will be enhanced by an enabling environment under the globalization and technological boom, consequently facilitating achievement of economic and development objectives (World Bank, 2006acd).

Foreign direct investment can potentially play a key role as an enabling driver (Black, 2003). However, it may be rather difficult in the future to attract such investment because of competition among the countries, although investment policies of national governments will tend to change and adjust to the situation. Proper investment policies provide a springboard to the development agenda since trade, investment, economic growth and development are closely linked. Poverty in the CWANA region will be alleviated to some extent with these changes in policies, especially with an increase in AKST investment, and thus will help realize development goals in the medium to long term.

With the forceful trend of globalization, national economies are bound to integrate with the international economy, sooner than later (FAO, 2006). This can lead, in the short term, to adjustment pains, before actual gains are realized in the long term. For example, agriculture will undergo structural reforms, including a shift from a self-sufficiency paradigm to export-led agriculture or a shift from subsistence crops to cash crops and from low-tech to high-tech practices. This, on the one hand, may provide better market opportunities for the production surplus that farmers generate. But at the same time, food security may be compromised and joblessness created among farmers, especially if compensatory policies and social safety nets are not in place (World Bank, 1997; WRI, 2005).

The need to establish social safety nets to cope with this situation will increase in these countries. If governments do not act swiftly in the short term, small-scale farmers will leave their farms and thus may become victims of the onslaught of globalization. Emerging democratic thinking and awareness of human rights will, however, compel governments to address social safety issues in the medium to the long term. While trying to adjust to the global policy framework, national governments will have relatively less policy space in which to maneuver, through price or other domestic support mechanisms such as subsidy. Thus they will have to focus on efficiency and higher productivity, for which they will need a strong AKST base to realize development and sustainability goals.

A lot, therefore, will depend on government capacity to adjust to the changing environment and deal with the complex issues outlined above. AKST will certainly have a central role in developing the required capacity and in making agriculture efficient and sustainable. Today ventures in the trade–investment–growth paradigm in many developing countries will continue, and the more progressive ventures will capitalize on this paradigm.

For the near future, the fate of World Trade Organization (WTO) negotiations is uncertain, especially the Doha Development Agenda with its built-in development dimension that makes trade liberalization development friendly.

Regional protectionism and support such as through the Organisation for Economic Cooperation and Development (OECD) continues to be a hurdle to overcome in free trade. Although WTO negotiations suspended in July 2006 were recently resumed, the risk of protectionism remains. However, mushrooming free trade agreements and preferential treatment agreements will hopefully be a starting point for integration, initially at regional and then at international levels.

Accordingly, proactive engagement of developing countries in multilateral negotiations will be important for securing their trade interests. If the issues of tariff peaks, tariff escalation and subsidies are objectively addressed, countries of the region will take advantage of emerging market opportunities in agriculture, including export of high-value, processed products. AKST will then be instrumental in developing the required capacity to make the region competitive, especially in agricultural processing (Johnson, 2005; Juma and Yee-Cheong, 2005). Future growth in trade-to-GDP ratio will depend to a great extent on both negotiation outcome and management ability of national governments. Integration of science and technology with national economic development will help solve such problems. Knitting agricultural research institutions into a coherent national agricultural research system (NARS) and integrating national agricultural research plans with the development agenda will be realized in the region with time (Kemmis, 2001; Adato and Meinzen-Dick, 2002; Thirtle et al., 2003; Ryan, 2004).

Agriculture and food markets in the CWANA region will hopefully be reorganized so as to provide access to both domestic and international markets. It will take some time for countries to solve such problems as poor market infrastructure and means of communication, lack of a cold chain and adequate storage facilities, inadequate transportation infrastructure, and not enough vertical linking of producers, industry and consumers (USAID, 2004). As a result, countries will continue to produce primary commodities and incur high postharvest losses, which will have deleterious consequences on trying to deal with food security, poverty alleviation and amelioration of hunger.

According to the current trend in investment in farm sectors and infrastructure, it will not be possible for CWANA countries to penetrate markets across borders in the near future. Integration of local markets with regional or international markets will be delayed. Limitations such as high tariffs will hamper the realization of the potential benefits of globalization, but these limitations may be overcome to a great extent by developing AKST (Foster and Welch, 2002).

Rehabilitation of sea, air and dry port infrastructure—basic for transporting goods and services—will continue to be a weak link in most CWANA countries in the near future. Legislation related to markets, especially for border control like quarantine facilities and the movement of goods and services within the country and across borders, will follow the same trend.

Collective regional efforts and multilateral initiatives will be helpful in evolving the required market system in CWANA. Success stories of the region in terms of market structure, like the one in the United Arab Emirates (UAE), a regional trade hub, will be replicated elsewhere, given political will and investment opportunities. Growth in trade will help countries attain development goals, alleviate poverty, and ameliorate hunger in the CWANA region.

In some countries of the region, nontransparent market practices hinder the development of competitive marketing systems, because of hoarding practices, and strong lobby groups with political support and monopoly. Effective policy regulations related to competition will help fight such monopolies.

Linking farmers with industry and consumers through contract farming will help evolve a better marketing system and cope with problems related to value-chain management, including post-harvest losses. Cooperative settings can also evolve by mobilizing and organizing farming communities. This will help overcome problems associated with the cyclical nature of agricultural markets, such as in poultry and horticulture. Another challenge will be the potential risk of exotic and transboundary diseases like avian flu and mad cow disease, which are likely to emerge movement of goods and services increases under globalization.

Some of these problems will be managed by applying standard management practices, especially those related to sanitary and phytosanitary (SPS) issues, and by adopting good agricultural practices. These practices will build confidence and certainty in the global marketplace that will help sell regional produce. AKST will play an important role in SPS management, in terms of both standardization and quality infrastructure development. Countries of the region therefore need to participate proactively in the international standardization process, through international standard-setting bodies like the International Plant Protection Council (IPPC), the World Health Organization (WHO), and the Codex Alimentarius Commission.

3.3.1.2 Food demand and consumption patterns

In the past two decades, average per capita incomes have increased around the world, more than doubling in many countries. In addition, the world population is expected to grow by more than one billion people in the next decade, most of whom will reside in low- and middle-income countries. This growth, combined with rising income levels in developing countries, is expected to increase and change the composition of global food demand over the next couple of decades (World Food Conference, 1974; UVIN, 1995).

Direct per capita food consumption of maize and coarse grains will decline as consumers shift from wheat and rice to foods such as meat, fruit, vegetables and dairy products, as their incomes increase. Growth in income in developing countries will drive strong growth in per capita and total meat consumption, which in turn will induce strong growth in consumption of cereals for feed, particularly maize (Pretty and Hine, 2001, 2003). These commodities will be procured from supermarket chains and fast food establishments, controlled by multinational and transnational corporations (MNCs, TNCs) (Jordan, 2000). These trends will lead to an extraordinary increase in the importance of developing countries in global food markets (Rosegrant et al., 2001).

Many developing countries are currently undergoing a rapid nutrition transition. Falling real prices for food enable a growing number of consumers to move swiftly toward greater calorie intake and allow them to embark on consumption patterns that had hitherto been reserved for consumers in industrial countries with, at least nominally, much higher income. FAO's long-term outlook suggests that the shift toward a greater supply of energy will accelerate and that it will encompass a growing number of countries.

In addition to falling real prices of food, rapid urbanization has and will continue to affect consumption patterns. Essentially the entire population growth over the next 30 years will be urban. Urbanization creates a new, improved marketing and distribution infrastructure, attracts supermarkets and their sophisticated food handling systems (cold chains, etc.), and brings about better roads and ports. It thus improves access for foreign suppliers, imports become important in the overall supply of food, and all in all urbanization globalizes dietary patterns.

Most importantly from a nutrition perspective, these changes include not only a shift toward higher food-energy supplies but also a shift toward more fats and oils and more animal-based foodstuffs, and thus higher intakes of saturated fat and cholesterol.

As discussed above, growing population and declining agricultural productivity are likely to create more demand for foods that most CWANA countries may not be able to meet locally, at least in the short to medium term (Boserup, 1965). However, with AKST, more efficient practices will evolve, in both planning for parenthood and agriculture production, that may in the long term help satisfy food demand locally (Santaniello, 2003).

3.3.1.3 Natural and agricultural resource management

Responsible management of natural resources is the key to attaining sustainable agricultural and rural development. Seeds of various plant varieties are basic to agricultural development, especially for crop breeding programs. Access to high-yielding varieties and a more diversified seed base will potentially increase production efficiency. However, local seed production in the CWANA region is insignificant, and there is extensive reliance on exogenous sources, which are mainly controlled by MNCs and TNCs in this field (CDE, 2002; Buck et al., 2003; Shyamsundar et al., 2005). This will continue at least for the short to the medium term. Moreover, indigenous crop varieties will be dislodged. Some may even eventually become extinct, with adverse implications for agrobiodiversity in the region, implying weaker social and food safety nets for generations to come (Warren and Rajasekaran, 1993; FAO, 1996; Thrupp, 1998; Huang et al., 2002; Howard, 2003).

Productivity will be constrained as a result of the reduced access to indigenous sources of seed caused by scarcity of seed preservation measures, that is, gene banks. In the same vein, the lack of AKST and R&D capacity will limit the plant breeding needed to develop better seed varieties. Furthermore, promoting AKST will help preserve indigenous seeds through gene banks and local knowledge and practices (Ellen et al., 2000; Ellis, 2000; Evenson and Gollin, 2003; UN Millennium Project, 2005).

Commercial fertilizers have made it possible in the twentieth century to dramatically increase the quantity and quality of food produced on agricultural land. The ability of agriculture to produce far greater quantities of food than in previous centuries can be attributed to four factors: advanced plant breeding techniques, intensive irrigation, availability of fertilizers on a commercial scale and development of plant protection products. With mounting pressure to increase production, more fertilizer will be needed to replenish the organic base of the land resource being depleted. However, no rehabilitation plans are in place to help restore soil fertility in the CWANA region (Scherr, 1999). Most countries of the region produce little fertilizer, and most of the requirements are met through imports, which will ultimately lead to monopoly. Under globalization, falling trade barriers will increase access to fertilizers, and through fair competition fertilizer prices will become cost effective in the long term.

Biodiversity provides not only food and income but also raw materials for clothing, shelter, medicines and breeding of new varieties; it also performs other services such as maintaining soil fertility and biota, and conserving soil and water, all of which are essential for human survival. The importance of agrobiodiversity encompasses sociocultural, economic and environmental elements. All domesticated crops and animals result from human management of biological diversity, which is constantly responding to new challenges to maintain and increase productivity.

The state of biodiversity in agriculture will change as a result of the effect of pesticides on wild species; farmers, the agrofood industry and government will all need to conserve biodiversity. Steps will include adopting integrated pest management (IPM) and making changes in government policies on crop and pesticide inputs and regulations to reduce risk from pesticides. A major challenge for CWANA countries will be to reconcile the desire to expand agricultural production with the obligation to meet national and international objectives and commitments to conserve biodiversity. CWANA countries also need to take into account what agricultural practices cause change in biodiversity in both negative (e.g., excessive farm chemical use) and positive (e.g., creating field margins as wildlife corridors) ways, in particular the effects of different farming practices and management systems (UN, 1992; OECD, 1998, 2005, 2006; Laird, 2004; CBD, 2006).

Deforestation, mostly occurring in developing countries, coupled with overgrazing has led to land erosion and ultimately to desertification in some areas of the region. It is also a major source of global emissions of greenhouse gases contributing to climate change. Lack of land rehabilitation plans together with meager investment in this area further exacerbates the situation. Efforts need to be made for integrated rangeland management, which AKST can potentially underpin.

Livestock will continue to be the mainstay of the farming community and will employ a sizeable agricultural workforce in many of the CWANA countries. Poor genetic makeup, limited feed resources and lack of effective animal health cover will continue to be the main constraints for livestock development in the region. Under globalization, and especially through transfer of technology, increased access to quality animal health services and upgrading of genetic livestock material will be possible. However, lack of capacity in CWANA countries will continue to be a limiting factor in the short and medium term (Inter-Academy Council, 2004b; International Council for Science, 2006).

Agrochemicals are some of the most important products used in agricultural production. They play an important role in increasing agricultural production and incomes. Being high-tech products, agrochemicals set certain requirements for the user. Pesticide research has provided compounds of progressively increased activity, and recently discovered insecticides, fungicides and herbicides are outstandingly potent. Consideration of the practical use of such chemical agents suggests that greater attention should now be given to methods of application and to the physiochemical properties that determine redistribution and biological availability following release.

The use of agrochemicals will increase as intensive agriculture expands. Runoff from intensively exploited areas will carry heavy loads of pesticides and herbicides as well as fertilizers. The chemicals will pollute water sources and overload them with nutrients. Such high chemical loads will lower water quality for such uses as drinking and irrigation or fish culture—compounding the water scarcity in the region. In particular, in countries where most of the population lives in villages that depend on ground or surface water for their own consumption, polluting this source can lead to extensive health problems in rural communities. Also, continuous and improper use of herbicides and pesticides will promote resistance to these chemicals in weeds, microorganisms and insects.

There is growing social awareness about pesticide-intensive pest control programs. This concern will be a driving force behind the adoption of modern AKST methods of pest control such as IPM, already popular in some countries (FAO, 2000).

3.3.1.4 Land and water resource management

Land resource management is the actual practice of how the local human population uses the land. Its use should be sustainable (FAO/Netherlands, 1991). In a broader sense management includes land-use planning, as agreed among stakeholders; legal, administrative and institutional execution; demarcation on the ground; inspection and control of adherence to decisions; solving of land tenure issues; settling of water rights; issuance of concessions for plant and animal extraction (timber, fuelwood, charcoal and peat, nonwood products, hunting); promotion of the role of women and other disadvantaged groups in agriculture and rural development in the area; and the safeguarding of traditional rights of indigenous peoples (FAO, 1995).

Improved land management that ensures better resource use and promotes long-term sustainability is basic to future food production and to the economic welfare of rural communities. Because of the dynamic aspects of land management, a flexible and adaptive "process" approach for monitoring the quality and quantity of the world's land resources (such as soil, water, plant nutrients) and for determining how human activities affect these resources is essential. However, systematic assessment of sustainability of current or planned land uses can be hampered by too much detailed data that are difficult to interpret, lack of baseline information from which to compare change, or data that are inconsistent over time or geographic area (USDA, 1994).

From the land management point of view, the major concerns are

- Decline in quality of the soil as a rooting environment
- Erosion and loss of topsoil by wind and water
- Loss of vegetative cover, including woody perennials
- Acidification, decline in soil fertility and depletion of plant nutrients
- Salinity and salinization, particularly in irrigated systems

Increased control over land resources by the powerful, especially large-scale farmers with political constituencies, leads to a monopolistic attitude, where land is a symbol of power and not productivity (Basu, 1986; Bonfiglioli, 2004).

In the long term, national governments will address issues of food security and rural development (Chambers and Conway, 1991). This may lead to better investment in AKST. As a result, an enabling environment will be provided for better management of land resources, e.g., through efficient soil conservation, water management, and salinity and waterlogging technologies and practices.

In most CWANA countries, traditional inheritance laws will continue to intensify fragmentation of landholdings and production systems. This too negatively influences the realization of economies of scale in agriculture in affected countries. The possibility of land consolidation and integration of the production system is the most promising scenario for overcoming the prevailing problems.

Water resource management is the integrating concept for a number of water subsectors such as hydropower, water supply and sanitation, irrigation and drainage, and environment. An integrated water resource perspective ensures that social, economic, environmental and technical dimensions are taken into account in managing and developing these resources.

Water is the most important natural input in agricultural development (UNWWAP, 2003; CA, 2006; IWMI, 2006). In some areas, ground water is used for irrigation, but because of factors such as drought and overdrawing of groundwater reserves, this important resource has a negative annual balance, with dropping water tables. Besides climatic changes and drought, the scarcity of water resources will be intensified because of pollution from industrial, agricultural and urban sources (WMO, 1997; Pearce, 2006).

Better investment in water resource schemes and better management of water resources will lead to sustainable increases in the productivity of water—and better livelihoods for poor people in rural areas. As a result of these smarter investments, over a 20-year time horizon, we expect less environmental degradation and less poverty.

Integrated water management will achieve positive results by using a three-pronged approach:

- Significantly influence how investments in irrigation development, improvement and management are made, by feeding results of relevant research into the global debate on water for food and environmental security.
- Develop and disseminate research tools to enhance the understanding of the most critical issues in managing irrigation water.
- Provide tools, processes and knowledge that allow water resource managers to adapt and respond to new and changing needs and expectations.

Water-use efficiency in CWANA countries is poor, because of several factors including poor on-farm water management, land slope, inadequate land grading and hydraulic structures. Rational use of water will not be practiced in the short to medium term, so water-use efficiency will deteriorate. A different technology mix will be required to optimize the efficiency, which will be possible by developing AKST (Penning de Vries et al., 2003; IWMI, 2006).

3.3.1.5 Climate change

Climate change in the past century has already had a measurable effect on ecosystems. The earth's climate system has changed since the pre-industrial era, in part because of human activity, and it is projected to continue to change throughout the twenty-first century.

The many challenges global climate change poses, from increased temperatures and extreme weather events to rises in sea level, are now widely recognized in both scientific and policy circles (Smit, 1993; McCarthy et al., 2001). The global mean surface air temperature is projected to increase from 1990 to 2100 by 1.4-6.4°C, accompanied by more heat waves (IPCC, 2007). Precipitation patterns are projected to change, with most arid and semiarid areas becoming drier but with an increase in heavy precipitation events, leading to increased incidence of floods and droughts. The Millennium Ecosystem Assessment scenarios project a sea level rise of 9-88 cm.

In this respect, the fourth report of the Intergovernmental Panel on Climate Change (IPCC) estimates that a global temperature rise during the century of between 2°C and 4.5°C is almost inevitable. Ominously, however, it also says that much higher increases, of 6°C or more, cannot be ruled out. IPCC's latest report makes it clear that climate change may be far worse than previously thought because of potentially disastrous positive feedbacks, which could accelerate rising temperatures (IPCC, 2007).

Furthermore, a recent report from Britain's Sir Nicholas Stern warned of the devastating economic effect global warming could have on the world in coming years (Stern, 2007). Stern, former chief economist at the World Bank, cautioned that if greenhouse gas emissions were not significantly reduced, by 2050 the global economy would shrink by up to 20%, millions of people would be permanently displaced and droughts would plague the earth.

So far, the main response both nationally and internationally has been to focus on initiatives aimed at reducing the potential size of these effects. Most industrial countries have sought to do this by committing themselves through signing the United Nations Framework Convention on Climate Change (UNFCCC) and the Kyoto Protocol, to stabilize or reduce greenhouse gas emissions and enhance carbon sinks (Klaey, 2000; Baumert et al., 2005).

Climate change will potentially lead to such eventualities as drought and famine, which some of the CWANA countries have already experienced. The capacity of national governments and communities to mitigate disasters will be limited in the short to medium term, rendering them still vulnerable to the adversities of climate change. Climate change is a global issue with regional implications. Many multilateral environmental agreements address these issues, and some countries of the region have ratified some such agreements (WCED, 1987; UNEP, 2002).

Too much or too little rain can be a matter of life or death in some countries in the CWANA region (UNEP, 2002; WRI, 2005). At different times and in different places across the region, climate change poses the threat of both. Prolonged and severe climatic desiccation, coupled with intensive exploitation of soil, pasture, forest and other natural resources, as well as the huge increase in human and livestock populations will lead to extensive degradation and result in an inherently fragile environment in some parts of the region to the extent that conflicts caused or catalyzed by these compounding ecological factors are bound to take place (De Wall, 1989).

In fact, ecological degradation, caused mainly by climate change, has been so severe that traditional means for preventing and managing interethnic disputes have been rendered virtually unworkable. Indeed, many of the current disputes are being fought not along traditional political borders but along ecological borders that divide richer and poorer ecozones (Bachler and Spillmann, 1992).

To continue to treat the conflict in Darfur and many other parts of Africa as purely ethnic, tribal, political or religious and to ignore the growing impact of ecological degradation and depletion of the resource base will ultimately lead to a distorted understanding of the real situation. This will consequently limit the possibility for genuine conflict resolution (El-Nour, 1992; Suliman, 1993, 1996, 2000). Therefore, a new model of development is called for, in which strategies to increase human resilience in the face of climate change and the stability of ecosystems are central. Above all, the challenge calls for new flexibility and not a one-size-fits-all approach to development (De Wall, 1989; Suliman, 2000).

The aforementioned issues and other environmental factors precipitating climate change, like industrial pollution, can only be addressed through efficient technologies. Yet the available technology mix (AKST) can have both positive and negative effects on the climate, and at times may not be efficient or environmentally friendly. National governments will remain complacent with what already exists, at least in the short term. But this will compromise their attaining their environmental health agenda in that time frame.

Regionally, countries sharing borders or that are members of regional cooperation or custom unions could address cross-border environmental issues. The coping strategies proposed could at times be costly, and countries of the region might not be able to handle the issues on their own. At present, as issues of environmental goods and services are being negotiated in WTO, national governments and communities may not be able to engage in them proactively. Finally, considering the current weak AKST infrastructure and scarcity of human resources, countries will not be able to give the support needed to mitigate climate change.

In the absence of effective mitigation efforts, climate changes will potentially lead to major crises, which will encourage stronger alliances between selected countries to attempt to reduce the incidence and damage of further natural disasters. Development of AKST will counteract the adverse fallouts on the region (Röling and Engel, 1991; Pardey and Beintema, 2001; Pingali and Traxler, 2002; Byerlee and Alex, 2003; Byerlee et al., 2005; NEPAD, 2005).

A key component of all the strategies is to link early warning and risk management systems to regional and local practitioners such as health-care providers and farmers

(WGCCD, 2006). Government departments and the media will also play an important role, particularly in getting information to rural areas. Present media gaps in communicating meteorological information make uninformed farmers more vulnerable to climate change.

3.3.1.6 Energy

Perhaps the greatest challenge in realizing a sustainable future is energy consumption. Energy is ultimately the basis for a large part of the global economy, and more of it will be required to raise living standards in the developing world. Today, we are mostly dependent on nonrenewable fossil fuels that have been and will continue to be a major cause of pollution and climate change. Because of these problems and our dwindling supply of petroleum, finding sustainable alternatives is becoming increasingly urgent (Starss, 2006).

Scientists are warning that the end of oil is coming sooner than governments and oil companies are prepared to admit. The London-based Oil Depletion Analysis Centre says that global production of oil is set to peak in the next four years before entering a steepening decline that will have massive consequences for the world economy and the way that we live our lives.

Under globalization, prices of oil will further increase. The present situation in the Middle East supports this scenario. Access to energy will be further hindered by the increased monopoly of associations such as OPEC. This state of affairs will have implications on evolving efficient technologies for agricultural development in the short to medium term.

Development of alternative means of energy such as solar, wind and biofuels needs strong underpinnings, from both a strong natural resource base and AKST capacity. With development of AKST capacity in the region, coupled with soaring petrol prices, alternative means of energy will become available in the long term. In oil rich economies, agricultural development is being accomplished by income-generated oil resources without integration with AKST. This will lead to unsustainable development in the long term. Tapping renewable energy resources such as solar energy, biomass and alcohol from plant residues is a more efficient way to achieve sustainable development. However, developing countries will find it difficult to produce biofuels without sacrificing food availability, as subsidized production for biofuel will compete with food production.

Biofuel production will increase demand for agricultural land at the expense of natural ecosystems. Perhaps more critically, it will also require large quantities of water—already a major constraint to agriculture in many parts of CWANA. Pursuing biofuel production in water-short countries will put pressure on an already stretched resource and will turn green energy into a major threat to resources.

3.3.1.7 Human resources and feminization of agriculture

Building human resource capacity directly implies agricultural education. By definition the term *capacity building* (and the process) has education, both formal and nonformal, at its core. In its broadest interpretation, capacity building encompasses human resource development (HRD) as an essential part of development. It is based on the concept that education and training lie at the heart of development efforts and that without HRD most development interventions will be ineffective. It focuses on a series of actions directed at helping participants in the development process to increase their knowledge, skills and understanding and to develop the attitudes needed to bring about the desired developmental change.

Human resources and capital along with natural resources are essential for development. Many dimensions of human resource development are end objectives of development, e.g., literacy, better health and nutrition, economic well-being. It is generally recognized that a country's human resource capacity for productivity is a prerequisite for social and economic development. However, the problems of development, and in particular food security and poverty, are complex, and improved HRD is only one of several necessary conditions for social and economic progress.

Sustainable development, with its management, technological and institutional aspects, clearly encompasses human resource development, and in particular HRD in agriculture. Unfortunately, the term *HRD* has been applied to such a wide array of activities that its meaning is often ambiguous. To be meaningful, *HRD* needs to be carefully defined.

Competitiveness in agriculture and food systems demands improvement and development of the human resource through investment in education, training, health, and information and communication technology (ICT) (Jaffé and Rojas, 1994; Rice et al., 2000; WHO, 2005; Stern et al., 2006). The establishment of agricultural schools and colleges will continue to present additional, sometimes alternative, knowledge systems for agriculture and related sciences. There is a threat of a vacuum of experienced and knowledgeable experts and professors in national educational and research institutions because of retirement regulations and an increased number of university graduates seeking job opportunities in NARS.

Capacity-building efforts should focus on institutional strengthening, including the design of new organizational structures to improve the "goodness of fit" between the policy context for sustainable development and enacting institutions in both public and private sectors. These institutions include agricultural education and training institutions as well as extension agencies, research institutions, NGOs and community organizations. A multiplier effect can be achieved if strong links among agricultural education institutions, NGOs, research organizations, public and private extension services and others are fostered. This approach recognizes that there are multiple sources of technology development and dissemination and that integrated institutional networking is needed to build capacity.

Building human resource and institutional capacity through agricultural education and training means enhanced investment, expanded international cooperation, improved quality and relevance of education, and broadened access to and participation in educational activities, especially by women. A wider financial base will be needed, including increased support from the private sector. This does not mean, however, that governments can detach themselves from the responsibilities of building human capacity. If there is not strong national commitment to sustained human resource development, the goal of sustained agricultural and rural development will not be realized.

In many parts of the world today the trend is increasing toward what is termed the feminization of agriculture. As men's participation in agriculture declines, the role of women in agricultural production becomes ever more dominant. War, and sickness and death from HIV and AIDS, have reduced rural male populations. Another major cause of this phenomenon is the migration of men from rural areas to towns and cities, in their own countries or abroad, in search of paid employment.

This will be in line with the present trend regarding the gradual increase of women's work over time in some countries of the region. Between the years 1960 and 1985, the percentage of women's participation in agriculture increased considerably in all countries (Narayan, 2002).

This trend is resulting in an increase in the proportion of households headed by women. Approximately one-third of all rural households in sub-Saharan Africa are now headed by women. Studies have shown that women heads of household tend to be younger and less educated than their male counterparts. They also generally have less land to work and even less capital and extra farm labor for working it.

With a shortage of labor and capital, women heads of household will often be forced to make adjustments to cropping patterns and farming systems. These adjustments will result in decreases in production and, in some cases, shifts toward less nutritious crops. Not surprisingly, these households will often suffer from increased malnutrition and food insecurity.

Nor has the fact that the participation of women will increase translate into their ownership of resources such as land, labor, credit and capital. As an example, traditional inheritance laws may be cited whereby the rights of the male heirs are twice those of the female.

Nevertheless, through higher education for women and change in attitude, society will develop institutions of governance, legal systems and social or gender-sensitive policies that will decrease disparities between women and men (UNDP, 2005).

3.3.1.8 AKST investment, indigenous knowledge and transgenics

The visible signs of science and technology are most evident in food production. Much of the increase in agricultural output over the past 40 years has come from an increase in yields per hectare rather than an expansion of area under cultivation.

Investment in agricultural research has a net payoff; in many developing countries, the share of agriculture in the GDP ranges from 25 to 70 percent (Hurni et al., 2001; FAO, 2003). However, the share of investment in agriculture and agricultural research will continue to be low, compared with that internationally recommended (FAO, 2004). This is one of the reasons that new technologies and better agricultural practices have not taken root in CWANA countries, and traditional agricultural practices, which are relatively less efficient, continue to dominate. Adoption of the new and more efficient agricultural technologies will require urgent and increased investment in AKST (Inter Academy Council, 2004a; CGIAR Science Council, 2005).

Agricultural technologies and knowledge have, until recently, largely been created and disseminated by public institutions. But over the past two decades, biotechnology for agricultural production has developed rapidly, and the world economy has become more globalized and liberalized. This has boosted private investment in agricultural research and technology, exposing agriculture in developing countries to international markets and the influence of multinational corporations. But the public sector still has a role to play, particularly in managing the new knowledge, supporting research to fill any remaining gaps, promoting and regulating private companies, and ensuring their effects on the environment are adequately assessed.

However, the relatively weak institutional and human capital base in the region has negatively affected agricultural development. With the establishment of globalization, CWANA countries will have opportunity to interact across borders with nations and communities that have advanced production systems and practices. Exchange of knowledge and experience across regional and international boundaries will let scientific information flow in, to the advantage of agricultural development in CWANA. The role of internationally recognized scientific and research organizations and institutions in the region will be of great benefit in this respect.

Countries of the region will get involved in investment treaties with transfer of technology. Considering that 90 percent of the genetic resources of the world are in the South, of which CWANA is a part, these countries will expand claims for intellectual property rights for providing access to genetic resources and indigenous knowledge. With good management and development, these resources will accrue monetary gains to the advantage of their custodians. But in the immediate future, organizational fragmentation of agricultural research will continue to prevail in most CWANA countries.

Indigenous knowledge is local knowledge, unique to a given culture or society. Indigenous agricultural and environmental knowledge gained global recognition through the United Nations Conference on Environment and Development (UNCED) in 1992. However, indigenous knowledge systems have never been systematically recorded in written form and therefore are not readily accessible to agricultural researchers, development practitioners or policy makers.

There is no standard definition of indigenous knowledge; however, there is general understanding as to what constitutes it. It is variously regarded as ethnoscience, folk knowledge, traditional knowledge, local knowledge and people's knowledge. The role of indigenous knowledge in preserving agrobiodiversity is essential to human development.

Small-scale, resource-poor farmers have good reasons for sticking to their local knowledge and the farming practices they have always used. Modern technologies can be successful and sustainable only if the local knowledge interplay of cultural, social and ecological systems are taken into consideration. In so stating, we suggest that, given the pervasive scenario of rapid population growth and the attendant domestic food demand deficits, the need has emerged to balance sustaining the indigenous knowledge production system with modern technology, through a systematic hybridization strategy (Titilola, 1990, 1994).

Indigenous knowledge of plant genetic resources is an invaluable tool in the search for new ways to conserve and use these resources to benefit local communities. Agenda 21, one of three nonbinding environmental agreements signed at UNCED, emphasizes that local governments and intergovernmental organizations should respect, record and work toward incorporating indigenous knowledge systems into research and development programs to conserve biodiversity and sustain agricultural and natural resource management systems.

It is impossible to predict exactly which new modern biotechnology derived from plants or animals will be ready for the marketplace over the next decade. Some possibilities:

- Genetically engineered plant varieties that provide improved human nutrition (e.g., soybeans enriched in omega-3 fatty acids)
- Products designed for use in improved animal feeds, providing better nutritional balance by increasing the concentration of essential amino acids often deficient in feed components, increased nutrient density, or more efficient use of nutrients such as phosphate that could provide environmental benefits
- Crops resistant to drought and other environmental stresses such as salinity
- Crops resistant to pests and diseases (e.g., *Fusarium*-resistant wheat; chestnut-blight resistant chestnut; plum-pox resistant stone fruit; various insect-resistant crops)
- Additional crops containing a number of transgenic traits incorporated in the same plant (stacked traits)
- Crops engineered to produce pharmaceuticals, such as vaccines and antibodies
- Crops engineered for particular industrial uses (e.g., crops with improved processing attributes such as increased starch content, producing useful enzymes that can be extracted for industrial processes, or modified to have higher content of an energy-rich starting material such as oil for improved use as biofuel)
- Transgenic animals for food, or for production of pharmaceuticals or industrial products (e.g., transgenic salmon engineered for increased growth rate to maturity, transgenic goats producing human serum factors in their milk, and pigs producing the enzyme phytase in their saliva for improved nutrient use and manure with reduced phosphorus content)

TNCs and MNCs have an aggressive trade interest in biotech products and GMOs (Atanassov, 2004). GMOs will lead to monocrop culture with adverse implications for bio- and agrodiversity (Benbrook, 2004; Brookfield et al., 2003). But given the low level of AKST in the CWANA region, coupled with religious and other social factors, the trend for not accepting GMOs and seed crops will continue at least in the near future (Cohen et al., 1999; World Bank, 2002; Tollens et al., 2004).

Nevertheless, better science and tougher regulations are needed to police the future growing of genetically modified crops in the CWANA region. Current genetically modified foods appear safe to eat but there are doubts about future products and the environmental effect of transgenic crops. For reasons of expense, concerns for safety, and inadequate understanding of basic biology, the use of transgenics in livestock production is likely to be minimal for many years to come. Substantial gaps in scientific knowledge remain that must be addressed. The prospect for genetically modified and organic farming coexisting in future is full of uncertainty and poses difficulties.

3.3.2 Indirect drivers

3.3.2.1 Demographics and human health

The population dynamics of the CWANA region are witnessing high rates of growth compared with other regions. These high rates coupled with problems like poor social safety nets, education, immunization, and child and mother care negatively affect human productivity and longevity. Children and women, the worst victims of the skewed demography, will continue to be marginalized in the absence of social safety nets (Ratner, 2004).

In the densely populated regions of CWANA, agriculture will remain a key source of livelihood, although over the past few years, its capacity to employ people as a dependable livelihood has declined. Migration will continue as people seek off-farm jobs not available in villages. This will lead to mounting pressure on urban satellites. The United Nations Millennium Project found strong links between rapid population growth, high fertility and ill-timed pregnancies—which add up to a demographically related poverty trap. Demographic trends affect both development prospects and security.

The interaction between health and agriculture operates in two directions—agriculture affects health, health affects agriculture. Agricultural production and its outputs can contribute to poor health, depending on the system of production and consumption. Agricultural policies clearly affect the quantity, quality and price of food, all of which play important roles in diet change in developing countries.

The greatest challenge now is the burden of nutritionally related diseases. Agricultural policy should take into consideration the whole spectrum of these diseases, to help strengthen the "one agenda" that will gradually break the cycles of both poverty and hunger in a sustainable way.

Under globalization, because of expanding tourism, cross-cultural contacts, and trade liberalization, diseases such as avian flu and HIV/AIDS are expected to rise in the region. Some countries in the region are hard hit by both of these disease problems. HIV and AIDS also have long-term implications for the nutritional health of affected communities and families, which further compound the risk of food insecurity under the business-as-usual scenario discussed earlier. This in turn adversely affects the rehabilitation of those infected with HIV (ILO, 2005; UN, 2005).

It is expected that population growth over the next several decades will be concentrated in the poorest urban communities in sub-Saharan Africa, South Asia and the Middle East. Populations in all parts of the world are expected to age substantially during the next century. While industrial countries will have the oldest populations, the rate of aging may be extremely fast in some developing countries in the CWANA region.

3.3.2.2 Sociopolitical drivers

Sociopolitical drivers encompass the forces influencing decision making and include the amount of public participation in decision making, the makeup of participants in public decision making, the mechanisms of dispute resolution, the role of the state relative to the private sector, and levels of education and knowledge.

Over the past 50 years, there have been significant changes in sociopolitical drivers. There is a declining trend in centralized authoritarian governments and a rise in elected democracies.

Good governance is a general term for the way in which a government acts and functions for the benefit of a society. Good governance should therefore not be seen as a sector in itself. If there is good governance in a country, we expect the government of that country to be transparent (clear and complete financial accountability, accountability concerning its policies); to play a facilitating and encouraging role toward civil society on economic, political and social grounds; to be a neutral arbiter in society; to be accountable for its rules and laws and to be cost effective and service-oriented toward its citizens.

All these elements tell us something about the way a country and government operates and the type of processes and dynamics that are going on within a country. The success of these elements often depends on political, cultural, social and economic aspects that can be found within the context of a country.

To promote good governance, specific attention is paid to democratization and decentralization, gender, human rights, strengthening of the rule of law, promotion of independent media, institutional strengthening, public finance management and civil service reform.

Citizens in many developing countries are demanding better performance on the part of their governments, including increased transparency and accountability. Institutional development activities focus on strengthening the ability and capacity of countries to design and implement development policies on their own and in a sustainable way. It is a multidimensional concept not limited to training activities and organizational capacity development; it also includes the overall environment of institutional performance: budgeting procedures, formal and informal patterns within organizations, organizational culture, social structures, etc.

The trend toward democratic institutions has helped empower local communities, women and resource-poor households. The need for good governance, democratization, decentralization and other sociopolitical reforms will remove the stumbling blocks and will speed up development. Pressures from globalization can contribute to adoption of democratic thinking in some countries, consequently reducing corruption in developing economies.

There is room for improvement in coordination at all levels, public, private and civil society, to facilitate the development process. Better coordination and collaboration will enhance development objectives, especially poverty alleviation and amelioration of hunger (Hemmati et al., 2002).

References

Adato, M., and R. Meinzen-Dick. 2002. Assessing the impact of agricultural research on poverty using the sustainable livelihoods framework. FCND Disc. Pap. 128, EPTD Disc. Pap. 89. IFPRI, Washington, DC.

Alcamo, J. 2001. Scenarios as tools for international assessments. Prospects and Scenarios No. 5. European Environ. Agency, Copenhagen.

Atanassov, A. 2004. To reach the poor: Next harvest study on genetically modified crops. Public research and policy implications. EPTD Disc. Pap. 116. IFPRI, Washington, DC.

Bachler, G., and K. Spillmann. 1992. How to cope with environmental conflicts. ENCOP, Swiss Federal Inst. Tech., Zurich.

Basu, K. 1986. The market for land: an analysis for interim transactions. J. Dev. Econ. 20(1):163-177.

Baumert, K.A., T. Herzog, and J. Pershing. 2005. Navigating the numbers: Greenhouse gas data and international climate policy. World Resources Inst., Washington, DC.

Benbrook, C. 2004. Genetically engineered crops and pesticide use in the United States: The first nine years. BioTech Infonet Tech. Pap. No. 7. Available at http://www.biotech-info.net/technicalpaper7.html.

Black, J. 2003. Dictionary of economics. Available at http://www.oxfordreference.com/views/. Oxford Reference Online, Oxford Univ. Press, Oxford.

Bonfiglioli, A. 2004. Land of the poor: local environmental governance and the decentralized management of natural resources. UN Capital Dev. Fund, NY.

Boserup, E. 1965. Conditions of agricultural growth: the economics of agrarian change under population pressure. George Allen and Unwin, London.

Brookfield, H., H. Parsons, and M. Brookfield (eds) 2003. Agrodiversity: Learning from farmers across the world. United Nations Univ., Tokyo.

Buck, L.E., J. Anderson, and D. Chandrasekharan-Behr. 2003. Strengthening natural resource institutions in Africa: applying social learning to reconciling poverty reduction and environmental management. International workshop on reconciling rural poverty and resource conservation: Identifying relationships and remedies, 3 May 2003. Dep. Agric. Managerial Econ., Cornell Univ., Ithaca.

Byerlee, D., and G. Alex. 2003. Designing investments in agricultural research for enhanced poverty impacts. ARD Working Pap. No. 6. World Bank, Washington, DC.

Byerlee, D., X. Diao, and C. Jackson. 2005. Agriculture, rural development, and pro-poor growth: country experiences in the post-reform era. ARD Disc. Pap. 21. World Bank, Washington, DC.

CA (Comprehensive Assessment). 2006. Comprehensive assessment of water management in agriculture. http://www.iwmi.cgiar.org/assessment/. IWMI, Colombo.

CBD. 2006. Global Biodiversity Outlook 2. Convention on Biological Diversity, Montreal.

CDE (Center for Development and Environment). 2002. NRM—natural resource management [Online]. Available at http://www.cde.unibe.ch/Themes/NRM_Th.asp. Center for Dev. Environ., Berne.

CGIAR Science Council. 2005. Science for agricultural development: changing contexts, new opportunities. CGIAR Science Council Secretariat, Rome.

Chambers, R., and G. Conway. 1991. Sustainable rural livelihoods: practical concepts for the 21st century. IDS Disc. Pap. 296. Inst. Dev. Studies, Brighton.

Cohen, J., C. Falcón, and J. Komen. 1999. Research policy and management issues. *In* G.J. Persley and J.J. Doyle (eds) Biotechnology for developing-country agriculture: Problems and opportunities. 2020 Focus No. 02, Brief 08. IFPRI, Washington, DC.

Conway, G. 1997. The doubly Green Revolution: Food for all in the 21st century. Cornell Univ. Press, Ithaca.

Das, S., and R. Laub. 2005. Understanding links between gendered local knowledge of agrobiodiversity and food security in Tanzania. Mountain Res. Develop. 25:218-222.

De Wall, A. 1989. Famine that kills. Darfur, Sudan, 1984-1985. Clarendon Press, Oxford.

DFID. 2005. Growth and poverty reduction: the role of agriculture. Available at http://www.dfid.gov.uk/pubs/files/growth-poverty-agriculture.pdf. DFID, London.

Dixon, J., A. Gulliver, and D. Gibbon. 2001. Farming systems and poverty: improving farmers' livelihoods in a changing world. FAO, Rome, and World Bank, Washington, DC.

Ellen, R., P. Parkes, and A. Bicker (ed). 2000. Indigenous environmental knowledge and its transformations. Harwood Academic Publ., Amsterdam.

Ellis, F. 2000. The determinants of rural livelihood diversification in developing countries. J. Agr. Econ. 51(2):289-302.

El-Nour, I. 1992. Displaced and refugee studies in the Sudan, an annotated bibliography. Univ. Juba, Sudan.

Evenson, R.E., and D. Gollin. 2003. Assessing the impact of the Green Revolution: 1990-2000. Science 300:758-762.

FAO. 1995. Planning for sustainable use of land resources: toward a new approach. Background Pap. to FAO's Task Managership for chapter 10 of Agenda 21 of the United Nations Conference on Environment and Development (UNCED). FAO Land and Water Bull. 2. FAO, Rome.

FAO. 1996. Global plan of action for the conservation and sustainable utilization of plant genetic resources for food and agriculture. [Online]. Available at http://www.fao.org/ag/AGP/AGPS/GpaEN/gpatoc.htm. FAO, Rome.

FAO. 2000. The state of food and agriculture: Lessons from the past 50 years. FAO, Rome.

FAO. 2003. FAOSTAT database. Available at http://apps.fao.org/. FAO, Rome.

FAO. 2004. State of food and agriculture 2003-2004: Agricultural biotechnology—meeting the needs of the poor? FAO, Rome.

FAO. 2006. Glossary of the Food and Agriculture Organization of the United Nations [Online]. Available at http://www.fao.org/glossary/. FAO, Rome.

FAO/Netherlands. 1991. Conference on Agriculture and the Environment, 15-19 April 1991. Vol. 2. S-Hertogenbosch, Netherlands.

Foster, P., and C. Welch. 2002. Poverty. p. 7-11. *In* Marketing the earth: The World Bank and sustainable development. Friends of the Earth, Washington, DC, and the Halifax Initiative, Ottawa.

Fresco, L. 2002. The future of agriculture: challenges for environment, health and safety regulation of pesticides. From a presentation to the OECD Working Group on Pesticides, Paris, 4 February 2002. Available at http://www.fao.org/ag/magazine/oecd.pdf. FAO, Rome.

Global Monitoring Report. 2006. Millennium Development Goals: Strengthening mutual accountability, aid, trade, and governance. World Bank and IMF, Washington, DC.

Hardi, P., and T. Zdan (ed). 1997. Assessing sustainable development: principles in practice. Int. Inst. Sustainable Dev., Winnipeg.

Hemmati, M., F. Dodds, J. Enayati, and J. McHarry. 2002. Multi-stakeholder processes for governance and sustainability: Beyond deadlock and conflict. Earthscan, London.

Howard, P.L. (ed). 2003. Women and plants: gender relations in biodiversity management and conservation. ZED Books, London.

Huang J., C. Pray, and S. Rozelle. 2002. Enhancing the crops to feed the poor. Nature 418:678-684.

Hurni, H., U. Christ, J.A. Lys, D. Maselli, and U. Wiesmann. 2001. Challenges in enhancing research capacity. Pp. 50-58. *In* KFPE (Commission for Research Partnerships with Developing Countries). Enhancing research capacity in developing and transition countries: experience, discussions, strategies and tools for building research capacity and strengthening institutions in view of promoting research for sustainable development. Geographica Bernensia, Berne.

ILO (International Labour Organisation). 2005. HIV/AIDS and employment. AIDS Brief, Sept. ILO, Geneva.

Inter-Academy Council. 2004a. Realizing the promise and potential of African agriculture: science and technology strategies for improving agricultural productivity and food security in Africa. IAC Secretariat, Amsterdam.

Inter-Academy Council. 2004b. Inventing a better future. A strategy for building worldwide capacities in science and technology. IAC Secretariat, Amsterdam.

International Council for Science. 2006. Priority area assessment on capacity building in science. ICSU Report of the CSPR Assessment Panel on Capacity Building in Science. ICSU, Paris.

IPCC. 2007. Climate change 2007: The physical science basis. Summary for decision makers. Contribution of the working group I to the fourth assessment report of the IPCC. Cambridge Univ. Press, Cambridge.

IWMI. 2006. International water assessment. IWMI, Colombo.

Jaffé, W., and M. Rojas. 1994. Human resources for biotechnology: Latin America. Biotech. Dev. Monitor 21:21-22.

Johnson, I. 2005. Science in the service of poor farmers: The CGIAR-JIRCAS partnership. Trop. Agric. Assoc. Newsl. 25:25-27.

Jordan, M.C. 2000. Privatization of food: Corporate control of biotechnology. Agron. J. 92:803-806.

Juma, C., and Yee-Cheong Lee. 2005. Innovation: Applying knowledge for development. UN Millennium Project Task Force on Science, Technology, and Innovation. Earthscan, London and Sterling, VA.

Kemmis, S. 2001. Exploring the relevance of critical theory for action research: Emancipatory action research in the footsteps of Jürgen Habermas. Pp. 91-102. *In* P. Reason and H. Bradbury (eds) Handbook of action research: Participative inquiry and practice. Sage Publ., London.

Klaey, A. 2000. The Kyoto Protocol and the carbon debate: a plea for an international policy on sustainable management of renewable natural resources. Centre Dev. Environ., Berne.

Laird, S.A. (ed) 2004. Biodiversity and traditional knowledge: Equitable partnerships in practice. Earthscan, London.

Lipton, M. 2005. Family farm in a globalizing world. the role of crop science in alleviating poverty. 2020 Disc. Pap. 40. IFPRI, Washington, DC.

McCarthy, J.J., O.F. Canziani, N.A. Leary, D.J. Dokken, and K.S. White (ed). 2001. Climate change 2001: Impacts, adaptation and vulnerability. Third Assessment Report of the Intergovernmental Panel on Climate Change. Cambridge Univ. Press, Cambridge.

MEA (Millennium Ecosystem Assessment). 2005. Ecosystems and human well-being. Vol 2: Scenarios: findings of the Scenarios Working Group, S.R. Carpenter et al. (eds) Island Press, Washington, DC.

Narayan, D. (ed). 2002. Empowerment and poverty reduction: A sourcebook. World Bank, Washington, DC.

NEPAD (New Partnership for Africa's Development). 2005. Needs assessment for science policy capacity building for African agriculture. AU-NEPAD. Department of Science and Technology, Government of South Africa, Pretoria.

OECD. 1998. Reaching the goals in the S-21: Gender equality and the environment. Working Party on Gender Equality, OECD-DAC, Paris.

OECD. 2005. Multifunctionality in agriculture: What role for private initiatives? OECD, Paris.

OECD. 2006. Challenge of capacity development: working toward good practice. DAC Network on Governance. DCD/DAC/GOVNET (2005)5/REV1. OECD, Paris.

Pardey, P.G., and N.M. Beintema. 2001. Slow magic: Agricultural R&D a century after Mendel. ASTI Initiative. IFPRI, Washington, DC.

Pearce, F. 2006. When the rivers run dry: Water—the defining crisis of the twenty-first century. Beacon Press, Boston.

Penning de Vries, F.W.T., H. Acquay, D. Molden, S.J. Scherr, C. Valentin, and O. Cofie. 2003. Integrated land and water management for food and environmental security. CA Res. Rep. 1. IWMI, Colombo.

Pingali, P.L., and G. Traxler. 2002. Changing locus of agricultural research: Will the poor benefit from biotechnology and privatization trends? Food Policy 27:223-228.

Pretty, J., and R. Hine. 2001. Reducing food poverty with sustainable agriculture: a summary of new evidence. Final report from the Potential of Sustainable Agriculture to Feed the World Research Project. Centre Environ. Society, Univ. Essex, Colchester.

Pretty, J., and R. Hine. 2003. Reducing food poverty by increasing agricultural sustainability in developing countries. Agric. Ecosyst. Environ. 95:217-234.

Ratner, D. 2004. Equity, efficiency, and identity: grounding the debate over population and sustainability. Population Res. Pol. Rev. 23:55-71.

Rice, A.L., L. Sacco, A. Hyder, and R.E. Black. 2000. Malnutrition as an underlying cause of childhood deaths associated with infectious diseases in developing countries. Bull. WHO 78(10):1207-1221.

Röling, N., and P. Engel. 1991. The development of the concept of agricultural knowledge information systems (AKIS): implications for extension. *In* W.M. Rivera and D.J. Gustafson (eds) Agricultural extension: Worldwide institutional evolution and forces for change. Elsevier, Amsterdam.

Rosegrant, M.W., M.S. Paisner, S. Meijer, and J. Witcover. 2001. Global food projections to 2020: emerging trends and alternative futures. IFPRI, Washington, DC.

Ryan, J. 2004. Agricultural research and poverty alleviation: some international perspectives. ACIAR Working Pap. No. 56. Australian Centre Int. Agric. Res., Canberra.

Santaniello, V. 2003. Agricultural biotechnology: implications for food security. Proc. 25th Int. Conf. Agric. Economists, 18-22 Aug. 2003. IAAE, Durban.

Scherr, S.J. 1999. Soil degradation: A threat to developing-country food security? 2020 Vision Disc. Pap. No. 27. IFPRI, Washington, DC.

Shyamsundar, P., E. Araral, and S. Weeraratne. 2005. Devolution of resource rights, poverty, and natural resource management: a review. Environ. Econ. Ser. No. 104. World Bank, Washington, DC.

Smit, B. (ed). 1993. Adaptation to climatic variability and change: Report of the Task Force on Climate Adaptation. Canadian Climate Prog., Univ. Guelph, Ontario.

Starss, T. 2006. Fossil fuel consuming our future. Available at http://www.ecoliteracy.org/publications/rsl/tom-starrs.html. Center for Ecoliteracy, Berkeley, CA.

Stern, N. 2007. The economics of climate change: The Stern review. Cambridge Univ. Press, Cambridge.

Stern, E., L. de Vaccaro, and J. Lynam. 2006. Evaluation and impact of training in the CGIAR. CGIAR Sci. Council, Rome.

Suliman, M. 1993. Civil war in the Sudan: From ethnic to ecological conflict. The Ecologist 23(3).

Suliman, M. 1996. War in Darfur. *In* G. Bachler and K. Spillmann (eds) Environmental degradation as a cause of war. ENCOP. Vol. 2. VR, Zurich.

Suliman, M. 2000. Sudan civil wars: new perspective (in Arabic). Cambridge Acad. Press, UK.

Thirtle, C., L. Lin, and J. Piesse. 2003. The impact of research-led agricultural productivity growth on poverty reduction in Africa, Asia and Latin America. World Dev. 31(12):1959-1975.

Thrupp, L.A. 1998. Cultivating diversity: Agrobiodiversity and food security. World Resources Inst., Washington, DC.

Titilola, S.O. 1990. The economics of incorporating indigenous knowledge systems into agricultural development: a model and analytical framework. Studies in Tech. Social Change Program. Iowa State Univ., Ames.

Titilola, T. 1994. Indigenous knowledge systems and sustainable agricultural development in Africa: essential linkages. Indigenous Knowl. Dev. Monitor 2(2):18-21.

Tollens, E., M. Demont, and R. Swennen. 2004. Biotechnology in developing countries: North–South partnerships are the key. Outlook Agric. 33:231-238.

UN. 1992. Convention on Biological Diversity. Concluded at Rio de Janeiro on 5 June 1992. Vol. 1760, I-30619. United Nations, New York.

UN. 2005. World population prospects: The 2004 revision. Vol. 1: Comprehensive tables. Dep. Econ. Social Affairs, Popul. Div. ST/ESA/SER.A/244. United Nations, NY.

UNDP. 2005. Women's empowerment: biodiversity for development—the gender dimension. Available at http://www.undp.org/women/mainstream/index.shtml. UNDP, NY.

UNEP. 2002. Global environment outlook 3: past, present and future perspectives. Earthscan, London.

UN Millennium Project. 2005. Innovation: applying knowledge in development. Task Force 10 on science, technology and innovation. Earthscan, London.

UNWWAP (United Nations World Water Action Programme). 2003. Water for people: water for life. UNWWAP, Paris.

USAID. 2004. USAID agricultural strategy: linking producers to markets. Available at http://www.usaid.gov/our_work/agriculture/ag_strategy_9_04_508.pdf. USAID, Washington, DC.

USDA. 1994. Agricultural resources and environmental indicators. Pp. 25-33. Agricultural Handbook No. 705.ERS, Nat. Resources Environ. Div., USDA, Washington, DC.

Uvin, P. 1995. The state of world hunger. Pp. 1-18. *In* E. Messer, and P. Uvin (eds) The hunger report: 1995. Gordon and Breach Sci. Publ. SA, Amsterdam.

Warren, D.M., and B. Rajasekaran. 1993. Putting local knowledge to good use. Int. Agric. Dev. 13:8-10.

WCED. 1987. Our common future. Oxford Univ. Press, NY.

WGCCD (Working Group on Climate Change and Development). 2006. Africa: Up in smoke 2. Second report on Africa and global warming. New Economics Foundation, London.

WHO. 2005. The world health report 2005: Make every mother and child count. WHO, Geneva.

WMO. 1997. Comprehensive assessment of the freshwater resources of the world. WMO, Geneva.

World Bank. 1997. Expanding the measure of wealth: indicators of environmentally sustainable development. ESDS Mon. Ser. No. 17. World Bank, Washington, DC.

World Bank. 2002. Biotechnology for poverty alleviation and economic growth: Challenges and options for the World Bank. World Bank, Washington, DC.

World Bank. 2006a. World development report 2006: Equity and development. Oxford Univ. Press, NY.

World Bank. 2006b. Enhancing agricultural innovation: How to go beyond the strengthening of research systems. World Bank, Washington, DC.

World Bank. 2006c. 2006 world development indicators. World Bank, Washington, DC.

World Bank. 2006d. Capacity development in science and technology for development. World Bank, Washington, DC.

World Bank and IFPRI. 2006. Agriculture and achieving the Millennium Development Goals. ARD Rep. No. 32729-GLB. Available at http://siteresources.worldbank.org/INTARD/Resources/Ag_MDGs_Complete.pdf. World Bank, Washington, DC.

World Food Conference, 1974. Report of the world food conference. Rome, 5-16 Nov. 1974. FAO, Rome.

WRI (World Resources Inst.). 2005. World resources 2005: The wealth of the poor—managing ecosystems to fight poverty. WRI, Washington, DC.

4

Looking Forward: Policies, Institutional and Organizational Arrangements for AKST Development and Application

Coordinating Lead Authors:
Ruba Al-Zoubi (Jordan), Mohamed Moussaoui (Morocco)

Lead Authors:
Roy Abijaoude (Lebanon), Mohamed Abo El Wafa Gad (Egypt), Mukhtar Ali (Pakistan), Nour Chachaty (Syria), Alessandra Galié (Italy), Ashot Hovhannisyan (Armenia), Sanginov Rajabovich (Tajikistan), Ahsan Wagha (Pakistan), Lokman Zaibet (Tunisia)

Contributing Authors:
Sandjar Djalalov (Uzbekistan), Alisher Tashmatov (Uzbekistan)

Review Editors:
Gulcan Eraktan (Turkey), Fahri Yavuz (Turkey)

Key Messages

1. Natural resources will continue to become limited. As globalized trade expands and markets liberalize, CWANA's competitiveness in agriculture will rely more and more on increased productivity and higher product quality. Degraded land, depleted water resources and expanded deserts imply that agriculture will take place in less favorable environments. Further trade liberalization, implementation of which is expected to come into effect after the Doha development round closes, will make trade barriers, production support and export subsidization obsolete when trying to compete in international and domestic markets. All these future prospects are calling for AKST as a means to sustain CWANA agricultural competitiveness.

2. Applying AKST advances is crucial if we are to meet the challenges for sustainability and development in the CWANA region. CWANA agricultural research systems must adjust to the context of new challenges such as land degradation, water scarcity, migration, loss of biodiversity, increase in population growth rates and climate change. At the same time, with support from the Consultative Group for International Agricultural Research (CGIAR) and the Food and Agriculture Organization of the United Nations (FAO), they have to orient themselves toward new directions of research such as biotechnology, agrobiodiversity, GIS technology, IPM, water and soil conservation, rangeland and drought management, value chains and market research. Applying advanced AKST may bring new varieties of crops, breeds of livestock, and advanced technologies that are suitable to tackle the problems of biotic and abiotic stresses and to meet the challenges for sustainable development.

3. Agricultural productivity improvements will depend on substantial public and private investments in agricultural research and extension. The national agricultural research systems (NARS) are generally weak, and investments in agricultural research and extension are low. This situation is not likely to improve, considering current dismissal of agriculture as an engine of economic development and the lack of constituency for stronger NARS. Increasing public investments and providing incentives to the private sector to engage in research and extension to complement public efforts will likely help acquire adequate capacity to contribute to poverty alleviation, food security and economic progress. Moreover, a sustained public sector role in agricultural research will be essential, particularly for production areas in less favorable environments, unlikely to be served by the private sector.

4. Private ownership of intellectual property rights (IPR) is increasing, making it likely that developing countries will find more barriers preventing their access to international research spillovers. A self-reliant research policy is required to build domestic AKST capacity, with research directed toward identifying biodiversity and variety of species. Ways to achieve such an objective include identifying CWANA agricultural resources and biodiversity and establishing CWANA-based IPR (e.g., Arab IPR League) and forums for equitable exchange of IPR-based research results.

5. Food safety and quality standards are important for trade, access to industrial-country markets and domestic consumers' health, as outbreaks of food illness are expected to increase. The cost of assuring quality in food will increase due to intensive use of chemicals, transformation of traditional systems, and large-scale production structures and trade. Compliance with food safety regulations and quality assurance in CWANA has been relatively slow and is mostly driven by government laws made to secure traditional export markets, responding to provisions of importing countries. In local markets as well, it is important to safeguard the right to food safety for all consumers. Good agricultural practices at the farm level with stringent veterinary controls along the supply chain are required to ensure the safety of both fresh and processed foods. Institutions in charge of protecting public health and of promoting the adoption and implementation of standards have to be strengthened. Legislation needs to be enacted and strictly enforced. Prioritizing local consumer awareness, private enterprise commitment, and risk assessment and laboratory infrastructure will ensure good traceability of food.

6. AKST in the CWANA region has too often used a nonholistic approach with little involvement of stakeholders. As a consequence, it is lagging behind international trends in innovativeness and effectiveness. Adopting a participatory and integrated approach can support AKST in CWANA to face the fast pace of its population growth and find a relevant role vis-à-vis agricultural needs and trends at national, regional and international levels.

7. Developing and applying AKST in CWANA is not truly geared toward the goals of alleviating poverty and promoting sustainability. These goals, however, are expected to be major paradigms of agricultural development in the next decades. Transparent, participatory and accountable mechanisms for setting AKST priorities at the institutional level can enhance the implementation of policies that are able to tackle poverty in CWANA while also addressing the sustainability dimension.

8. Proper and well-established links among agricultural education, research and extension are important if AKST is to work efficiently. The source of knowledge for education and extension is directly connected to the results of scientific research, and the research is driven by collaborating with extension workers who are well aware of local problems and are in close touch with farmers. The curricula of agricultural education and the content of the courses must be up to date and applicable to the needs of the market. These needs can mainly and correctly be determined by extension activities, which are the best way to tap into local knowledge. In CWANA countries these links are not well established because legitimate interaction among agricultural education, research and extension sectors is lacking. To enhance AKST effectiveness, links among agricultural education, research and extension are to be strengthened so that all links, including farmers, can be included in the system. Policy options for forging well-established links are to put these institutions under one authority such as the land-grant universities in the U.S. or to ensure legitimate horizontal and vertical interaction among them.

9. If persistent needs for national agricultural technology are to be effectively met, NARS in the CWANA region should be structurally empowered and their activities supported by regional and international resources. AKST development in CWANA suffers from lack of an enabling environment. The educational and research infrastructure is poor; policies and institutions place only limited emphasis on domestic and regional efforts toward developing AKST. This situation requires a radical policy shift to favor strengthening educational infrastructure and to adopt a policy framework that provides human capacities and offers incentives for AKST development.

10. Policies that promote agrobiodiversity and use of traditional knowledge lead to sustainable development of agriculture, despite the intensification of farming systems. Green Revolution practices and intensification of farming systems with the introduction of new varieties of crops, livestock, mechanization and aggressive use of chemicals cause us to lose traditional knowledge and biodiversity. Formulating new agricultural policy to protect and enhance agrobiodiversity has to become an important part of the agroenvironmental objectives and actions for many CWANA countries. Developing these policies will be in response to growing public concern over the increasing pressure on natural and existing ecosystems brought by agricultural activity. Actually, government policies toward biodiversity should balance the tradeoff between benefiting the economy and conserving biodiversity.

11. Future AKST in the CWANA region is to be visualized as transitional, to benefit from local knowledge and incorporate and transform local agricultural practices into scientific ones. Options in that line include recording, preserving and researching local knowledge, devising new AKST models to take advantage of local agricultural practices and considering traditional knowledge as a base of every international attempt for modernization in a CWANA country.

12. The CWANA region continues to lack appropriate technologies that could help effectively address key concerns like desertification, low productivity and loss of biodiversity. One reason for this is that the available technologies are not necessarily appropriate, as they are not based on indigenously developed or documented AKST. This problem can be addressed by national, regional and international initiatives aimed at strengthening research, technology development and extension capacities within the CWANA region.

13. Higher, stable and continuous economic growth substantially encourages farmers to make better use of AKST. In CWANA, the per capita consumption of food is low, especially of meat and milk products. Higher demand for agricultural products would mean more cash flowing in to the farmers, who in turn would most probably seek out and use appropriate AKST.

14. Continuing population growth is likely to increase rural-to-urban migration. This will result in small-scale farming employing those sectors of the population that have limited opportunities of movement and choice, mainly women. These sectors will have to rely on agriculture to support their livelihoods and at the same time face harsh environmental conditions due to climate change and difficult market access because of global trade. Technological innovations in agriculture, which are not designed or applied to meet the needs and conditions of women, carry the risk of further increasing the burden on women as providers for their families, and may impair women's productivity. The negative consequences of these trends can be contained if institutions adjust to the changing circumstances in rural areas. This means, for example, that women's role as farmers is acknowledged together with their role as food providers and, often, as heads of household. If their rights are redefined accordingly, their access to resources will be facilitated, their agricultural work supported and their livelihoods and lives of their families enhanced. Also, policies aimed to build capacity may be put in place so that rural livelihoods can depend on diversified sources of income, thus reducing the vulnerability of disadvantaged sectors. If AKST addresses the needs and priorities of these new farmers and adopts a more participatory model of development, it may boost the role of agriculture and sustain the livelihoods of the sectors that increasingly rely on it.

4.1 Implications of Future Challenges for AKST-related Policies

4.1.1 Market and trade issues

Markets and trade are important factors in determining the access to and adoption of AKST. Rising demand for agricultural products and more competitive markets are likely to result in higher demand for AKST. For example, protectionist policies would not encourage the adoption of certified seeds, while liberalization and appropriate marketing policies may be accompanied by the adoption of more productive technologies, higher efficiency and economic growth. We are interested to find the best ways and options to develop AKST based on our assessment of market and trade developments in the region.

4.1.1.1 Trade arrangements

Although trade liberalization globally represents the goal of multilateral trade negotiation under the auspice of the World Trade Organization (WTO), most WTO members have engaged in regional or bilateral agreements due to the relative ease of forming regional blocs. The CWANA region is not an exception. It has seen emerge many regional and bilateral trade agreements among neighboring countries. For instance Egypt has concluded about 40 agreements (ESCWA, 1998). In 1981 the gulf countries established the Gulf Cooperation Council to enhance intraregional trade and cooperation. In February 1989 the Arab Maghreb Union was established in Marrakech. The Customs Union between the European Union (EU) and Turkey is a unique event in the region; it has increased trade volume between the two partners and has been particularly profitable to Turkey. These arrangements are to be fostered to facilitate AKST adoption in the region.

WTO trade negotiations, however, also create threats for developing countries and for the CWANA countries in particular. Not only benefits are expected. For example, WTO blue box payments for reducing production and setting land aside will be reduced according to a tiered formula. Under this formula, members having higher levels of trade-distorting domestic support will make greater overall reductions to achieve harmonious results. The same approach will be valid for the total aggregate measure of support and market access. So developing countries, which need more support for their agricultural sector, will be affected by these developments (Zaibel et al., 2003).

Intraindustry trade is also growing among regional trading groups. Such a trend is an indication of economic integration and economic diversification and development. Intraindustry trade within the regional trading blocs occurs mostly between neighboring countries with similar demand structure. Transportation and transaction costs are among the constraints that hamper its development within the region. Policy and institutional changes are required to follow these developments and overcome the current constraints.

Since the EU is an important partner for many CWANA countries (e.g., the Mediterranean countries), its enlargement with the entrance of the Central and Eastern European countries (CEECs) will bring benefits but also threats to the region. There would also be benefits if Turkey were to enter the EU; it would bring the EU boundary closer to CWANA and be an excellent opportunity to increase mutual trade.

We do not expect that enlargement of the EU to the east will divert foreign investments to the newly added countries instead of the countries in the CWANA region, as the incentives to invest in these regions are dissimilar and the foreseen investments in CEECs had begun to be realized even before the expansion.

4.1.1.2 Trade negotiations: more integration

Since the inception of WTO in 1994 efforts have focused on launching a new, comprehensive round of multilateral trade negotiations. From the Seattle ministerial meeting up to the Doha Declaration there have been advances on a number of trade and nontrade issues. The ministerial conference at Cancún, Mexico, set a milestone toward achieving the Doha Development Agenda round of trade negotiations as mandated by ministers at the 2001 Doha conference. However, given the few achievements in past negotiations, observers remain skeptical that a new comprehensive round can be completed as planned (Miner, 2001). The big players are expected to make additional policy reforms (e.g., trade legislation in the USA and European Common Agricultural Policy [CAP] reforms in the EU) before undertaking strong concessions and commitments in the upcoming negotiations.

It is somewhat disappointing that benefits from agricultural trade liberalization have not materialized as was predicted. There are at least two reasons why trade benefits were only partial. First, negotiations on agriculture alone do not consider the comparative advantage principle. As a result, the Doha Declaration made provisions for broad-based negotiations extending trade negotiations to further liberalize trade for the industrial products and services of which nations may take advantage (Ingco, 2002). Second, national policies and legislation are creating additional cross-national boundary transaction costs and limiting liberalization efforts. Trade relations remain far denser within nations than between nations and a lot of trade does not occur according to predictions of the neoclassical model (Gerber, 2000). Accordingly, "deep" economic integration requires that not only border barriers but also domestic policy barriers be removed. More integration is needed to achieve regional cooperation to develop AKST.

4.1.1.3 Recurrent and newer issues

The main issues already identified in the General Agreement on Trade and Tariffs (GATT) on agriculture embodied market access, export competition and domestic support. However, a body of new trade and nontrade concerns are emerging and attracting growing public interest. The agreement on agriculture already included issues of food security, food safety and quality, environment concerns, resource conservation and rural development (Miner, 2001). Additional issues raised in the last negotiation meetings included animal welfare, biotechnology, species preservation, landscape safeguards, poverty reduction and preservation of rural culture (Miner, 2001).

Newer border-trade topics embodied items such as the rules of origin, standards and technical barriers, intellectual property rights, sanitary and phytosanitary (SPS) standards, dispute settlement and the role of small countries (Gerber, 2000). Among the nontrade domestic policy issues are foreign investment, competition policies, and labor and environmental standards. All these issues affect AKST; more investment is required, CWANA seeks more aid in the area of SPS and in general there is a need for research and capacity building.

4.1.1.4 Regional links: the EU-CAP reform

Traditional regional links are shaping export markets and observed trade flows. According to Diao et al. (2002), export markets for many developing countries are concentrated in a few countries in the North because of geographic proximity and historical links. As a result trade negotiations will be shaped by regional blocs. North African and Middle Eastern countries are thus more interested in the EU agricultural markets and consequently in EU agricultural reforms under the 2003 CAP reform.

Indeed, the work program annexed to the Barcelona Declaration cites the following objectives with regard to the countries that have signed the declaration, which are options for AKST development as well (Chioccioli, 2002):

- Integrated rural development
- Support for policies implemented by Mediterranean countries to diversify production
- Reduction of food dependency
- Promotion of environment-friendly agriculture

4.1.1.5 Food safety and product quality

With the decline in the use of traditional trade barriers such as tariffs and quotas, there is evidence that technical and regulatory barriers are increasingly used instead. In industrial countries many firms are moving toward adopting international standards. This move is relatively slow in CWANA countries and might therefore represent an obstacle to international trade. Food safety and quality standards are important for trade and access to markets in industrial countries but also for domestic consumers' health, with a view to reducing food-borne morbidity and mortality and improving nutritional and hygienic quality. Food-borne diseases such as *Salmonella* and *Escherichia coli* infections remain responsible for high levels of morbidity and mortality in the general populations of CWANA, but particularly for at-risk groups, such as infants and the immunocompromised. Many zoonoses such as brucellosis and tuberculosis that are associated with handling diseased domestic and wild animals are also prevalent in CWANA countries. Because of intensive use of chemicals, transformation of traditional systems, large-scale production structures and trade, the cost of maintaining quality in foods will increase. Organic agriculture is an alternative to traditional farming systems and greatly appreciated by consumers, mainly in industrial countries (import markets). In many countries, including CWANA, products are registered with country of origin designated to assure consumers of the assumed high quality. Compliance with food safety and quality assurance in CWANA has been driven by government laws to secure traditional export markets. In recent years, several CWANA countries such as Bahrain, Morocco and Pakistan have planned and implemented extensive reviews of their food safety systems, updating their legislation and generally improving their systems as a whole (WHO, 2001). In local markets as well it is important to safeguard the right of food safety to all consumers, protecting their health from unsafe or potentially unsafe food by preventing health hazards associated with microbiological and chemical contamination and additives. Good agricultural practices at the farm level with stringent veterinary controls along the supply chain are required to ensure the safety of fresh and processed foods. Highly useful preventive and cost-effective approaches to food safety (such as the Hazard Analysis Critical Control Point System or HACCP) exist and CWANA countries should adapt and adopt them. Institutions in charge of promoting the adoption and implementation of standards have to be strengthened, and strategic partnerships between the multiple concerned disciplines (such as health, agriculture, and food industry and trade) encouraged. Consumer education is key to preventing food-borne diseases. Donor support for building capacity in the area of food safety is to be called upon, and legislation needs to be enacted and strictly enforced.

4.1.2 Pricing policies

Pricing policies for agricultural products ought to follow the rules of a free market. Further, strategic planning is needed to shift toward market-oriented agriculture policy closely integrated with national development objectives, without compromising food security or food sovereignty. This however depends on the prevailing local market structure and the engagement in multilateral and regional economic cooperation and negotiation toward establishing free markets. If the conditions of a free competitive market are prevailing, this will lead to efficient price formation, which in turn influences positively the development and adoption of AKST.

In most CWANA countries, however, agricultural markets are not competitive. Small-scale farmers in particular are facing problems of scale, with market power in favor of the middleman. Marketing conditions and marketing margins are changing as a result of evolving supermarket requirements, mostly affecting small farmers. Under these conditions pricing policies will be developed in parallel with the development of coordination strategies. Vertical coordination will guarantee stable prices and markets. Farmers' associations are also an effective way to create market power for small- and medium-scale farmers. Vertical coordination and farmers' associations are more likely to favor the adoption of AKST in response to new requirements of the supermarket phenomenon that characterizes the new marketing scene. For instance, supermarkets are adopting private quality schemes. Farm enterprises need to adopt these private standards if they want to stay in business.

The pricing policy, when coordinated by bureaucratic mechanisms through administered prices, does not reflect marginal production costs. Under this scenario, for administrative convenience, monopolies are created that lead to prices that are distorted when compared with product quality. It should be noted that in this scenario there is no market-based price formation and no possibility to compensate for seasonal deficiencies and overstocks. Prices set by the government are rarely revised and do not reflect the opportunity cost on the international market, which brings negative added value for some producers if evaluated on the basis of international market prices. Because of government intervention, entrepreneurs will find it more profitable to trade on the basis of barter or mutual agreements, as the transaction costs will be too high. In this case, producers see no necessity to seek alternative resources or adopt newer techniques, because they have no incentive to improve their work processes.

Most CWANA countries have made significant progress toward establishing free market conditions. Negotiations are under way with major trading partners to enter into trade relations based on WTO rules. At the national level, agricultural production is no longer centrally planned and is now in the hands of private sector farmers who are free to choose what crops to grow. Agricultural incomes have risen significantly as a result. Government policy toward trading inputs and outputs, including processed goods, is steered toward creating a liberal market, although some interventions that cause distortions and inefficiencies remain in some countries. In these countries, governments are undergoing reform programs to completely liberalize the sector and redefine the relationship between government agencies and the private sector. This will create a more favorable environment with freer markets and prices. Liberalization will likely be accompanied by better access to AKST, first to meet international markets' requirement, second to be competitive in the marketplace; third, international markets will have access to AKST.

The private sector must be prepared to assume the role of market regulation and to serve as an engine of growth for the whole agricultural sector. Working directly with farm associations, private enterprise will improve marketing conditions by changing traditional concepts of how to market and by creating useful information systems and fostering business links. Useful information will be needed about prices, but also about quantities and the quality of products as required by the supply chain actors. This will improve price formation mechanisms. While helping the industry to process high-quality products efficiently and create better conditions to foster processing capacities through transferring technologies, this will also lay the foundation for sustainable growth in the industry and provide the agricultural sector with the means to respond to ever-changing market conditions. The private sector may also be involved in AKST development through involvement in joint ventures with research institutions to make AKST available as a public good to smaller farms.

Changes in price-formation policies will occur mainly as a result of shifts in the demand curve and as a consequence product prices will be affected differently. What factors will cause this shift?

- Demographic—growth in population normally brings equal growth in demand for all types of goods. However, concomitant changes in the age structure may affect the demand and consequently the price for certain goods. For instance, an increase in the percentage of children in CWANA countries population may cause a higher demand for milk.
- Economic—changes in per capita income levels may affect the degree of demand for most goods. Increase in income will change the food patterns, with expensive meat and sea products dominating. The demand for less tasty foods containing starch will decline as incomes increase. If the income level falls, less expensive necessity foods, such as bakery products, will prevail.
- Socio- and psychological—these factors have recently emerged because of growing concerns about human health. Thus recently there was a decline in the demand for beef, especially in western Europe, resulting from fears that mad cow disease could cause mortal disorder in the human brain. Avian flu caused a drastic decline in the demand for poultry. But the demand for olive oil grew in view of the belief that it reduces the risk of cardiovascular diseases as compared with adipose or other vegetable oils.

While the above factors will directly influence the demand for final agricultural products they also indirectly influence the derived demand for AKST. Demand- or market-oriented production will focus more on the adoption of AKST.

Price disparities have been most visible at the producer level, where prices for agricultural products increased much less than prices for means of production. Calculations indicate that the rise in price for means of agricultural production is 40% faster than for agricultural products. Purchasing prices set by monopolistic processing industries are below world market prices, and farmers have no option but to accept them.

Notwithstanding government support to producers in the form of subsidies, most means of production such as agricultural machinery, fertilizers, pesticides, and veterinary services are inaccessible for producers. Also, food-pricing policies, based on an extensive system of food subsidies, have a negative effect on macroeconomic variables such as the rate of inflation, the balance of payments and the exchange rate. Moreover, the subsidy system has destabilized industrial output and investment. Restricting the benefits of subsidies only to those most deserving would lower the inflation rate, reduce the volume of imported food (thus the government deficit), and increase industrial output and investment. Subsidies to producers will come through public services such as research and extension and may be a more effective way to diffuse and adopt AKST.

To overcome the negative consequences of the transition to free market conditions it is necessary to take several measures to improve agricultural policy:

- Prices and agricultural trade should be liberalized. Unless prices for agricultural products are harmonized with world market levels and payments are made directly to producers, we cannot expect significant growth in the agricultural sector, and productivity will remain low.
- The primary task is to improve price-formation policies through increased competition at the level of farms. Antimonopoly legislation should be developed. Creation of a more competitive environment in the sphere of purchases will increase farm income and encourage farmers to improve productivity, marketing and trade, and quality of agricultural processing.
- It is necessary to abolish the system where production requirements are based on government order and production scheduling is done by the state. For products that in government's opinion represent national interests, a price policy should be introduced that would stimulate their voluntary production based on profitability.
- Public purchases should be based on market prices. Productivity could be improved using a contract-based system. In future, the state and farmers will buy and sell futures contracts in response to changes in market conditions and generate income before harvest.
- It is necessary to undertake thoroughgoing reforms in agricultural and trade policies. Trade barriers should be removed and a system of customs duties established. Export and import licensing should be abolished; private companies should be allowed and encouraged to take part in international trade on condition that only the above-mentioned customs duties are collected from them.

4.1.3 Research policy for NRM

One of the most challenging issues is the emerging expansion and diversification of the research portfolio. In addition to conventional topics, AKST is called upon to cover a variety of new research and innovation domains.

Pasture management. Agricultural land for the whole CWANA region is mainly devoted to permanent pasture and rangeland. The proportion of rangeland to total land is 83 percent, the lowest proportion being in Southwest Asia (55%), the highest in the Arabian Peninsula (98%). In many CWANA countries, rangeland carrying capacity is decreasing because of overgrazing.

Research and technical options for improved rangeland management are available, e.g., practicing rotational grazing, corralling to rehabilitate degraded spots, seeding and planting possibly supported by fertilization and water harvesting, practicing agroforestry, maintaining livestock biodiversity and reducing the number of artificial water points. However, these practices have often been developed in completely different ecosocial regions, and adaptation of these technologies in other countries is important. Further research in the area of rangeland will contribute to solving environmental problems and to developing a livestock industry in the region, and it will mitigate the climate change problems.

Soil and water management. Research in soil salinity management will be essential for the region. Research priorities in this area include development of measures to prevent soil salinity; land reclamation by using low-cost technologies to improve the properties of saline soils; assessment of soil salinity through GIS technologies; biological reclamation of saline soils; biodrainage systems (tree plantation) in saline and waterlogged soils; selection of salt-tolerant crop species and varieties; and development of halophyte agriculture.

In the field of irrigation and drainage management: deficit irrigation, conjunctive or drainage water use, irrigation scheduling, irrigation, drainage-water quality management, identification of optimum furrow length, water discharge, development and adoption of advanced water-saving technologies, selection of promising irrigation technologies. In rainfed areas: supplementary irrigation, water and soil conservation technologies, diversification of cropping patterns, crop residue management, land leveling, integrated plant nutrition management, irrigation, wind and water erosion control, GIS technologies and erosion control, traditional and introduced soil conservation technologies in mountain areas, slope land management, and watershed management.

Integrated pest management. Integrated pest management (IPM) is an effective and environmentally sensitive approach that relies on a combination of common-sense practices. IPM programs use current, comprehensive information on the life cycles of pests and their interaction with the environment. IPM takes advantage of all appropriate pest management options including, but not limited to, the judicious use of pesticides. Use of biological alternatives instead of harmful chemicals, research on soil biota, especially on nematodes, for soil-borne diseases and on parasitic weeds could alleviate many problems farmers of the region face.

Organic farming. Industrial countries have developed markets for organic products and there are today numerous opportunities for organic agriculture. Organic crops include cotton, cereals and potatoes. Beef, dairy, and sheep and goats are the focus for livestock. The Swiss Agency for Development and Cooperation has already implemented a project for organic cotton in Central Asia. Research and implementation activities aimed at adopting organic agriculture can potentially meet the challenges CWANA farmers face.

Conservation agriculture and reduced tillage research. Promotion of research in the field of conservation agriculture could save water, labor, fertilizers and pesticides and fuel; it could solve many problems connected with the degradation of natural resources in the region. The management of cover crops and crop residues will be closely related to carbon emission issues, and will increase nutrient and soil organic matter content.

Livestock. For many poor households in CWANA countries, livestock is an important asset. In this region where many challenges impose themselves in the field of natural resource degradation, integrating crops and livestock production is a promising agricultural system for low-income small-scale farmers. AKST therefore needs to find ways to ensure that crop and livestock resources will be developed sustainably with enhanced output per unit while increasing area productivity. Lack of feed resources, poor genetic makeup and lack of effective cover for animal health are the main constraints hampering livestock development in the region. Conservation of local livestock breeds would be required to sustain development, and nontraditional feed resources need to be developed to contain rangeland losses. Development and access to quality animal health services and genetic material for upgrading of livestock should be possible. Embarking on such initiatives in the region could be made possible through AKST.

Crop management. The region needs to enhance germplasm and take advantage of its genetic resources. Advanced plant breeding may help achieve productivity gains, introduce resistance to pests and diseases, reduce pesticide use, improve crop tolerance for abiotic and biotic stress, improve the nutritional value of some foods, and enhance the durability of products during harvesting and shipping. Raising productivity could increase smallholder incomes, reduce poverty, increase food access, reduce malnutrition, and improve the livelihoods of the poor.

Drought tolerance. CWANA countries are classified as drylands, susceptible to desertification and mostly drought prone (UNEP, 1997). These countries vitally need drought management and mitigation. Thus CWANA governments have to make difficult tradeoffs between short-term benefits and long-term solutions. Droughts always require immediate attention because they threaten human lives, but long-term solutions are also necessary. From this perspective, it is essential to note that drought-tolerant crops, varieties and hybrids are essential for countries of the region.

High-value crops. The main objective of the research system in the region during the process of commercialization and diversification remains to generate new technologies that improve productivity and farmer income. In addition to the productivity objective, research should focus on providing farmers with the flexibility to decide on crop choices and to move relatively freely to growing the crops they choose. Gearing farmers to meet more exacting safety and quality standards ought to be an essential part of the strategy.

Postharvest methods. Many times, large shares of food produced are lost after harvest. Reducing postharvest losses has been an important focus of AKST and development programs in the past. But on several occasions technical innovations have faced sociocultural or socioeconomic problems like low profit margins, additional workload, or incompatibility with the existing production or postproduction system. The divergence between technical recommendations and the realities of rural life translated in many cases into a low adoption rate. Now the rationale for improvement in postharvest systems has been shifting from preventing loss to opening new market opportunities. Making markets work for the poor is emerging as the new rationale for development, reflecting a shift away from governmental operation of postharvest tasks to frameworks that enable private sector initiatives in this field.

Biotechnology. Agricultural biotechnology will contribute to poverty reduction and food security if scientists can develop technologies to increase quality and yield of food crops, and if small-scale farmers adopt the technologies. Research has to focus on crops, livestock and fish. Major crops are rice, maize, wheat, sorghum, millet, oilseed and potato. Biotechnology should also focus on high-value cash crops: cotton, soybean and vegetables that can increase the incomes of small-scale farmers through crop diversification. Fish and livestock—cattle, sheep, goats, pigs and chickens—are also important. The technology must be simple, low cost, and carry little or no risk to human health and the environment.

Genetic engineering could be widely used as a breeding technique. Genetic engineering involves the transfer of one or more precisely selected genes into the genome of the host organism. The ability of genetic engineering to transfer genes across the species barrier or indeed across kingdoms is precisely what gives the tool such power and what attracts such controversy. Gene banks and genetic engineering can also be used to speed up the breeding process by inserting a specific gene into an otherwise desirable genetic background without requiring multiple generations of backcrossing to eliminate unwanted change, as is necessary with conventional breeding. Biotechnology can also be used to develop vaccines for animals. Bio-information could support molecular research, e.g., breeding and GMO activity.

Value-added technologies and market analyses. Value addition to primary goods offers a major income opportunity and is not being achieved in many countries of the region. Research and development of value-added products and markets could increase income of poor farmers and be used to generate income in rural areas. For value chains and market analyses, this type of research is essential: analysis of constraints of access to market information; development of better methods to communicate price and quality information; new technology to reduce postharvest losses; role of production for different markets; availability of external and domestic markets for the poor; improved access to financial capital and markets; input markets and services; and capacity building in marketing.

Given existing research capacities and capabilities in the CWANA region, it is unlikely that such an overwhelming agenda will be met under the business-as-usual scenario.

4.1.4 Investment and funding policy

During the twentieth century, highly accelerated improvements in agricultural productivity have significantly contributed to poverty alleviation, food security and economic progress. These productivity improvements have been achieved as a result of substantial and deliberate investments in agricultural research and development (R&D). Because of associated high returns, it is recognized worldwide that a minimum target of spending on investment in agricultural R&D should be set by developing countries, in addition to ensuring larger share gains from international public spillovers.

Historical trends of investments in agricultural R&D show, however, that government spending slowed in the Middle East, North Africa, and Central Asia and the Caucasus. Meanwhile, international technology spillovers and corresponding knowledge have also decreased. Taking into account the low density and poor to medium performance of NARS, these trends currently pose critical challenges to AKST development and application. Business-as-usual prospects show that higher investment will be of great value to ensure a critical level of AKST self-reliance. This is vital in light of persistent signals that developing countries are not likely to benefit, as they used to do, from international spillovers from the North and from the CGIAR centers.

It is suggested that under globalization, countries would still have opportunities to benefit from investment spillovers by interacting with nations and communities who are well equipped in agricultural science, technology and information. However, risks are likely to be faced regarding the availability, price and quality of needed new technologies. Research agendas and investment structures are changing in the direction of diverging research objectives between industrial countries and developing ones, and of the emergence of private corporate bodies providing AKST (Alston et al., 2006).

As a result, only substantial self-reliance in agricultural R&D will ensure developing efficient agriculture production systems that are able to successfully compete in price and quality in domestic and international markets (Pardey et al., 2006ab). This is of particular importance for the future of small-scale farmers who cannot generate or do not have access to the AKST needed to improve their livelihoods.

Therefore, it seems that business as usual will not, under all circumstances, ensure a continuous flow of affordable AKST. An increase in national spending of CWANA countries will still be needed to counter increasing monopoly building in the AKST system that may be detrimental for agricultural development and sustainability objectives by excluding developing and less-developed countries from AKST benefits.

4.1.5 Intellectual property rights policy

Growing IPR protection as one endorsed by WTO members is intended to promote innovation and technology transfer and dissemination to the mutual benefit of both the producer and the user of the technology. This is why all countries are called upon to establish and enforce appropriate IPR-related regulations to help innovation take place in sectors vital for socioeconomic and technological development. As a result if required regulations are adopted, technology transfer toward less-developed countries can occur (Abbot, 2003). Such cooperation comprises assistance in preparing law texts related to IPR promotion, enforcement and protection; prevention of their abuse; implementation of institutions and agencies serving this aim; and last but not least, personnel and technical training.

However, in developing countries, regulations protecting IPR can be perceived as a means of principally serving rich countries since they are the technology generators. Holding IPR, AKST producers will invest only in industrial countries with established and functional laws that comply with international standards. It is true that developing countries are more and more present, but the technology generated in these countries comes either from multinational companies that relocate their production plants or from small national companies.

In addition, perfect compliance with trade-related intellectual propriety rights (TRIPS) does not guarantee that poor countries will have access to new top technologies. Often infrastructures are insufficient and professionally qualified personnel are lacking to make use of them. Furthermore, technology patenting is not always followed by use in production, which prevents consumers from taking advantage of technological progress.

Abolished trade barriers and globally protected IPR may be antagonistic, if effective holding and use of IPR is not rightly controlled. According to the WTO report on interactions between trade exchanges and competition policy, IPR protection and competition policy are seen as complementary notions aimed to promote competition and consumer welfare. But in some cases, IPR protection might threaten competition (Drexl, 2003). To avoid such a negative outcome, we might suggest that IPR protection be placed under the control of a global competition law. But, should harmonized competition laws include a sensitive concept like IPR protection? If yes, what would be the effect of that extended law on high technologies?

At present, few CWANA countries have established IPR protection laws and hence are not likely to take advantage of accessible new technologies to strengthen their own capacity for innovation. While working toward establishing a domestic legal environment (market competition and IPR protection laws), developing countries can consider

- Abolishing barriers, for better access to innovation
- Supplying adequate engineering and managing skills
- Promoting an adequate national marketing environment
- Reducing the technology gap
- Implementing IPR standards for dynamic competition

These suggestions are acceptable if the imported technology is relevant and if the importing country has adequate capacity, policy, regulation and institutions to optimally exploit IPR provision.

In a fair-competition environment with protected IPRs, innovation, consumer welfare and development are evident consequences. In other words, competition enhances dynamic efficiency, which through protection can give access to an exclusive right to innovation through appropriation in respect to patent law while diverting free riders and misappropriations. This gives the consumer better access to innovation and encourages information dissemination. Monopoly ownership, resulting from IPR protection, may not be harmful to innovation in given applications (scientific research, computer licenses, etc.).

However, often an optimal mix of competition policy and patenting laws is required to effectively induce a productive equilibrium between innovation and IPRs, as mentioned above, creating stronger markets.

4.2 Implications of Future Challenges for AKST-Related Institutions and Organizations

In such a rapidly growing world with tremendous challenges, CWANA has a lot to worry about while striving for a better and sustainable future. CWANA countries share complex situations, beginning with their harsh climate and scarce resources. These factors are compounded by high population growth rates; they pass through wars and natural disasters, and end with the newly emerging issues of globalization and trade liberalization. All these factors have significant implications on the ability of CWANA countries to achieve development and sustainability goals, and more specifically to reduce hunger and poverty and improve livelihoods. Institutional arrangements and partnerships are major actors in developing and applying AKST. Their effect varies, reflecting different levels of involvement and maturity across the region.

4.2.1 Cooperation

Institutional and organizational arrangements of interest comprise regional and international conventions (Framework Convention on Climate Change, biodiversity, etc.), regional organizations (e.g., the Arab Center for the Studies of Arid Zones and Dry Lands—ACSAD), national institutions, local and community-based arrangements to enhance technology generation, transfer and adoption, access to new technology and better technology management.

These arrangements affect directly (as direct drivers) the generation, access, dissemination and use of AKST in achieving development and sustainability goals. If the CWANA region is to attain development goals, member countries need to cooperate and coordinate their efforts.

CWANA countries need to coordinate and collaborate within and across the region to deliver the development objectives, especially with reference to poverty alleviation, amelioration of hunger, and socioeconomic and sustainable development. Also they need to establish networks to preserve and develop natural resources and human capital, to mitigate natural disasters such as droughts and floods, and to resolve conflict over natural resource management.

Global cooperation. Institutional arrangements within developing countries are needed to conform with and provide input into overall government reform, particularly into restructuring their economic, social and related fields. Cooperation principles should be based on an action- and results-oriented approach and be consistent with the principles of universality, democracy, transparency, cost-effectiveness and accountability. These institutional arrangements should elaborate strategies and measures to increase national and international efforts to promote sustainable and environmentally sound development in the CWANA countries and to promote economic growth.

To be effective, these efforts need to be coordinated and implemented by private or public organizations in relation to international organizations in the form of networks to support and facilitate the transfer and adoption of technology. The involved organizations include

- National research centers
- Nongovernmental organizations (NGOs)
- Trade associations (chambers of commerce, associations of enterprises)
- State and parastatal institutions for converting economic and policy approaches (ACSAD)
- Private service providers, active NGOs

These institutions and international networks contribute to development, diffusion and adoption of AKST. They should be enabled by financial funds, strong networking capabilities, continuous learning and assessment, explicit incorporation and voice of producers in the AKST process, business management and planning approaches, and clear and transparent priority-setting mechanisms to achieve significant success in realizing the development and sustainability goals.

Regional cooperation. Regional and subregional cooperation includes the regional development banks, NGOs, and regional economic and technical cooperation organizations. Within their respective agreed mandates, these organizations can contribute to AKST development and adoption by

- Promoting regional and subregional capacity building
- Promoting the integration of economical, social and environmental concerns in regional and subregional development policies
- Promoting regional and subregional cooperation, where appropriate, regarding issues related to sustainable development

In particular regional organizations for technology generation, evaluation, diffusion and study will need to be developed. It is likely that new AKST will flow toward the region from all around the globe, promoting R&D in this field. This will be further enhanced by the increased pressure on natural resources associated with increased population. Countries of the region may be encouraged to share resources (water, energy, gas), which would help stabilize the price of such goods. Nevertheless, a basic assumption for stronger regional cooperation is the high level of commitment for institutional development and reform from various countries, especially from industrial countries and donors.

Such cooperation is more effective if an outward liberalization policy is adopted. If an inward-looking and protective approach in dealing with development issues is adopted, it is not likely to enhance the development and application of AKST to achieve development goals and reduce poverty in CWANA.

In the latter case, increased prices and the monopoly of some associations will prevent poorer—or nonoil-producing—countries from developing and applying AKST. Under this scenario, countries will continue to have inward-looking policies that will hinder any potential cooperation across borders. In addition, links with R&D institutions will be weak and thus access to new technology and innovation will be limited. This will likely have long-term implications on reducing poverty and achieving development goals.

Given these expected negative results, CWANA countries will more likely take a proactive role in going through a transitional phase to enter global markets. Such a phase will be enhanced through developing the regional trading blocs that are already emerging, making it easier to develop and apply AKST at national, regional and international levels. Regional cooperation will be enhanced in the fields of research and AKST. It will mainly target product processing, storage and marketing—ultimately providing food security and protecting human health and the environment. This will contribute strongly to poverty alleviation and will improve the quality of life in the region. As a result, investment in science and technology in general and in agricultural R&D in particular will be enhanced on national and regional levels—thus contributing to achieving the development goals.

National cooperation. Of equal importance to CWANA countries are what arrangements national institutions will make and the effects they will have on developing and applying AKST in their efforts to achieve the development goals.

It is assumed that CWANA will adopt knowledge-driven economic development in which AKST is the key factor. CWANA countries will enjoy needs-based decision making integrated within countries of the region and across regions, leading toward achieving the development goals and improving livelihoods.

While CWANA countries struggle for integration within the global market through regional trade areas, they need to face major challenges including developing AKST infrastructure at different levels such as academia and research as well as developing and planning for transformation and change management under globalization. Overall, the process of change should feed into enhanced well-being of nations and improved health, education, and use of natural resources and infrastructure.

Policy and institutional reform in various sectors will be a major feature of this storyline. National policies, plans and legislation will be improved to support integration into the global market and to meet all the required criteria and conditions for promoting investment and facilitating trade. In parallel, public institutions will need to be developed to accommodate the changes. Local producers will strive to meet the conditions for entering the global market. The role of the private sector and other national stakeholders will be enhanced through better cooperation, and strengthened public–private partnership will be witnessed with more emphasis on gender equality and empowerment of local communities. CWANA countries will live in an era flourishing for development institutions, especially those working on AKST and other relevant issues including natural resources and property rights. Farmers' organizations will emerge as a major player to support research and technology transfer and application and protection of farmers' rights. Civil society organizations promoting the conservation of natural resources will advocate land conservation and rehabilitation. Education and capacity building for various players will be integrated into various activities. Sustainability will become a culture and way of living for the people of CWANA, leading the countries and the region toward more accomplishments on the scale of development goals.

As stated earlier, inward policies would contribute to increased prices. Also, the monopoly of some associations would prevent poorer—or nonoil-producing—countries from development and application of AKST. CWANA countries would suffer from focusing on food security from the local perspective, and not in the global context. Research and development would focus on adaptive research, but investment in basic and applied research might not get priority. As a result, the capacity to innovate would be limited. The media would continue to be under central control, sifting the information, and thus agricultural informatics and the flow of scientific information would be blocked to a greater extent.

Consumers would have to rely on limited information and because of the limited role civil society would have, consumer activism would not take root. The human resource quality would remain at low ebb and agriculture would continue to be complacent with unskilled or low-skilled labor, with scarce capacity to transform agriculture and thus increase its productivity. Over-controlled governance would prevent agriculture and its relevant institutional arrangements from responding to the change out of and across borders.

As governments embark on more people-caring and outward-looking policies, they become more proactive to provide equitable access to education, health and information. Thus AKST development will be enhanced, focusing mainly on processing, storage and marketing rather than agricultural production.

Local organizations will receive more support from local and national governments. Governments will become more proactive to provide equitable access to education, health and information. The aim will be to improve knowledge about the environment and to ensure an optimal national natural resource management (NRM) system. In addition, new actors will engage in agricultural production. The goal of achieving a better quality of life as opposed to generating income will get prominence. Higher awareness and responsibility levels will help fight problems like environmental pollution and public health hazards on national and regional levels, and thus achieve sustainability goals.

Affected by WTO negotiations, environmental problems will be solved through technology and market-oriented institutional reform. People will pay for the pollution they create. Under these policies with expanded property rights, people providing ecosystem services will be paid. Ecotechnologies for managing ecosystem services will be demanded as interest in increasing economic values of property rights grows and benefits of ecosystem services increase.

In addition, countries will be encouraged to produce and sell products tailored to diversified market niches. This is applicable to both regional and global markets. Problems of agriculture in CWANA will be addressed holistically and efforts made to align agriculture with WTO negotiations aimed at global reduction of subsidies and removal of barriers to agricultural trade. Markets for ecosystems services and relevant technologies will be created and developed as a result of agricultural multifunctionality and diversification. New companies and cooperative institutions will evolve to provide these services. These companies, however, requiring large amounts of capital and knowledge, will develop in rich countries and operate as multinationals in poor countries, imposing their own fees and operating under less control from local governments or institutions. Poor countries will be at a disadvantage and may not approve of such institutions.

4.2.2 Capacity building for innovation

Public research organizations. CWANA countries do not possess the institutional, managerial or financial capacity to absorb current levels of project aid or to sustain project activities after foreign aid is phased out. The challenge for donors is to continue moving beyond the resource-transfer model of financing the construction of buildings and purchase of equipment and vehicles for NARS and pursue a model based on human capability and institutional building that is geared to the specific needs of CWANA countries at this stage of their development. The following constraints face most NARS of developing countries in their institutional development: weak research management, institutional instability (donor driven), human resource instability, funding instability, research program instability, limited relevance of research and deficiency in priority setting, defective linkage with the world knowledge system; insufficient links within the NARS themselves with universities, the private sector and NGOs, and with outside partners such as international agricultural research centers, regional institutions and advanced research institutions in industrial countries; and weak monitoring and evaluation of research. Generally speaking, the role of foreign assistance has been prominent in developing NARS in the region. Building agricultural research capacity means developing the capacity to design organizational rules that will help people organize, support, conduct and monitor agricultural research. Research management capacity development measures may involve

- Setting medium- and long-term research plans and strategies to serve as a frame for priority research programs and projects, in light of integrated sustainable development priorities and policies
- Identifying appropriate research instruments for achieving research objectives
- Transforming human, physical and financial resources of research institutions into research outputs and practical technologies
- Upgrading and executing a research agenda consistent with minimum environmental degradation
- Monitoring, evaluating and revising the agricultural research system

The agricultural research agenda must respond to the challenges of the world food supply. It will be influenced by the choices of research investments and strategies made by governments and institutions in both industrial and developing countries.

It is now recognized that a rigid borderline between public and private sector roles cannot be established, and there are many gray areas where public–private partnerships are needed, often in conjunction with civil society and producer and community organizations. In some least-developed countries, the withdrawal of the public sector from markets has left a vacuum that the private sector has not adequately filled, because of high transactions costs and risks. This means that the public sector needs to take a more active role in coordinating activities, jointly financing and building the capacity that the private sector needs to fill its role. In addition it must finance core public goods, especially infrastructure. Many responsibilities are also being devolved to local or state governments for decentralized program implementation, providing additional challenges and opportunities. Strategies such as contracting out to the private sector, providing targeted matching grants to support activities within the public interest, expanding collaborative action in the context of development of market supply chains and trade associations, various types of consultations and coordination forums with the private sector are all important. CWANA countries, while signing free trade agreements and proceeding with trade liberalization, are facing tremendous direct and indirect challenges that will need to be addressed carefully; among these is the capacity of local public and private entities as well as regulatory and institutional maturity.

Farmers need to recognize that agriculture is the key to sustainable development, food security and biodiversity conservation; it is central to international action in trade and investment. It has been the main user of freshwater resources and central to producing bioenergy. Thus farmers have begun—but not yet sufficiently—to form partnerships, covering such areas as how to manage water, land, genetic resources and energy. Farmers have also strengthened partnerships in research and technology. Such partnerships have been good, but they must be supported by capacity building and good governance. Successful development of agriculture requires democratic, consultative processes that involve farmers' organizations. On the other hand, indigenous communities should continue to seek partnerships and associations with governments and transnational bodies to maintain access to traditional lands, based on principles of good faith and equity.

Public–private partnerships. When discussing partnerships, we should note that sustainable development requires partnerships among all stakeholders and at all levels. In particular, the regional aspect has been stressed as crucial, if implementation is to achieve the stated goals. Despite the fact that many encouraging partnerships toward implementing the declarations and conventions have emerged following Rio, real implementation has been less satisfactory due to the lack of resources and political will. Implementation has also been hindered by structural and institutional failings, such as questionable government policies and incentives associated with trade and agriculture. The international community has a responsibility to consolidate the multistakeholder dialogue by establishing an institutional structure to facilitate the building of partnerships.

Recent approaches adopted by some international entities, such as the World Bank's strategy in rural investment to promote agricultural growth and poverty reduction, are founded on the fact that the public sector, the private sector and civil society can work together to enhance productivity of the agricultural sector and promote its competitiveness in ways that reduce rural poverty and sustain the natural resource base. These actions involve a rich mixture of science, technology, people, communication, management, learning, research, capacity building, institutional development and grassroots participation.

4.2.3 Governance and information

It is essential in striving for sustainable development to seek and maintain transparent democratic institutions capable of protecting the environment and natural resources while providing basic needs and economic opportunities. In communities where people came together to protect their ecosystems, they also had better schools, health care and economies. Hence, developing institutional capacity has been the core of the recent national and global attempts to achieve the Millennium Development Goals. Moreover, and with continuous globalization, sustainable urbanization that covers environmental, social, economic and institutional sustainability should be based on the proposition that transformation from rural to urban life requires a change in the institutional framework.

4.2.3.1 Governance principles

While rapid technological advances may in many cases help achieve economic growth without harming the environment through what is known as "green economics," real cases have raised the question: How can the international community guarantee that it will not continue to fail? The answer lies in emphasizing that greater overall sustainability goes hand-in-hand with less institutional constraints on decision-making powers, greater openness of political competition, and more widespread civil and political rights. Inevitably, national efforts to achieve sustainable development must focus on productive capacity and the institutions that are its key determinants, as well as human and natural resources. Moreover, capacity must be strengthened to be able to monitor performance where the results would feed into the process of influencing policy at the highest level.

It is essential to stress that all types of institutional setups could play a role in achieving the IAASTD sustainability goals. For CWANA, on the political level, the democratic deficit in decision making, both nationally and internationally, had to be overcome. Far too many governments and institutions in positions to act focused only on narrow interests without special focus on the will of the people. Parliaments had been working, at national and international levels, to provide a parliamentary dimension to the work of intergovernmental organizations working on sustainable development issues.

Local governments, on the other hand, could show leadership through increasing the coherence and integration of their own policies, including integrating sustainable development concerns across ministries and ensuring that existing policies have not worked against each other.

Trade liberalization has been a means to an end, not an end in itself. Each of the international regimes and institutions should be judged on its contribution to eradicating poverty and maintaining a viable natural resource base. The new perspective must build the bridges between trade and environment, between investment and development, and between finance and sustainable development.

4.2.3.2 Transparency and accountability

The poor state of governance and weak protection of rights of vulnerable communities, including smallholders, is attributed to lack of transparency and accountability in government as well as corporate activity, which restricts the ability of citizens, civil society groups and public representatives to effectively monitor the performance of various public and private institutions. Access to information is the first step toward promoting and institutionalizing public accountability at various levels; while its absence or lack of it often results in arbitrary and nonparticipatory decision making, weak monitoring, inefficient project execution, human rights violations and rampant financial corruption in public bodies (Transparency International, 2006). Lack of access to information also contributes to sustaining excessive bureaucratic controls, eliminating stakeholder participation and weakening democratic institutions.

Currently, almost all government activity in CWANA takes place in a pervasive culture of official secrecy, manifest in both official attitude and various pieces of legislation (e.g., Official Secrets Act 1923 in Pakistan). Any disclosure or sharing of information, if and when it takes place, is on a "need to know" basis, as determined by official authorities, and not in recognition of the "right to know" as a fundamental human right. As a result, whether information is made accessible or not and at what time or in what manner it is disclosed is determined by the government. Citizens and communities have hardly any say or control over it, even though the information and records held by various government departments may have direct implications for their environment, health, safety and well-being as well as their ability to make political or economic choices. It particularly affects the weaker members of society, as the powerful find ways to access the information they require by using their contacts and influence.

The culture of secrecy is so predominant that it has seriously undermined almost all mechanisms created for providing access to government information. Official statements and press releases often provide one-sided information and lack credibility. Annual reports are either not published or lack details and appropriate analyses, which could help in determining the credibility of data presented and in assessing the year's performance of related departments. Parliaments either do not exist or parliamentary proceedings do not provide adequate mechanisms for maximum disclosure of information about public policies and plans, participation of farming communities, transparency and accountability. Information could also be made accessible through Web sites but most government Web sites provide little that is useful. All of this is, partly or wholly, because there are no comprehensive policies that recognize the right to information as a fundamental human right and that provide an efficient legislative and institutional framework to assure this right.

The few countries in CWANA that have enacted and implemented right-to-information laws include Pakistan, Tajikistan, Turkey and Uzbekistan. Even where such laws exist, they do not conform to best international practices and hence offer little opportunity to promote a culture of transparency and accountability. This situation has adverse implications across the board but especially in relation to AKST, which is the mainstay of the economy of many CWANA countries. This lack of transparency and access to information explains, at least partially, the grave nature of the problem of corruption. On the Corruption Perception Index (CPI) of Transparency International in 2005, not even one country from CWANA is among the top 20 better-performing countries. Among the first 50 best-performers, only 7 are from CWANA. Almost all the major countries in CWANA are among the poor performers on CPI. For instance in 2005, out of 158 countries, Turkey ranked 65, Egypt, Saudi Arabia and Syria 70, Morocco 78, Iran 88, Algeria 97, Uzbekistan 137 and Pakistan 144.

4.2.3.3 Information technology

New information and communication technology (ICT) potentially will have a profound effect on transmitting information and knowledge on agriculture and natural resource management. New systems will be emerging to provide up-to-date market, weather and extension information to rural producers, processors and shippers (USAID, 2005). Geographic information systems (GIS) will be increasingly used in linking geographic information to agriculture and NRM to help decision makers. GIS will allow more efficient use of inputs, which will not only save money in materials but will also make labor available for other activities (World Bank, 2007). Innovations in biological and information sciences have resulted in several emerging fields that hold promise for the development of future agricultural technologies. The new fields of bioremediation, nanotechnology, genomics and bioinformatics will increase knowledge that can be shared and used to improve sustainable agricultural production and protect ecosystem functions in industrial and developing countries alike (USDA, 2003).

We will need to facilitate the exchange of scientific information and knowledge among all stakeholders in the CWANA region, and between them and the outside world. The goals of facilitating sustainable development and developing a global partnership for development can only be realized in cooperation with the private sector to make available the benefits of new technologies, especially information and communications (World Bank, undated). To meet the need to exchange information and knowledge, it is highly essential to improve and enhance ICT in the region. ICT will help bring together the scientific strengths and talents available in the region to collectively tackle the formidable challenges and tasks ahead (World Bank, 2007).

There is great potential to improve the access to information necessary for boosting production, using traditional communications technologies (such as radio) to disseminate information and ideas on agricultural technologies, markets and investors (USAID, 2005). For information without proprietary constraints, national and international agencies are increasingly using modern communication technologies, such as the Internet, to disseminate information. While such technologies are important bridging mechanisms for sharing information and experience between various sections of society and across countries (Juma and Gupta, 1999), and their use is likely to grow in the future, excessive reliance on them with the presence of the digital divide could prevent those CWANA countries with the least capacity and the greatest need for information (such as on biosafety and other risk-related fields) from having timely access to the latest knowledge they need. Measures should be taken to complement information dissemination through the Internet, including establishing information clearinghouses to act as bridges for sharing information and experience, and disseminating the lessons learned between various sections of society and across countries (Roozitalab, 2000). Appropriate specialized Web sites organized and managed by international organizations should play a more prominent role in spreading AKST. To deliver solutions for the poor in CWANA, biotechnology and information technology should be actively linked so that new scientific discoveries worldwide can be accessed and applied to the problems of food security and poverty in a timely manner (IFPRI, 1999). In addition to the growing challenge of facilitating and regulating access to information and information technologies, CWANA countries will need to harness modern science and skills for pro-poor growth, in a world in which agriculture is becoming more knowledge and information intensive. The challenges here require global efforts to reach agreements on access for the poor to proprietary information and technologies. In addition, a modernizing agricultural sector requires harnessing new skills and capacities to use modern science and technology (World Bank, 2007), a formidable task ahead for CWANA countries.

To summarize, considerable advances in Internet and electronic commerce and their application to the needs of CWANA countries present great opportunities to provide new cost-effective knowledge systems. They offer much potential to make agricultural growth more pro-poor, but at the same time they are often controversial. The challenge will be how to use these new advances together with developments in biotechnology and other agricultural technologies to make the complex agricultural systems of CWANA more productive and sustainable.

4.2.4 Social factors

Market and trade. Competitive global markets in the past years have favored corporate farming to the detriment of small-scale economies, diversity in agricultural products and farming systems. Small-scale farmers, semi-, low-skilled or informal laborers are likely to suffer most from purely market-oriented agricultural production. Women, who constitute the majority in these categories, are likely to suffer more from liberalization policies in agriculture (Baden, 1998).

Since "markets are not abstract, neutral entities but are real processes of exchange embedded in social institutions, including gender relations" (Baden, 1998) a number of policies can be adopted to balance their negative effects. These include providing credit to initiate new businesses, information on new market possibilities and requirements, and training on compliance with new production standards. Also, constructing infrastructures to facilitate movement between rural areas and the markets, and storing, transporting and preserving agricultural produce could be effective ways to integrate farmers from the most remote areas and enhance female participation in the market. This would increase the control by farmers over the returns for their agricultural work and eventually empower them, particularly the female farmers. This might also positively affect the general economy of many rural households.

Agricultural market liberalization has generally reduced state intervention. A different approach might assign a new positive role to the state to support fair globalization of the market. Alternative systems of agricultural production that favor locally produced and organic products of quality can support small economies, help preserve local systems of agronomic management and benefit the environment. They can also help diminish the marginalization of the most vulnerable rural sectors.

Climate change. Addressing climate change has recently become an urgent concern. Pollution and unsustainable development megaprojects in the past have mostly affected dwellers of marginalized areas, which are, for example, often chosen as sites for dams (McCully, 1996). Displacement, worsening health standards and general impoverishment are among the related consequences. Unpredictable changes in the ecosystem can cause droughts and other ecological disasters that will affect the most vulnerable people—mainly poor women, children and the old.

Some see paid ecosystem services as the solution to pollution. These, however, like other market-based solutions to environmental degradation, could have a negative effect on the poor, who, unable to pay for these services, will have to cope with increasing pollution that negatively affects their health. Development of ecofriendly technology, on the contrary, such as less harmful substitutes for pesticides and fertilizers, and alternative sources of energy, will partially limit environmental pollution and will primarily benefit rural users.

Agricultural policies and AKST development. Focusing AKST development on discovering alternative sources of energy, improving agricultural production and optimizing the use of available natural resources can be the first step toward sustainability. A sustainable approach to development of agricultural technology will not aim at agricultural production per se but will integrate a number of concerns such as environmental, sociocultural and economic ones. It will also include the needs of all stakeholders in establishing priority areas, research performance and technology development by adopting a gender-sensitive participatory approach.

For many years now, including end users in development intervention is considered a premise for sustainability (Zuger, 2005). This is because AKST developed with a top-down, non-participatory method is unlikely to address the diverse conditions, needs and preferences of end users. Technology developed under a purely market-driven system also is likely to focus on profitable topics, marginalizing the needs and interests of those who lack the financial means to support research or influence its development. Technology can become a tradable good available to the most affluent countries and people only.

The exclusion from technology development of the actual doers of agricultural duties leads to ineffective results. In CWANA countries women contribute significantly to agriculture and should participate in AKST development. A gender-blind approach to AKST can produce inefficient results, is likely to improve only the agricultural work of men and also can disempower the overlooked end users. Because men's work is considered productive, as opposed to women's domestic work, which is regarded as unproductive, it is generally considered more worthy of investment. As a result, research and social spending are directed to irrigation infrastructures more than to safe drinking water, with women and children usually being the ones to fetch water (WEDO, 2003). Cash crops, mainly cultivated by men, often receive more attention than subsistence crops, generally grown by women (Chambers, 1983). Agricultural machinery is mainly designed for male users and their needs. Because engagement with mechanized agriculture often corresponds to more powerful positions in intrahousehold or community dynamics (Boserup, 1970) a gender-blind AKST can disempower women.

In a truly participatory approach to agricultural technology, both men and women farmers will develop AKST and produce machinery with technical characteristics that make it easier for smaller and weaker persons to use. This could help limit the gender division of labor. Historically, men in CWANA have been assigned the use of machines, leaving the manual and time-consuming jobs to women and children (Rassam and Tully, 1988). A gender-sensitive AKST development will also expand the range of crops on which to focus, by including subsistence crops and local varieties as well as cash crops. It will take into consideration all phases of agronomic management plus postharvest duties and related domestic activities that are often neglected. By integrating local and gender-differentiated understanding of seeds and the cultural values connected to food preservation, preparation and storage, AKST could enhance the success of technological adoption and eventually be more effective in enhancing rural livelihoods. This is particularly important in the case of ethnic minorities, who connect dietary habits and the preservation of landraces to their culture.

A gender-sensitive approach to agriculture development is particularly important in areas characterized by feminization of agriculture. In countries like Syria male farmers often migrate to urban areas in search of work, and women are in charge of the agricultural work (Abdelali-Martini et al., 2003). Nonetheless, women are not considered farmers, and policy makers and development planners overlook their needs and preferences, negatively affecting agricultural production, women's daily labor and rural livelihoods. Furthermore, laws and policies rarely adapt to these changing circumstances. Entitlement and access to land, water and seeds rest with absent husbands or fathers, upon whom women must depend to get access to the basic means for their daily work. Control over key economic resources can determine intrahousehold distribution of benefits from increased agricultural productivity (Tipilda et al., 2005). Also, labor laws rarely protect the rights of women farmers or those of the informal workers, whose number is constantly growing in the agricultural sector of CWANA. Policies should be formulated that reflect the changes in social composition of rural areas and deal with emerging issues. Moreover, policies aimed to build the capacities of the rural population can help diversify the sources of household income, thereby decreasing their vulnerability. This is particularly important in light of the agricultural sector being increasingly marginalized.

Technology. Literacy rates in the CWANA region have recently risen. However, the gender gap in education is still wide. According to the Economic and Social Commission for Western Asia (ESCWA), in 2002 almost half of the female population in the Arab countries was illiterate (ESCWA, 2005). Policies should be pursued that support women's school attendance and completion, and training in fields that are usually male dominated. This is particularly important in rural and agricultural areas where access to education is limited by poor infrastructure. Cultural norms disfavor female education; poverty causes high school dropouts because children are needed to help provide income for the household. Many girls abandon school after puberty because both the trip to school and the lack of proper toilet facilities in the buildings jeopardize their modesty and honor. Safe means of transport to reach the school and proper facilities could improve their attendance. Finally, the quality of education could be improved by removing stereotypical gender images in school texts. By training students in the latest agricultural technologies or other skills the labor market requires, education could become a path in the rural areas towards better employment.

Technologies can be developed and applied to meet the needs of women in particular. For example, biofortification and foods enriched to supply the nutrients that women in CWANA tend to be deficient in, such as calcium, iron and zinc, should be considered (Gender Advisory Board, 2004). Currently in several CWANA countries, technologies suitable for women farmers are lacking, particularly labor- and energy-saving farm and household technologies. This lack of suitable technology impairs women's productivity (Kasnakoglu, 1997). Agricultural technology developed with close attention to alleviating some of the labor constraints experienced by rural women potentially can improve not only the well-being of the woman farmer but also of others in her household who are dependent on her care (World Bank, undated). Alleviating the labor burdens of rural women is an important dimension in empowering them. Technology targeted at men and implemented with men's goals and situations in mind may put women at a disadvantage by leading to an increase in the amount of labor they must expend to attain the same level of production (Gender Advisory Board, 2004).

There are no easy answers to the question of what kind of technology will promote the autonomy of women in rural societies of CWANA. These women may indirectly—but drastically—be affected by technological innovations. Technologies, as seen in many instances during the Green Revolution, may displace women and actually decrease their income (Gender Advisory Board, 2004). Such biotechnologies carry the risk of increasing the burden on women as providers for their families: increasing competition and lowering world market prices as a consequence of applying modern technologies could lead to the migration of men from rural areas. Such migration will cause an increase in the heavy burden that women must carry alone, and consequently lead to their impoverishment (Pingali and Rajaram, 1998). If the present global power structure and the current bias towards males in agricultural research, extension and development policies persist, modern agricultural technologies will most likely further widen the gap between men and women, and between rich and poor. Public research also generally bypasses women and their needs. However, since women rapidly take up technologies that improve efficiency, researchers assume they will adopt practices such as modern biotechnology (Zweifel, 1995). If women's situations, concerns, technological skills, use of technologies and knowledge continue to be overlooked, women will be displaced and marginalized by technology development, with many of their activities becoming sidelined or taken over by men. This will have resulting implications for the health and well-being of women and children, environmental sustainability, and income levels in developing countries (Gender Advisory Board, 2004).

The need is urgent in both research and priority setting to ensure that women will be in a position to benefit from modern agricultural technologies, rather than being disadvantaged as has often occurred in the past (Gender Advisory Board, 2004). Women's participation both before and during the introduction of new technology is of central importance. Their participation should go beyond consultation, aiming to implement outside innovation more easily, and to include shared responsibility, trust and cooperation. The exchange among women of technology and knowledge they have developed would be a more sensitive step in improving women's autonomy than expensive and advanced technology. If women are involved in the whole innovation process, they can set their own priorities collectively for appropriate capital-intensive technology (Pingali and Rajaram, 1998). This implies the need to change national and international research and agricultural policy in favor of women's possibilities and capacities (Zweifel, 1995). There is also need to increase the participation of women in the biotechnological sciences and other modern sciences, especially at senior levels, as well as their representation in regulating and making policy for biotechnology (Gender Advisory Board, 2004).

To conclude, technologies developed and implemented to meet the needs of women and women farmers have the potential to contribute to the IAASTD development goals, mainly through alleviating their labor burdens. Measures should be taken to ensure that modern agricultural technology will not undermine women's autonomy but will rather help women gain more autonomy. Acknowledgement of this autonomy leads to the logical conclusion that women must play a key role as decision makers in designing the direction of research and in agricultural policy-making processes and governance in the CWANA region.

4.3 Options for Strengthening AKST Future Effectiveness

The previous chapter identified the most prominent challenges that agriculture might face in the CWANA region over the next 50 years should agricultural practices continue to operate according to a business-as-usual scenario. In this section, agriculture-related technology is presented as a key tool to address these challenges, and the role of various technology management practices (and support systems) in reaching development goals is examined.

4.3.1 Options to improve AKST generation

Options exist for improving the generation of AKST:
- Research policy and funding
- Intellectual property rights
- Farmers' innovation capacity

At present most CWANA countries do not adequately invest in agricultural research to generate AKST. The research capacity required to generate appropriate environmentally friendly technologies to increase agricultural production has not yet been fully built. And agricultural research has been largely concentrated in public agricultural research institutions (Roozitalab, 2000).

Developing effective and efficient research systems to generate agricultural technology in the region will require CWANA countries to set research priorities that are well attuned to the needs of farmers and preferences of consumers, and to mobilize all partners to generate technology.

The future outlook for generating agricultural technology

In developing countries, where growth in food production had relied heavily on plowing up new land and on irrigation development, technology was increasingly responsible for production growth after 1965 (Oram, 1995). However, since shrinking land and water reserves are placing a greater burden on technology, increase in food production during the next 25 years will have to be achieved using less water, labor and cultivated land. This can be done only if scientists can develop new crop varieties with high water-use efficiency and high yield potential (ADB, 2001).

Concerted and systematic efforts should be made to develop priorities for technology generation and R&D that are consistent with socioeconomic, cultural, agricultural, environmental and political realities and goals (Gender Advisory Board, 2004). Agricultural technology generation needs to be directed to the needs of the poor. It needs to focus on the problems of marginal dry areas and to emphasize simple, low-cost technology appropriate for smallholders and resource-poor farmers. For example, biotechnology can address the issue of poor-quality seed and introduce improved materials into the local seed sector (USAID, 2005). Inappropriate technology could radically change local employment patterns; although it could increase production, it could cause greater unemployment and hence poverty.

Technology generation in the following areas should receive high priority (Oram, 1995):
- Technologies to improve natural resource management
- Technologies to protect crops from biotic stresses without heavy reliance on pesticides
- Genetic improvement of key crops

Need for increased public-private sector collaboration

Public investment in agricultural technology is crucial for achieving future food security and reducing poverty. Accelerated public investments are needed to develop technology applications that address difficult problems in rainfed and marginal areas (ADB, 2001). However, most governments in CWANA have limited resources to finance technology research, and few CWANA countries or even international public-sector institutions have the resources to create an independent source of modern technology, such as biotechnology innovations. Additional private investment is required. Currently, it is the private sector that has the knowledge, skills and capital to solve the problems of small-scale farmers, even though comprehensive data on private sector biotechnology research in developing countries are not available (FAO, 2004). Meanwhile, the private sector is unlikely to undertake much of the R&D needed by small-scale farmers because it sees little potential for return on investment. Hence financial incentives or policy initiatives are essential for increased collaboration in technology generation and R&D between public and private sectors.

The leading role of the public sector in technology generation cannot be overemphasized. Agricultural research is often long term, large scale, and risky, and while returns to generated technologies are often high, the firm responsible for developing the technology may not be able to appropriate the benefits accruing to the innovation. The benefits of agricultural research often accrue to consumers (through reduction in commodity prices due to increased supply), rather than to the adopters of the new technology, so social returns may be greater than private returns to research. Therefore, a sustained public sector role in funding agricultural research will be essential, particularly for production areas in less favorable environments that the private sector is unlikely to serve (World Bank, undated).

The challenge for CWANA countries is not to develop new agricultural technologies (such as plant breeding techniques or disease diagnostics) but to design and implement the capacity-building programs and regulatory systems needed to facilitate the sustainable transfer and adaptation of these technologies to the relevant farming systems (Dhlamini, 2006).

Technology transfer

One of the lessons of the Green Revolution was that agricultural technology could be transferred internationally, especially to countries that had sufficient national agricultural research capacity to adapt the imported high-yielding cultivars to suit local production environments (FAO, 2002). Advances in agricultural technology hold great promise, but the full benefits of scientific breakthroughs will not be realized unless the new technologies are properly disseminated and CWANA farmers adopt them successfully. Concerted and systematic efforts to transfer new technologies should incorporate participatory approaches as well as a clear assessment of users and beneficiaries.

Public engagement

Evidence has shown that public engagement is identified as an important precondition for the appropriate and successful transfer of new, modern technology (Gender Advisory Board, 2004), as farmers in resource-poor areas are innovators and adapters (Chambers et al., 1989). Indeed, technology transfer strategies that have proved successful in CWANA countries have used a community approach and direct farmer participation. Thus technologies can be transferred through extensive programs of on-farm demonstrations, where local extension services play a vital role (Haddad, 2004).

Knowledge transfer

Because new technologies are more demanding for both the farmer and the extension agent, they require more information and skills for successful adoption than did the initial adoption of modern varieties and fertilizers. A bottom-up information flow combined with adaptive, location-specific research is particularly important in transferring complex crop-management technologies (World Bank, undated). While transferring new technologies, it is important to recognize and take into account the social status of recipients as well as employment patterns and cultural norms in the community. In the context of transferring "controversial" technologies, such as biotechnology, it should also be recognized that farmers and communities may have knowledge that will affect decisions on how the technology is used in the local context (Gender Advisory Board, 2004). At present, there is widespread distrust of biotechnology, and the public needs to be engaged in dialogue before it is disseminated widely (ADB, 2001). The public should be made aware of the potential risks, harm and benefits of new technologies being transferred and given opportunity to discuss them. It is also important to inform communities about the service level they would require for the technology they may want to use. Communities should especially have a clear understanding of long-term costs and maintenance implications, so that they can choose what is most appropriate for them under their budget constraints (World Bank, undated). All these aspects should be clearly explained to the public at all levels in terms that are understandable and relevant to local farmers (USAID, 2005).

Technology adoption

Experience has shown that a number of key conditions help maximize the benefits of a growing agriculture sector for poor people by facilitating the adoption process of modern agricultural technology.

Good governance

Good governance is crucial to ensure that new agricultural technology reaches the poor (ADB, 2001). In each CWANA country, successful local adoption of innovations from others will depend on incentives and barriers producers face. In addition to investment in technology generation and transfer, significant policy and governance reform is required to ensure that the poor in CWANA benefit most from greater investment and higher agricultural productivity. The increasing importance of new, knowledge-intensive technology requires a market-friendly environment for adopting and adapting new technologies and removing restrictions on technology imports, which must be encouraged through continued progress in economic liberalization. Alongside favorable macroeconomic and trade policies, good infrastructure and access to credit, land and markets must be in place. Equitable conditions give farmers incentive to adopt new and sustainable technologies and diversify production into higher-value crops—actions that raise incomes and lift households out of poverty (World Bank, undated). Decentralization of existing extension service structures that encourage a bottom-up flow from farmers to extension and research will also help farmers cope with the additional complexity of efficiency-enhancing technology, as local governments are usually more knowledgeable about rural agricultural needs and adept in dealing with them.

Dissemination in a package

Exploiting the growth potential of staple crops from dissemination of modern technology requires not only investment but also changes in farm management and a transition from current farming traditions to more modern systems. Since the returns to technology adoption are low if modern inputs are used in isolation and not supplemented by other technologies, modern technology needs to be disseminated in a well-defined package of technologies and services to be successful in the field (World Bank, undated). Farmer surveys of successful technology adoption experiences from Jordan indicate that farmers prefer accepting new technologies as packages, rather than accepting only one component at a time (Haddad, 2004). However, in practice, components of a promising package could be taken up in a piecemeal, stepwise manner, where the sequence of adoption would be determined by factor scarcities and the potential cost savings achieved, as was often the case in disseminating and adopting Green Revolution technologies (FAO, 2002).

Training farmers
Informal training in understanding modern technology is necessary for farmers, local communities and the public to participate effectively in development programs. Extension services in CWANA are often weak in this respect, and knowledge in the labs often does not reach farmers or the beneficiaries of technologies (Gender Advisory Board, 2004). In addition, if rural people are to obtain relevant technical, entrepreneurial and management skills, they need adequate training. In an increasingly technical and communications-oriented world, specialized training schemes (in computing and accounting, for example) are needed, including programs for women, who dominate many service and trading activities (World Bank, undated).

Constraints to adoption
Modern technologies that increase agricultural productivity do exist, but many factors prevent farmers from adopting them. As researchers want to maximize production, this can lead to technologies that require a high degree of management and a high level of precision. But farmers tend to want robust technologies and are prepared for lower potential returns if their risk and vulnerability are reduced (CGIAR, 2002). Evidence so far suggests that technologies that are embodied in a seed, such as transgenic insect resistance, may be easier for small-scale, resource-poor farmers to use than more complicated crop technologies that require other inputs or complex management strategies. On the other hand, some biotechnology packages, particularly for livestock and fisheries, require a certain institutional and managerial environment to function properly and thus may not be effective for resource-poor smallholders (FAO, 2002). Farmers should assess the different management options available and adapt them to fit their own circumstances and production goals.

Given the technical and financial constraints that they face, many resource-poor farmers in CWANA countries are unable to adopt the very technology that is intended to reduce their poverty. An example is an agricultural technology designed to produce nontraditional crops that will expand nontraditional exports, which can result in agricultural growth and overall economic growth. However, the nontraditional export sector's contribution to poverty reduction may be relatively small because nontraditional export growth is often concentrated around cities where there is greater access to transportation and other market facilities (World Bank, undated)

4.3.2 Options to improve AKST access and use

Access to natural, physical and financial resources. The world has the technology to feed a population of 10 billion people. However, access to such technology is not assured (FAO, 2002). New technologies—particularly biotechnology and information technologies—require new approaches to facilitate access to CWANA countries, especially for the poor (World Bank, 2007). The range of potential barriers includes issues related to civil society and governments accepting the technology; it includes financial, informational, educational and technical barriers that keep poor farmers marginalized and unable to adopt new, unaffordable technology; it includes intellectual property rights.

Farmers in developing countries experience a need for support services and resources such as access to credit, infrastructure, and services such as transportation and market facilities (Gender Advisory Board, 2004). The high cost of modern technologies is one of the most serious setbacks to their use, slowing adoption. High cost of a technology can be further exacerbated by poor investment of both public and private sectors (USAID, 2005).

The policy bias against agriculture in developing countries and the trade barriers put up by industrial countries is well known (USAID, 2005), and a situation that calls for reform in both camps. Particularly important are government policies to enhance access for smallholder producers and other agricultural and NRM entrepreneurs to regional and world markets (domestic and international trade policies), as well as to build the capacity of developing country governments in these areas. Improving national macroeconomic policies is critical (World Bank, 2003, in USAID, 2005) to support agricultural trade and market access as well as markets for agricultural inputs and services, and to facilitate entrepreneurship.

Information on agricultural technologies, markets and investors affects decisions on adopting the technology. Therefore, to boost production, it is crucial to improve access to this information using communication technologies (Gender Advisory Board, 2004). Policies that improve access to global knowledge and technology should be identified and introduced to reduce the gap between knowledge systems and technologies available to agronomists, plant breeders and farmers in developed and developing countries.

4.3.3 Options to activate enabling factors of AKST generation and application

In its report, the UN Task Force on Science, Technology, and Innovation concludes that to achieve the MDGs, developing countries must strategically embrace the role of science and technology in their development efforts. Then they must begin "improving the policy environment, redesigning infrastructure investment, fostering enterprise development, reforming higher education, supporting inventive activity, and managing technological innovation." These components are part of the enabling environment that will encourage the generation, transfer and adaptation of agricultural technologies, leading to greater productivity and sustainable development (USAID, 2005).

CWANA countries will have many crucial decisions to make in meeting their sustainable agricultural goals. These decisions need to be made and implemented based on decision makers' knowledge of the unique environmental, social and economic characteristics of their country. Those CWANA countries with strong research, health and education capacity will offer a supportive environment for technology development and investment (USDA, 2003). The role of the civil society (consumer groups and small-scale farmers' societies) in technology assessment cannot be overemphasized.

The European Commission notes a growing recognition that technology development needs to be "meshed with social, economic and policy dimension to have impact on beneficiaries" (EIRAD, 2004, in USAID, 2005). Modern biotechnology, for example, will not solve all the problems of food insecurity and poverty in CWANA. But it could provide a key component to a solution if given the chance, and if steered by a set of appropriate policies that would guide an increased public investment in R&D, foster regulatory arrangements that inform and protect the public from any risks arising from the release of GMOs, implement intellectual property management to encourage greater private sector investment, and introduce appropriate regulation to protect the interests of small farmers and poor consumers (IFPRI, 1999).

Policy research is needed to better understand the political and institutional factors that promote or inhibit the use of new ideas in CWANA at all levels (local, national and regional), and specifically those factors that encourage institutions and mechanisms for effectively articulating science and technology policies. Other areas for policy research:

- Identification of policies that improve access to global knowledge and technology
- The interface between technological change, institutional change and policy environments
- Formulation of and education about appropriate policies for biotechnology and biosafety
- Investigation of policies on intellectual property

New technologies, particularly biotechnology, will require new approaches to regulate their use. Some of these technologies remain controversial because of potential health and environmental risks (World Bank, 2007). It will be necessary to have biosafety procedures in place to ensure that the benefits of these modern technologies are realized (FAO, 2003). It will also be necessary to provide appropriate regulatory mechanisms to ensure that modern biotechnology products such as GMOs that might interact with the environment are as safe as the products of traditional biotechnology (ADB, 2001), and that the benefits of GMOs outweigh their risks. The sociopolitical ramifications of new agricultural biotechnologies should not be overlooked. These include the potential widening of the prosperity gap between the North and the South; the exploitation of indigenous genetic resources without appropriate compensation to indigenous populations; and an increased inequality in the distribution of income that biotechnology might create since the privileged classes derive earlier and greater benefits from the introduction of powerful technologies than do the socially disadvantaged (Leisinger, 1996). The good news is that sound domestic policies and international cooperation can go a long way toward reducing the sociopolitical risks of new technologies.

Technology regulation should be science based (FAO, 2002), and the regulatory framework should not be regarded in isolation from the broader policy context of agriculture and the contribution that technology might make in the particular economic, social and environmental context of individual countries (ADB, 2001). In addition, the establishment of an effective regulatory capacity must go hand-in-hand with investment in technology appropriate for farmers (USAID, 2005). For example, to realize the goal of facilitating environmental sustainability, appropriate regulatory institutions should be in place to limit environmental degradation. Since the underlying driving force for environmental degradation through the harmful use of farming technologies is frequently poverty rather than factors inherent in agricultural technology itself, farmers should realize personal economic benefits from using environmentally friendly technologies and also recognize the social benefits from environmental protection. Only when sustainable agricultural technologies are profitable for farmers will they comply with regulatory requirements and employ environmentally sustainable production techniques (World Bank, undated).

It is the responsibility of national governments to ensure that national regulatory systems are applied, enforced and monitored (ADB, 2001). The regulatory capacity of the public sector in CWANA countries to address food safety and environmental issues will determine the success of modern technologies in individual countries. Without functioning regulatory systems, the private sector is unlikely to invest in modern technologies appropriate for CWANA countries. Effective intellectual property regulations are also important for any long-term investment in modern agricultural technology on the part of the private sector. Regional cooperation in intellectual property and biosafety regulations has great potential for simplifying both technology access and agricultural trade (USAID, 2005).

Judgment and dialogue are essential elements in any science-based regulatory framework (FAO, 2002). At present, there is widespread distrust of biotechnology, and modern biotechnologies such as genetic engineering in food and agriculture cannot succeed unless the public is engaged in dialogue and convinced of its safety and usefulness before these technologies are disseminated widely (ADB, 2001). It is important for CWANA communities to be informed about the technology they want to use, the service level they require, and especially to have a clear understanding of long-term costs and maintenance implications, so that they can choose what is most appropriate for them under their budget constraints (World Bank, undated). Public engagement will be a precondition for the appropriate and successful implementation of modern technologies in CWANA, since communities may have knowledge that will affect decisions on uses of modern technology in the local context (Gender Advisory Board, 2004).

In each country in CWANA, the successful local development of technologies or the transfer and adaptation of innovations from others will depend on the supportive environment faced by investors and producers alike. Only if countries have appropriate policy, regulatory and institutional frameworks in place to support science and technology can they contribute to the achievement of sustainable development goals by increasing agricultural productivity and stimulating economic growth (USDA, 2003).

Improved technologies alone cannot do the entire job of sustainable agricultural development. A combination of improved incentives and policies, reinvigorated institutions, and increased investments must occur in CWANA if agriculture is to develop and the benefits are to be spread widely (USAID, 2005). The potential value of modern science to agriculture and the environment in CWANA countries will not be realized without major additional efforts involving all stakeholders, including civil society, producers, consumers and governments (Serageldin and Persley, 2000).

References

Abbot, F.M. 2003. Competition provisions in the TRIPS agreement: Implications for technology transfer. Available at www.wto.org/english/tratop_e/trips_e/techtransfer_e.htm. WIPO-WTO workshop Intellectual Prop. Rights Trans. Tech., Geneva. 17 Nov 2003. WTO, Geneva.

Abdelali-Martini, M., P. Goldey, G. Jones, and E. Bailey. 2003. Towards a feminisation of agricultural labour in northwest Syria. J. Peasant Studies 30:2:71-94.

ADB (Asian Development Bank). 2001. Agricultural biotechnology, poverty reduction, and food security. ADB, Manila.

Alston, J.M., and P.G. Pardey. 2006. Developing-country perspectives on agricultural R&D: New pressures for self-reliance? Pp. 11-28. *In* P.G. Pardey et al. (eds) Agricultural R&D in the developing world: Too little, too late? IFPRI, Washington, DC.

Baden, S. 1998. Gender issues in agricultural liberalisation. Bridge No. 41. Inst. Dev. Studies, Brighton, UK.

Boserup, E. 1970. Women's role in economic development. Allen and Unwin, London.

CGIAR. 2002. Task force on integrated natural resource management: Putting INRM into action. 4th INRM Workshop, Aleppo. 16-19 Sep 2002. CGIAR.

Chambers, R. 1983. Rural development: Putting the last first. Longman, Harlow.

Chambers, R., A. Pacey, and L.A. Thrupp (ed). 1989. Farmer first: Farmer innovation and agricultural research. Intermediate Tech. Publ., London.

Chioccioli, E. 2002. Euro-Mediterranean cooperation in agriculture within the new international context. New Mediterr. 1:4-6.

Dhlamini, Z. 2006. The role of non-GM biotechnology in developing world agriculture [Online]. Available at http://www.scidev.net/en/policy-briefs/the-role-of-non-gm-biotechnology-in-developing-wor.html. Science and Development Network.

Diao, X., T. Roe, and S. Agapi. 2002. Developing countries interests in agricultural reforms under WTO. Am. J. Agric. Econ. 84:782-790.

Drexl, J. 2006. International competition law: A missing link between TRIPs and transfer of technology. Available at www.wto.org/english/tratop_e/trips_e/techtransfer_e.htm and www.wipo.int/documents/en/meetings/2003/wipo_wto/index.html. Joint WIPO-WTO Workshop: Intellectual property rights and transfer of technology, Geneva, 17 Nov 2003. WIPO, Geneva.

EIRAD (European Initiative for Agriculture Research for Development). 2004. Strategy for the European Initiative for Agricultural Research for Development (EIARD), 2005-2010. *In* Agriculture and natural resources management research priorities desktop review. First draft. Available at http://www.eiard.org/strategic_docs.html. Int. Resources Group, USAID, Washington, DC.

ESCWA (Econ. Social Commission for Western Asia). 1998. Challenges and opportunities of the new international trade agreements for the ESCWA member countries in selected sectors. United Nations ESCWA, Beirut.

ESCWA (Econ. Social Commission for Western Asia). 2005. Women's empowerment in the Arab region: Achievements, challenges and future actions. United Nations ESCWA, Beirut.

FAO. 2002. Impact of foreign assistance on institutional development of national agricultural research systems in sub-Saharan Africa. Available at http://www.fao.org/DOCREP/005/Y4349E00.HTM. FAO, Rome.

FAO. 2003. World agriculture: Towards 2015-2030. FAO, Rome.

FAO. 2004. Agricultural biotechnology: Meeting the needs of the poor—the state of food and agriculture 2003-04. FAO, Rome.

Gender Advisory Board. 2004. Regional consultative expert workshops: Understanding the gender dimensions of biotechnology, research and development, Pretoria. 26-27 Nov., 2004 and 30 Nov-1 Dec., 2004, Islamabad. UN Commission on Science and Technology for Development.

Gerber, J. 2000. National policies and the limits of international integration. Estey Centre J. Int. Law Trade Policy 1:11-22.

Gibbon, D. 2001. Global farming systems study: Challenges and priorities to 2030. Regional analysis: Middle East and North Africa. FAO, Rome.

Haddad, N.I. 2004. Technology development and transfer to small farmers in low rainfall areas. Jordan: Approaches and perspectives. Available at http://www.aaaid.org/pdf/magazine/2004/Jordan%20Approaches%2053-57.pdf.

IFPRI. 1999. Biotechnology for developing country agriculture: problems and opportunities, 2020. Focus 2. IFPRI, Washington, DC.

Ingco, M.D. 2002. Agricultural policy reforms in the new multilateral trade round. Am. J. Agric. Econ. 84:3:798-799.

Juma, C., and A. Gupta. 1999. Biotechnology for developing country agriculture: problems and opportunities. Safe use of biotechnology, 2020. Focus 2, Brief 6. IFPRI, Washington, DC.

Kasnakoglu, Z. 1997. Women and agricultural development in the Near East [Online]. Available at http://www.skk.uit.no/WW99/papers/Kasnakoglu_Zehra.pdf. Middle East Tech. Univ., Ankara.

Leisinger, K.M. 1996. Sociopolitical effects of new biotechnologies in developing countries. 2020 Vision Brief 35. IFPRI, Washington, DC.

McCully, P. 1996. Silenced rivers: The ecology and politics of large dams. Zed Books, London.

Miner, W.L. 2001. An overview of the issues and positions of the major countries in the WTO negotiations. Estey Centre J. Int. Law Trade Policy 2:1-23.

Oram, P. 1995. Potential of technology to meet world food needs in 2020. 2020 Vision Brief 13. IFPRI, Washington, DC.

Pardey, P.G., J.M. Alston, and R.R. Piggott (eds) 2006a. Agricultural R&D policy in the developing world: Too little, too late? IFPRI, Washington, DC.

Pardey, P.G., N. Beintema, S. Dehmer, and S. Wood. 2006b. Agricultural research: A growing global divide? Food Policy Rep. IFPRI, Washington, DC.

Pingali, P.L., and S. Rajaram. 1998. Technological opportunities for sustaining wheat productivity growth toward 2020. 2020 Vision, Brief 51. IFPRI, Washington, DC.

Rassam, A., and D. Tully. 1988. Gender related aspects of agricultural labor in northwestern Syria. *In* S.V. Poats et al. (eds) Gender issues in farming systems research and extension. Westview Press, Boulder.

Roozitalab, M. 2000. Strengthening partnership in agricultural research for development in the context of globalization. West Asia and North Africa (WANA) regional forum. *In* Proc. Congr. Global Forum on Agricultural Research, AARINENA Doc. GFAR/00/23, Session 3, 15 May 2000.

Serageldin, I., and G. Persley. 2000. Promethean science: Agricultural biotechnology, the environment and the poor. CGIAR and World Bank, Washington, DC.

Tipilda A., A. Alene, V. Manyong, and B.B. Singh. 2005. Intra-household impact of improved dual-purpose cowpea on women in northern Nigeria. Presented at the PRGA Impact Assessment Workshop, CIMMYT, Mexico City, 19-21 Oct 2005.

Transparency International. 2006. CPI 2006. Available at www.transparency.org/policy_research/surveys_indices/cpi/2006. Transparency Int., Berlin.

UNEP. 1997. World atlas of desertification. 2nd ed. John Wiley, New York.

USAID. 2005. Agriculture and natural resources management research priorities desktop review. First draft. Int. Resources Group, 31 May 2005. USAID, Washington DC.

USDA. 2003. 21st century agriculture: A critical role for science and technology. USDA, Washington, DC.

WEDO (Women's Environment and Development Organization). 2003. Untapped connections: Gender, water and poverty: Key issues, government commitments and actions for sustainable development. WEDO, New York.

WHO. 2001. Work of the WHO in the eastern Mediterranean region: Promotion and protection of health. Available at http://www.emro.who.int/Rd/AnnualReports/2000/chapter4-3.htm. Annual Rep. Reg. Director. WHO, Geneva.
World Bank. 2003. Reaching the rural poor: A renewed strategy for rural development. World Bank, Washington, DC.
World Bank. 2007. World development report 2008: Agriculture and development. Draft Prospectus 1. World Bank, Washington, DC.
World Bank. Undated. Agriculture and achieving the millennium development goals. Report 32729-GLB. ARD, World Bank, Washington, DC.
Zaibet, L., M.S. Bachta, and M. Chaffai. 2003. Concerns about WTO negotiations: A developing country perspective. Working Paper. Dep. Agric. Econ., ESA Mograne, Zaghouan.
Zuger, R.C. 2005. Participatory development projects in the Andes: Looking for empowerment with Q-methodology. Presented at the PRGA Impact Assessment Workshop, CIMMYT, Mexico City, 19-21 Oct 2005.
Zweifel, H. 1995. Modern biotechnologies in agriculture: Impact on women in the South. Biotech. Dev. Monitor 23:10-13.

5

Looking Forward: Role of AKST in Meeting Development and Sustainability Goals

Coordinating Lead Authors:
Numan Mizyed (Palestine), Rania Shatnawi (Jordan), Christoph Studer (Switzerland)

Lead Authors:
Sandjar Djalalov (Uzbekistan), Ahmed El-Wakeel (Sudan), Aggrey Majok (Sudan), Ahmed Rafea (Egypt), Alisher Tashmatov (Uzbekistan), Maria Wurzinger (Austria)

Contributing Authors:
Dawood AlYahyai (Oman), Azzam Saleh (Palestine), Ahu Uncuoglu (Turkey)

Review Editors:
Abid Qaiyum Suleri (Pakistan), Asha El Karib (Sudan)

Key Messages

1. Agricultural knowledge, science and technology (AKST) has an essential role in meeting sustainable development goals of reducing hunger, improving human health, reducing poverty, improving livelihoods, and attaining environmental, social and economic sustainability. Furthermore, it may help cope with scarcity of resources and food insecurity, which are major causes of conflicts. Increased agricultural productivity is a direct driver for reducing hunger and improving nutrition and human health in that sufficient and more nutritious and diverse food results in a healthier constitution and improved body defenses. Increased production may also help increase income, thereby reducing poverty and improving livelihoods of farming populations (economic sustainability). Livestock not only generates income for many poor families in CWANA but also increases their security by serving as "living banks." Adequate farming practices allow for sustainable and efficient management of natural resources and enhance ecosystem services. AKST may thus reduce pressure on scarce or disputed resources and thereby reduce conflict potential. Holistic approaches in AKST therefore are appropriate for the multifunctional role of agriculture.

2. Agricultural productivity in crop and livestock production and aquaculture may be substantially improved in many areas of CWANA. However, appropriate measures have to be taken so that increasing productivity does not compromise sustainability of production with regard to ecological, economic and social aspects. Through intensification of irrigated production, certain countries in CWANA achieve the highest crop yields worldwide. Substantial increases in crop production can result from increasing soil fertility with organic and inorganic fertilizers, protecting crops against pests and diseases, controlling weeds, developing and using high-yielding species and varieties (derived through both conventional breeding and biotechnology) adapted to site-specific conditions (participatory decentralized breeding) combined with locally adapted mechanization. Integrated crop management practices that include crop rotation, integrated pest management (IPM), regular soil fertility analysis, and use of buffer and compensation areas may reduce negative effects on the environment of such intensification. Increased livestock production in CWANA to meet the rapidly growing demand for meat and milk products will probably have to be based on intensified mixed systems since land degradation due to excessive stocking rates on rangelands is already widespread. However, increased inputs in intensified systems require monitoring to avoid soil and water pollution and to safeguard animal and human health. Removing policy distortions that promote artificial economies of scale (e.g., in livestock production), developing approaches to let poor producers capitalize on the benefits of production, and regulating environmental and public health concerns will represent important challenges for CWANA decision makers.

3. Capitalizing better on the wealth of locally developed and modern technologies for improving the productivity of scarce water resources will allow for substantially higher production or reduced water use in agricultural production, or both. Management factors that increase crop yield generally also increase water productivity. Optimal planting dates, appropriate soil management, mulches, windbreaks and protected production can reduce crop water requirements substantially. In rainfed production, maximizing infiltration of precipitation and reducing runoff, water harvesting, and using drought-tolerant varieties may further increase productivity of scarce water. Supplemental and deficit irrigation may increase water productivity massively. Water losses in irrigation and conveyance systems can be reduced by piping, lining and regularly maintaining the system; optimizing water distribution in the field using appropriate irrigation systems and scheduling irrigation properly can increase field application efficiency. Proper irrigation practices and the assurance of good drainage can avoid salinity. However, even the most sophisticated irrigation scheduling tools are of little value if systems for organizing, allocating and distributing water are deficient, and if the capacity to deal with these systems and the awareness about the importance of saving water are lacking.

4. Integrated water resources management (IWRM) aims to coordinate development and management of water and related resources. Involving all pertinent stakeholders in IWRM allows consideration of water demands in sectors other than agriculture. Major challenges of IWRM in CWANA include developing currently untapped water sources, preserving water quality, managing demand and handling transboundary collaboration. Potential for capturing currently untapped water resources exists through water harvesting, including dams and groundwater recharge, and using unconventional sources such as reclaimed, recycled, brackish, salty, desalinated water and fog collection. Preserving water quality is important for all water users; it may pose particular problems where agricultural productivity is pushed to use more agrochemicals. Managing water demand may include using water efficiently in irrigated agriculture, but also building awareness, handling incentives and disincentives with financial and economic measures such as water pricing, or trading in virtual water. IWRM aims at managing water and related resources efficiently in ways that will maximize the resultant benefits in an equitable manner for different uses in all sectors without compromising the sustainability of ecosystems. Participation in pertinent negotiations of all stakeholders involved helps avoid conflicts over water resources at various scales (watershed, canal, international).

5. The different forms and causes of land degradation in CWANA require specific approaches to reduce and reverse the degradation. Besides developing and disseminating sustainable land management practices that fit specific conditions, socioeconomic measures are required for widespread adoption of appropriate land management practices. Land degradation in CWANA ranges from nutrient depletion and erosion through waterlogging and salinization to rangeland degradation and loss of productive land to other sectors. Numerous practices and technological options fostering sustainable land management at the field, farm and community or watershed level, which are adapted to site-specific conditions, are available from traditional and modern knowledge. However, besides disseminating this information efficiently, a conducive environment must be created for these technologies to be adopted. Long-term land-use rights for owners and leaseholders, risk reduction measures that include safety nets and credit and saving schemes, and profitability of recommended technologies are prerequisites for their adoption. Participatory land-use and land-management planning that organizes access to and use of land and adequate pricing policies and employment opportunities outside agriculture may ease pressure on land and promote the investments required for more sustainable land use. Developing and implementing national action plans under the United Nations Convention to Combat Desertification (UNCCD) will help combat land degradation through coordinated approaches.

6. AKST can capitalize on the rich biodiversity existing in CWANA, but it also has to counteract the threat that agriculture poses to biodiversity. Given the global changes occurring, particularly climate change, the rich biodiversity in CWANA may gain importance in crop and livestock breeding. Furthermore, markets capitalizing on biodiversity as a source of food, herbal remedies and income are gradually emerging. Strategies for conserving biodiversity include different means of in situ and ex situ conservation. Agricultural practices in ecoagriculture such as agroforestry, compensation areas and biodiversity-enhancing landscape elements and adequate land-use planning including the creation and maintenance of protected areas help conserve biodiversity. Establishing and strengthening gene banks may simultaneously allow capitalizing on biodiversity by using genes from wild relatives of crop species and neglected landraces and supporting their conservation. Implementing National Biodiversity Strategy and Action Plans (NBSAPs) developed through the Convention on Biological Diversity (CBD) may facilitate biodiversity conservation as well as making use of this treasure in CWANA.

7. Numerous approaches to mitigate and adapt to climate change are available, but further research is needed to tackle the differing challenges that CWANA subregions will face. Conservation agriculture, improved rangeland management and adaptations in rice cultivation as well as improved feeding of ruminants and improved manure management may substantially reduce greenhouse gas emissions and possibly increase carbon sequestration in CWANA. Research regarding adaptation to climate change will need particular focus on pest and disease management (resistance, forecasting and modeling, IPM) and the introduction of adapted crops and varieties. Erosion control, floodwater management and ways to cope with saltwater intrusion will probably have to receive additional attention, and efficient management of scarce water becomes even more important. Developing capacity may be required to successfully face the challenges ahead and may also help in benefiting from the Flexible Mechanisms included in the Kyoto Protocol (e.g., the Clean Development Mechanism).

8. CWANA has great opportunities to strengthen and reorient capacity development. Agricultural education, research and extension will need to reorient the currently technology-focused approach to holistic and integrative systems approaches. Higher consideration of socioeconomic aspects and introduction of participatory approaches, including acknowledging the important role of women in agriculture, are required to respond to the real questions of farmers, markets and consumers, and may strengthen local ownership. Value-chain management and risk reduction are important topics for CWANA to strengthen in AKST. Blending local knowledge with modern science may bring novel technologies and approaches to the fore. Education, research and extension may greatly benefit from modern technologies (GIS, simulation modeling, expert systems, etc.) and from improved knowledge management. Links and collaboration between education, research, extension and farmers as well as the interaction with the private sector may make AKST more efficient and effective. AKST impact monitoring and evaluation allow for continued priority setting and sound strategy development.

9. Information and communication technologies (ICT) will allow capitalizing to a greater extent on the wealth of information and knowledge available for AKST. Besides improving information and knowledge sharing, exchange and dissemination through ICT infrastructure development and Internet connectivity, modern technologies such as geographic information systems (GIS), simulation modeling and expert systems make better use of existing information. Investments in ICT infrastructure and capacity development together with adequate information policies will allow sharing, exchanging and disseminating traditional and modern information and knowledge, thus strengthening links among AKST stakeholders in education, research, extension and production. This may reduce duplication of research activities and enable stakeholders in AKST systems to make use of the latest technologies. Access to up-to-date market information can assist decision making at various levels. Investments in ICT will narrow the digital divide gap between rich and poor people.

10. Improving market organization may help reduce postharvest losses and contribute substantially to poverty alleviation and development in CWANA. Only if market organization in CWANA countries is improved may the stakeholders in agricultural value chains fully capitalize on increased agricultural production. Producers, processors and traders need access to credits, markets (e.g., by closing the gap between rural areas and urban centers) and reliable market information, particularly in view of more diversified and market-oriented production. Appropriate technologies and infrastructure are required for well-functioning value chains. Processing facilities at different levels may substantially reduce postharvest losses, and together with the development of agribusiness provide additional income along the value chain, particularly if diversified production (with more focus on nonstaples) targets newly emerging market opportunities (organic products, supermarkets, etc.). Vertical integration and professional value-chain management facilitate quality and safety management at relevant levels and foster compliance with newly emerging standards.

11. Policy adaptations are necessary to realize technological options and develop the capacity required to achieve sustainable development goals (SDGs) to their full extent. Adaptations to policies focusing on risk reduction, functioning markets and value chains, trade, sustainable natural resource management, and strengthened capacity are particularly important to achieve SDGs. For producers, improved access to credits, loans and insurances, and adequate input and product pricing (considering the possibility of subsidies and direct payments) are particularly important with regard to economic performance. Appropriate policies regulating access to and use of land and water resources, possibly realized through decentralized participatory approaches, may promote investments and adoption of sustainable practices. Adequate import and export policies (including transit procedures) and trade arrangements (based on proactive engagement in trade negotiations) can greatly strengthen the position of CWANA producers and agrobusinesses in globalized markets. Food safety and biosafety guidelines and regulations, legislation for being compliant with and competitive in international trade, regulations and legislation on intellectual property rights (IPR), and liberal information policies stressing access and transparency are framework conditions required for a productive, efficient and competitive agricultural sector. Integrating AKST in national development strategies and plans may reconcile conflicting views and ambitions with regard to national goals such as national security, food sovereignty (trade in virtual water, etc.), economic growth and development, and quality of life.

5.1 Role of AKST in Meeting Sustainable Development Goals

5.1.1 Hunger, nutrition, and human health

Increased agricultural productivity is a direct driver for reducing hunger and improving nutrition and human health in that sufficient and more nutritious and diverse food results in a healthier constitution and improved body defenses (Rosegrant et al., 2006). Increased agricultural productivity also directly helps increase income, thereby reducing poverty and securing livelihoods of farming populations (IFAD, 2002; Rosegrant et al., 2006). Higher productivity may further allow for more diverse food production and thus more diverse and higher-quality diets, which not only provide sufficient protein and vitamins, but also help combat micronutrient deficiencies.

However, increasing productivity has to be approached cautiously; too often a narrow focus on productivity gains results by exploiting natural resources unsustainably through overuse and pollution (environmental degradation), causing problems related to food quality and safety with negative effects on nutrition, health or marketability, or by neglecting social aspects in trade-off with profitability (including abuse and social dumping). Therefore, AKST-related initiatives need to be guided toward sustainable agricultural productivity in achieving development goals.

5.1.2 Poverty and livelihoods

In CWANA an estimated 70% of the poverty is in rural areas, even though only some 43% of the total population lives there (El-Beltagy, 2002). Despite the large dependence of the rural population on agriculture, emphasis on agriculture and rural development is declining. In addition, the region is facing a number of converging trends that threaten the future livelihoods of the poorest sector of society (Thomas et al., 2003).

It is generally agreed today that poverty reduction requires a holistic perspective focusing on understanding root causes, removing constraints, and creating opportunities and choices for improving livelihoods. Sustainable livelihood approaches based on the principle of reducing poverty by empowering the poor to build on their opportunities are today considered more successful than sectoral approaches to development, which focused on resources rather than people (Carney, 2002).

AKST has to play an important role in various aspects: AKST may support farmers in managing assets, reducing vulnerability, and transforming structures and processes. AKST plays a key role in shaping the quality and quantity of and access to natural, human and other resources, as well as the efforts of those working at different levels (household, national, international) to reduce poverty and hunger in a sustainable manner (DFID, 2007). With regard to agricultural production, some studies indicate that a higher crop yield of just 10% may lead to 6-10% reduction in the number of people living below the poverty line (Irz et al., 2001). CWANA can improve crop production considerably through optimizing the use of production inputs such as water, fertilizers, pesticides and proper crop varieties in a sustainable approach. Agricultural education and better training of farmers may also accelerate development and improve livelihoods considerably.

5.1.3 Environmental sustainability

Environmental sustainability may be adversely affected by efforts aimed at increasing agricultural and economic development—efforts important for achieving the first Millennium Development Goal (Rosegrant et al., 2006). The pressure is to modernize and intensify agricultural systems to meet the food demands of an increasing population. It is unlikely that AKST can develop the agricultural sector in CWANA in such a manner that only benefits and no negative externalities accrue. Further expansion of agriculture dependent on inorganic fertilizers, pesticides and machinery applying agricultural practices like tillage, drainage, irrigation, and fertilizer and pesticide application will undoubtedly have their impact on the environment. In the light of this, it is imperative for CWANA countries to undertake, in their planning process for any agricultural development effort, a judicious, comprehensive and participatory assessment of environmental costs and benefits. A productive agricultural sector will reduce pressure on and contribute to ensuring environmental sustainability.

Since water is a major limiting natural resource in CWANA and agriculture is the leading consumptive user of water, AKST plays a key role in satisfying competing demands for this scarce resource and in raising public awareness of the effect agriculture has on the environment (Bonnis and Steenblik, 1998). Besides the focus on quantitative aspects, i.e., on increasing water supply and decreasing water demand, preservation of water quality will have to receive more attention in the future. In this regard, AKST will not only have to concentrate on protecting water resources against pollution—from agricultural activities as well as from other sources—but will also have to explore ways for using water of lower quality in agriculture. However, AKST will also have to foster the development and adoption of other options that are environmentally sustainable and economically and technically applicable, such as biological control, integrated pest management, integrated crop management, good agricultural practices and organic farming (Clay, 2004).

5.1.4 Economic sustainability

Economics is about using resources efficiently; usually this is expressed in monetary terms. In this sense, the theories regarding sustainable use of resources can be applied to economic sustainability, except that, in monetary terms, one resource can generally substitute for another (University of Reading, 2005). Economic sustainability is usually associated with the ability to maintain a given level of income and expenditure over time. Maintaining a given level of expenditure necessarily requires that the income or revenue that supports that expenditure also be sustainable over time. In the context of livelihoods of the poor, economic sustainability is achieved if a minimum level of economic welfare can be achieved and sustained (DFID, 2001).

With regard to agriculture, economic sustainability relates to the long-term ability of farmers to obtain inputs and manage resources. At the level of the individual farm this means that a farm business must remain financially viable while providing an acceptable livelihood for the farm family (Lien et al., 2007). Economic sustainability of a farm is therefore subject to the viability of, and markets for, an enterprise or product. Economic sustainability of an agricultural sector is subject to the whole economy—locally, nationally and internationally (University of Reading, 2005).

AKST therefore plays a crucial role with regard to economic sustainability of agricultural production. AKST can help in using resources and assets efficiently by developing appropriate technologies and practices that may reduce labor requirements. It can facilitate access to viable markets and lobby for adequate input and output prices. It can support the development of well-functioning saving and credit systems that allow for making necessary investments in agricultural production and value chains. Thus, AKST's role with regard to economic sustainability of agriculture has to focus on different levels, from field and farm up to the policy level.

5.1.5 Social sustainability

Social sustainability requires that the cohesion of society and its ability to work towards common goals be maintained. Individual needs, such as those for health and well-being, nutrition, shelter, education and cultural expression should be met (Gilbert et al., 1996). Social sustainability is therefore dependent on a particular set of social relations and institutions, which can be maintained or adapted over time (DFID, 2001).

For a socially sustainable community to function and meet the basic needs of its members, it must be able to maintain and build on its own resources and have the resiliency to prevent or address problems in the future. The community has two types or levels of resources available for building social sustainability:

- Individual or human capacity refers to the attributes and resources that individuals can contribute to their own well-being and to the well-being of the community as a whole. Such resources include education, skills, health, social values and leadership.
- Social or community capacity is defined as the relationships, networks and norms that facilitate collective action taken to improve upon quality of life and to ensure that such improvements are sustainable (Gate and Lee, 2005).

The role of AKST lies particularly in attaining social sustainability in rural communities. On one hand, AKST can foster social sustainability by improving farmers' capacity for enhancing their own well-being and, in accord with their environment, the well-being of communities through increasing their skills to achieve sustainable production and improved livelihoods. On the other hand, AKST can promote networks involving rural populations that facilitate achieving sustainable development. Contributing to more equity (e.g., regarding access to resources and key services) and participating within the society may represent a further task of AKST with regard to achieving social sustainability. By facilitating sustainable development in rural areas AKST contributes to reducing population growth rates, to decreasing differences between rich and poor and between genders, and possibly to reducing migration from rural areas to urban centers, which may further help in attaining social sustainability.

5.2 AKST Options to Overcome the Challenges in CWANA

5.2.1 Production systems management

Agricultural growth and increased farm productivity create wealth, reduce poverty and hunger, and may protect the environment, particularly in developing countries (CGIAR, 2006). Since crop and livestock productivity is generally low in rainfed production systems of CWANA, technologies facilitating increased agricultural productivity remain an important pillar toward achieving the development and sustainability goals of the region.

5.2.1.1 Crop production

Apart from the general management strategies and practices commonly applied to increase crop productivity, management strategies and practices to use water efficiently and productively are of utmost importance in the dry areas of CWANA. As a rule, any management factor that increases crop yield also increases water productivity because evapotranspiration generally increases less than yield production (Turner, 1986). Adequate soil fertility and fertilization, crop protection against pests and diseases (which reduce productivity and increase water use) and weed control (weeds compete for water, nutrients, light) therefore not only increase crop production but also increase the efficiency of water use.

Similarly, high-yielding species and varieties (developed possibly through hybridization, apomixis, or possibly genetic engineering) adapted to the specific conditions of a certain location (through participatory decentralized crop-breeding programs) may use water more productively if managed adequately than varieties or landraces with inferior yield potential.

The choice of optimal planting date can, in combination with short-duration varieties within suitable crop rotations, increase water productivity substantially by making the best possible use of limited precipitation and by moving the cropping season into a period of low evaporative demand ("seasonal shifting"; see van Duivenbooden et al., 2000). Improved meteorological forecasting and supplemental irrigation, possibly combined with mechanization, may greatly facilitate moving the cropping season for better water-use efficiency.

Appropriate crop rotations or relay and intercropping practices, including food legumes that fix atmospheric nitrogen, also make better use of limited precipitation; growing a legume crop instead of fallowing every second year has proved to increase water productivity substantially in cereal production in West Asia and North Africa (WANA) (van Duivenbooden et al., 2000). In addition, crop rotations may reduce weed, pest and disease pressure and positively influence soil fertility and structure.

Mulches of crop residues combined with appropriate soil management may not only reduce unproductive evaporative water loss from the soil surface but also enhance infiltration of scarce precipitation. Mulches may thus reduce wind erosion, soil temperature and surface sealing, and contribute to improved water productivity if their effect on soil temperature does not prolong the crop growing period into a dry season. A large soil volume that crop roots can explore is crucial for storing water for crops to use. Therefore, management factors that increase soil depth and give roots access to soil volume are important for making optimal use of scarce water. Such factors include breaking impermeable layers, building terraces and other structures to mitigate erosion, fertilizing for vigorous root growth and increasing soil water-holding capacity. Particularly in windy areas windbreaks may reduce evapotranspiration through an ameliorated microclimate and thus improve water-use efficiency provided that competition of the windbreak species (usually trees) does not limit crop production. Other options such as dense canopies, ground cover, shade, plastic tunnels and greenhouses can reduce evapotranspiration and increase the relative humidity of the ambient atmosphere, greatly increasing water-use efficiency.

In many cases, such as mulching or adequate soil management, some degree of mechanization may greatly support management practices fostering a more efficient use of limited water resources and precipitation. Furthermore, profitable markets and access to them are a general prerequisite that strategies and practices for increasing crop and water productivity can be implemented (profitability of investments required). Often access to credit at a reasonable interest rate is required to ensure that necessary investments can be profitably made.

Rainfed cropping. In addition to the above-listed general strategies and practices to increase crop and water productivity, certain management aspects are of particular importance in rainfed cropping. To store as much water as possible in the soil, maximizing infiltration of precipitation and reducing runoff is a major priority for improving the water supply to crops in environments where water is a limiting factor for plant growth.

Water-harvesting technologies of collecting, storing and concentrating precipitation at micro, meso and macro scale may not only increase crop productivity in dry areas but make it possible to produce crops in environments where cropping would not be possible without such technologies, because they minimize the risk of crop failure. Additionally, water harvesting may protect land from degradation and desertification. Developing and using drought-tolerant or drought-resistant plant material with high yield potential is a prerequisite in the drought-prone areas of CWANA if irrigation is not available or feasible. Recent evidence shows that prospects for improving yields and water productivity in rainfed agriculture are considerably more promising than previously assumed (Rosegrant and Cai, 2000; Rosegrant et al., 2002; Comprehensive Assessment of Water Management in Agriculture, 2007 vs. Seckler and Amarasinghe, 2000). A change in breeding strategy to directly target rainfed areas rather than relying on "spill-in" from breeding for irrigated areas seems key to this development (Rosegrant et al., 2002).

However, policy changes, such as providing a policy environment that does not discriminate against rainfed areas, and infrastructure investment in these areas, such as access to input and output markets, are required to increase productivity in rainfed cropping. Decision-making power will have to be increasingly delegated to communities and social groups, particularly considering the roles of women, who actually manage the natural resources and depend on their sound management for their livelihoods. Key strategies to follow for sustainable development thus include sustaining investments in agricultural research and extension; improving coordination among farmers, NGOs and public institutions; ensuring equitable and secure access to natural resources; empowering land users for effective risk management; and increasing investment in rural infrastructure (Rosegrant et al., 2002).

Irrigated cropping. In addition to the management factors discussed above, irrigation-specific options may be considered to render irrigated cropping more water efficient and productive. Irrigation and conveyance systems should be planned and improved to minimize water loss. Piping, lining and regularly maintaining conveyance systems are ways to reduce water loss through evaporation, percolation at the bottom of canals, seepage, overtopping, bund breaks, leakage through rat holes and runoff (Brouwer et al., 1989). Optimizing water distribution in the field is key for efficient water use in irrigation (University of Arizona, 1999). Field application efficiency may be increased by improving irrigation systems and scheduling irrigation efficiently (Solomon, 1988; Allen et al., 1998). Particular attention should be paid to exploiting the potential of supplemental and deficit irrigation, which may increase water productivity tremendously (Figure 5-1) and greatly reduce the threat of crop failure (risk reduction, stability) (Oweis et al., 1999). In Syria, for example, spreading a given amount of limited irrigation water supplementing precipitation on a larger area, hereby not fully satisfying crop water requirements, allows achieving considerably the higher total crop yields and water productivity as compared to using the water for full irrigation on a smaller area (Oweis and Hachum, 2001).

In some of the richer countries of CWANA, high-tech irrigation systems may be used to reduce water use for irrigation; however, low-cost irrigation systems are in general of greater need in the region. Besides improving the performance of gravity irrigation systems, such as basin, border or furrow irrigation, and low-cost sprinkler systems, in particular portable hand-move systems, great potential exists for low-cost micro-irrigation systems (SIMINetwork, 2006). However, proper irrigation practices (see Aillery and Gollehon, 2003) are prerequisite for efficient water use with any irrigation system; practices worked out in developed countries such as the United States may well be suited or adapted to conditions prevailing in the water-scarce environments of CWANA. Modern tools for improved irrigation scheduling such as crop simulation models fed with actual climatic data and coupled with GIS can greatly improve water productivity and help manage water resources for irrigation even at large scales, such as in the Jordan Valley or Egypt.

It is important to raise awareness that by using less water more efficiently high yields of good quality are achievable. Education and training are extremely important to raise understanding of how much water a crop requires and how water may be managed more efficiently, avoiding the negative effects of overirrigation. Socioeconomic factors such as awareness and experience, but also the financial capacity of the farmer, access to affordable inputs and knowledge regarding irrigation systems and their maintenance and scheduling are "soft" factors that are as important for productive water use in irrigated cropping as technological aspects.

However, even the most sophisticated irrigation and scheduling practices are of little value if water organization, allocation and distribution systems are deficient. In many CWANA countries water is allocated through public entities. Several excellent examples of water allocation systems, such as in the Jordan Valley, may show the way toward managing water more productively at the project or perimeter level. Establishing water user associations that jointly organize and manage water allocation and distribution may represent another way to facilitate productive water use considerably and at the same time empower local populations while relieving public institutions.

Salinity represents a major threat to irrigated agriculture in most areas of CWANA because evapotranspiration generally exceeds precipitation to a great extent—as early as in the third millennium BC, the Sumerian cities of Mesopotamia crumbled, partly because of salinization due to poorly managed irrigation. Proper irrigation management and availability and maintenance of suitable drainage systems are keys to avoiding land degradation due to salinity. Particularly in cases where unconventional water sources are used for irrigation, such as drainage of brackish water, management practices have to consider the considerable threat of salinity. Strategies and technologies to avoid salinization are sufficiently known.

Rendering water use more efficient in irrigated agriculture is not only important for increasing crop productivity in dry areas (and thus revenues and livelihoods of irrigating farmers; see Lipton et al., 2003). Saving water in agriculture frees up substantial water resources that may be used in other sectors and activities, which is important for overall development of a society. Therefore, IWRM has become a key principle, in particular with regard to irrigated agriculture.

In conclusion, numerous AKST options to use scarce water resources in CWANA more productively and efficiently in irrigated agriculture already exist. It is important to recognize that in addition to technical improvements, organizational and political considerations, plus increased

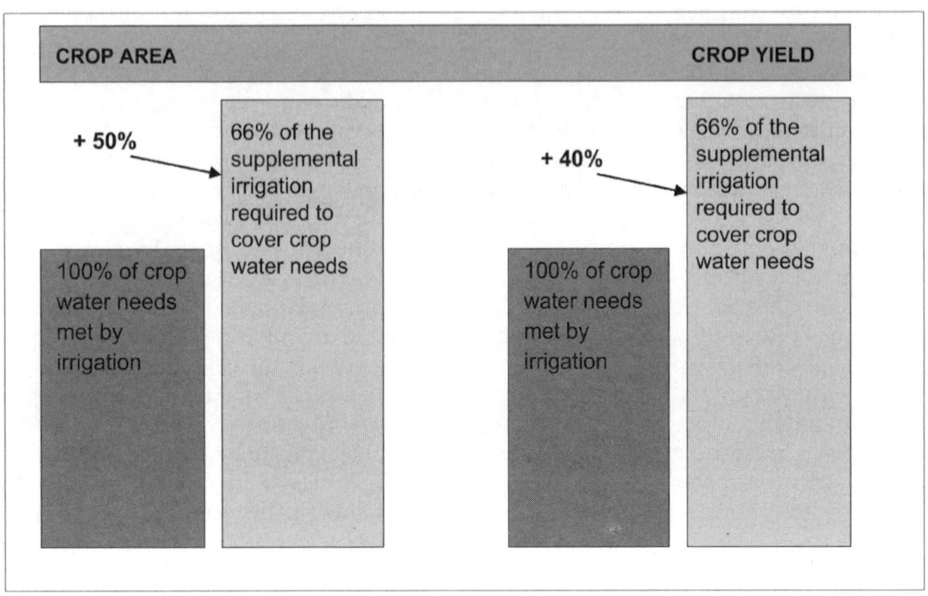

Figure 5-1. *Increasing wheat yields in Syria by reducing irrigation: The effect of deficit irrigation on yield and water productivity.* Source: Oweis and Hachum, 2001.

awareness and understanding are key to using water more productively in CWANA agriculture.

5.2.1.2 Livestock

Livestock is an integral part of most traditional production systems in dry areas, as it is in CWANA. Nevertheless, CWANA is importing large amounts of meat to keep up with the growing demand for livestock products, because of population growth, increased urbanization and higher incomes. Expansion of livestock production has so far been achieved mostly by increasing the number of animals—an approach that raises concerns about public health and environmental sustainability. Therefore, other options to increase productivity have to be explored to meet the expected increased demand (Delgado et al., 1999).

Livestock production systems are changing in CWANA as they are at a global scale with three linked factors at play: intensification of livestock production (in some places with stronger crop–livestock integration); increased commercialization of livestock production, particularly in peri-urban areas; and the gradual overcoming of animal diseases as a constraint on production (Morton and Matthewman, 1996).

Three major types of farming systems may be distinguished: grazing systems, mixed systems and industrial systems. These systems may be characterized by different stocking rates—grazing systems having the lowest rates and industrial systems the highest. In mixed systems livestock farming is combined with crop farming, where part of the crop or its by-products are used as feed resources (Chapagain and Hoekstra, 2003).

Traditional pastoralism (based mainly on sheep and goats) and to a limited extent mixed farming still exist in CWANA. However, in the grazing and small mixed-farming sectors of the region, little technological change has occurred (Delgado et al., 1999). Pastoralists are being driven into ever more marginal areas as arable terrain gradually expands. But these marginal lands are increasingly coming into focus as reserves of biodiversity, and thus pastoralism is likely to disappear in many regions where it competes with agriculture (Blench, 2000). In ecologically more favorable environments, notably the Nile Valley in Egypt, competitive dairy systems have emerged that use a mixture of domestic and imported feed resources and intermediate labor-intensive technology (Delgado et al., 1999).

Industrial production units, mainly for poultry and dairy, that have emerged based on oil revenues and the resulting economic expansion have state-of-the-art technology but require imported inputs and the domestic production of others, such as forage production for dairy cows. Most of these systems cannot compete with world markets but are maintained through protection as a matter of political choice. This is due to a certain extent to the fact that CWANA, as does most of Asia and sub-Saharan Africa, lacks the capacity to produce substantial amounts of feed grain at competitive prices. Given that many countries in the region cannot expand their crop area, two possibilities remain to increase feed grain availability: intensification of production on existing land resources and importation of feed. Because much of the gain from intensification will probably be used to meet the increasing demand for food crops (and possibly biofuel), many CWANA countries will have to import substantially more feed grains in the future (Delgado et al., 1999).

In the grazing and mixed farming systems in CWANA, productivity gains still seem possible. Animal nutrition can be improved through different technical interventions, such as dry-season supplementation, unconventional feeds or increased use of silvopastoral systems. Making better use of livestock manure may not only be of interest for mixed systems; by fertilizing fodder shrub plantings it may also improve fodder production and help conserve land in grazing systems. Another way of increasing productivity is through animal breeding, by improving local breeds or introducing high-yielding breeds for crossbreeding schemes. However, the latter approach bears the threat of progressively eliminating rare livestock breeds by genetic introgression, representing a corresponding loss of valuable genetic traits and biodiversity (Blench, 2001).

With regard to animal health, the control of serious diseases is becoming increasingly effective through better understanding of epidemiological aspects such as prevalence, risk factors and transmission mechanisms. Treatment is more easily accessible with easy-to-use control agents such as thermostable vaccines, and with the increased effectiveness, safety and ease of application of veterinary products. As farmers gain confidence that diseases can be controlled, reducing economic risk, they are prepared to invest more in animal production (Morton and Matthewman, 1996; Perry et al., 2005). This important fact will considerably increase livestock productivity in both grazing and mixed systems. Aspects of hygiene are important not only with regard to animal health but increasingly also for marketing livestock products, and they will thus receive more attention. However, institutional aspects related to delivering required animal health services in CWANA need to be evaluated in terms of (1) achieving a balance between public and private roles and (2) finding a mutually acceptable balance between regulatory standards to be maintained and the benefits accruing to those who keep livestock.

Preventing degradation and rehabilitating marginal and degraded land are possible with technical options for improved rangeland management: using rotational grazing, corralling to rehabilitate degraded spots, seeding and planting possibly supported by fertilization and water harvesting, practicing agroforestry with fodder shrubs such as *Atriplex* (saltbush), maintaining livestock biodiversity, reducing the number of artificial water points. However, experience to date suggests that technical inputs alone will have only a limited effect on rangeland productivity and conservation. This experience has been particularly so in completely different ecosocial regions such as the more stable environments of high-potential areas in North America, Australia or New Zealand (Sidahmed, 1996; ALAWUC, 2002).

Improving rangeland management to reduce or reverse land degradation in dry areas is a complicated and tricky issue. Most rangelands in CWANA are commonly owned, often by the state, and their profitability is rather limited, which discourages pastoralists from investing in pastures. Furthermore, traditional ways and authority systems regarding their management have been lost to a great extent (Blench, 1998). The traditional wisdom and knowledge of nomads and extensive livestock herders should thus receive more attention. Future research and technology transfer should be based on well-identified demand by the herders and should build on, complement, support or modify this indigenous knowledge (Sidahmed, 1996). Natural resources such as rangelands, forests, water and wildlife are best managed sustainably by local communities, who depend on them for their livelihood and food security, and not by the state or private sector. Thus state control and ownership of natural resources may have to be reconsidered and possibly returned to the local communities where traditional knowledge on sustainable management still persists (ALA-WUC, 2002). The role of the public sector should focus on providing an enabling environment for sustainable natural resource and rangeland management, and on monitoring rather than regulating (Seré and Steinfeld, 1996; Nasr, 1999; Ngaido et al., 2002).

Further aspects to be considered in reorienting policy include integrating crops and livestock more strongly by using pastoral outputs in mixed farming effectively, including intensifying the use of work animals; producing niche products from unusual species or breeds, or high-quality, value-added meat and milk products; developing interlocking strategies to link conservation of wild fauna and flora with pastoral production; and expanding ecologically sensitive, low-volume tourism, using pastoralists to provide services (Blench, 2000). Increasing productivity and profitability of grazing systems may finally allow for better management of rangelands. This is important because pastoralism on rangelands can indeed make efficient use of scarce and patchy resources, although its proportional contribution to the overall livestock product market in CWANA is continuously decreasing. Additionally, better rangeland management seems to offer great potential for increased carbon sequestration, particularly through an increase in soil organic matter (IPCC, 2000).

Mixed farming systems should still play an important role in the future, since integrating livestock and crop operations remains the main avenue for sustainable intensification of agriculture in many—particularly the drier—regions of the developing world (Delgado et al., 1999). Integrating crops and livestock has manifold advantages and benefits. Livestock uses land that is not suitable for crop production, provides manure for the crops, and may be used as draft power, which allows a certain extent of mechanization. Integrating fodder crops, particularly leguminous crops, in rotations may further improve crop productivity. Additionally, using rotations, including green manures, and integrating livestock into farming systems widens the range of outputs produced, thus reducing the damaging effect of failure or market collapse of any one crop.

Due to the increasing demand for livestock products, in particular milk and meat around urban centers, mixed and intensive peri-urban farming may become more and more profitable, as examples in East and South Asia demonstrate. However, to profit from this opportunity, farmers require access to sound information on markets and market prices. Increased livestock production in CWANA to meet the rapidly growing demand for meat and milk products will probably have to be based on intensified mixed systems since land degradation due to excessive stocking rates on grazing system rangelands is already widespread in the region. However, the potential threats of pollution (as well as of animal and human health and welfare) will have to be watched cautiously. Major problems to be considered and overcome in intensified systems include the threat of polluting water, soil and air through inappropriate waste management, causing environmental and public health dangers; animal health and animal welfare issues; and zoonotic and epizootic diseases and epidemics; and further human health aspects such as hormones and antibiotics in livestock products, and Creutzfeldt-Jakob disease (BSE—bovine spongiform encephalopathy, commonly known as mad cow disease) due to inadequate control of product safety (Delgado et al., 1999; Guendel, 2002). Important challenges for CWANA decision makers will be to remove policy distortions that promote artificial economies of scale in livestock production, develop approaches to let poor producers capitalize from the benefits of increased livestock production, and form regulations to address environmental and public health concerns.

5.2.1.3 Fisheries and aquaculture

The per capita consumption of fish in CWANA was 7.6 kg year^{-1} in 2004, whereas the world average consumption has increased from 13.1 kg year^{-1} in 1992 to 16.1 kg in 2003 (FAO, 2004). Fish consumption is expected to increase in CWANA. However, many fish stocks are under threat due to high fishing pressure. It will be difficult to expand production from capture fisheries at the current level of exploitation. It is therefore expected that production from capture fisheries will grow only slowly to 2020 (Delgado et al., 2003), even if fish resources are managed carefully.

Sustainable management of fish resources will have to include responsible use of fishing gear, reducing by-catch, and improving processing techniques. Developing infrastructure in fishing communities will allow fishermen to increase the quantity and quality of their fishing. Presently, large quantities of harvested fish are discarded due to wasteful postharvest methods. Education in coastal communities is important to introduce new techniques to local fishermen and increase their fishing abilities. This education should also increasingly reach women to increase their participation in fishing activities, particularly in postharvest activities. Having local communities participate in managing fishery resources is important. Local committees, which should include representatives from local fisher communities and government authorities, may facilitate such participation since fishers will participate in decision making in managing fish stocks.

Cooperation is needed among countries to provide more fish food for people and to alleviate poverty in coastal communities. This cooperation should include joint research and the exchange of information and data. More research should be directed toward substituting animal protein in the feed with non-animal protein sources; this will reduce the pressure on important fish stocks that are normally used for fish meal. International and regional organizations should also play their roles in providing assistance to the countries in this field. Such assistance may include financial funding, training local people, conducting research, and giving support for local fishery organizations. More research is particularly needed with regard to biological and economical aspects of fish stocks, to determine the optimum yields of these stocks and to ensure the suitability of production from certain fish species.

In conclusion, the following measures are important to sustainably manage fish resources:

- Provide more information about fish stocks—important for management. This information may include, but is not restricted to, size, structure and other biological parameters of the stocks.
- Develop a plan for sustainably managing fish stocks that allows their responsible use and rebuilding of depleted stocks.
- Develop infrastructure in fishery communities.
- Improve fishing techniques and reduce the amount of by-catch.
- Educate local people in new techniques on fishing and quality assurance.
- Increase the participation of women in postharvest activities.
- Increase the cooperation between countries and international organizations.

Aquaculture. It is expected that aquaculture will grow significantly in some CWANA countries since the growing demand for fish will probably be met mostly through increased production from aquaculture. This growth will differ from one aquaculture sector to another (marine aquaculture, aquaculture in integrated agriculture, production of non-food fish). Production may be increased either by expanding the cultivated area or by improving the production per unit area (Delgado et al., 2003). The increasing demand for fish, increasing domestic food supply and increasing export revenue are the main forces that will direct the expected growth of the aquaculture industry in the future. Many CWANA countries realize the importance of aquaculture as a valuable food source and an important source of employment, providing a development opportunity for rural communities. They will incorporate it in their development plans.

In poorer communities, small-scale integrated aquaculture can benefit local people and ensure food security. Growing fish along with agricultural crops can use scarce water in fish ponds for crop irrigation. In areas with saline groundwater that cannot be used for agriculture, fish species such as tilapia may be grown to increase the income of local farmers as well as to provide valuable food.

Challenges facing aquaculture differ from one country to another in CWANA, but in general they include limited availability of sites suitable for new aquaculture activities, uncertain provision of a continuous supply of fingerlings, limited availability of local species for aquaculture, risk of disease, and unreliable markets. In some cases, conflicting interests from different stakeholders, such as tourism, agriculture or other coastal uses, will have to be mediated to get access to and use suitable sites; this may often require collaboration between different authorities. In other cases suitable technologies for particular sites will have to be developed as some sites require special technology.

Constraints to developing aquaculture related to land, water and inputs may be faced through adapted technologies such as selective breeding, better health management, water control and modification of feed input (Delgado et al., 2003). However, more research regarding the sustainability of aquaculture production is certainly required; joint development and transfer of technologies between countries may speed up success. Improved harvesting and processing techniques will increase output of high-quality products. Biotechnology in aquaculture may improve production, but it will be important to use such new technologies in a sustainable manner, regulated through proper legislation and monitoring.

Using local fish species is important for sustainable aquaculture as many problems are associated with using exotic species, principally disease transmission. Therefore, research should focus on determining suitable local species for aquaculture. They should have a high growth rate, be resistant to environmental changes and diseases, and be in demand in the marketplace. If an exotic species is to be used for any reason, special procedures should be followed such as a detailed risk assessment of the effect of introducing this species on the local environment and local species, and rigorous quarantine procedures. Some countries do not allow fingerlings of exotic species to be imported, but they allow importation of brooders. This gives the country more surveillance over both the brooders and the hatchlings.

It is essential for CWANA to address environmental issues in developing sustainable aquaculture. Therefore, an environmental impact assessment should be carried out for every commercial project. These projects should promote environmentally sound technologies for managing production. Monitoring is important to study possible negative effects and the best methods to reduce them. Likewise, codes of conduct for best practices and methods in aquaculture should be prepared to ensure sustainability. These codes can be written jointly between the government and the private sector in any country.

Sustainable aquaculture requires effective policies, legal frameworks and institutions. In CWANA, some countries already have policies for developing aquaculture; others are lagging behind. Obstacles to overcome developing a legal framework include conflicting views between different governmental authorities, bureaucratic procedures and inconsistency among the various laws. Therefore, it is important to encourage cooperation among authorities to use suitable mechanisms to jointly develop a management plan for aquaculture.

5.2.1.4 Forestry and Agroforestry

The forested area in CWANA is limited—less than 5% in most countries—but it still harbors great plant biodiversity and offers considerable potential to serve as a carbon sink. Forest ecosystems in these countries are sources of timber, firewood and fiber as well as non-wood forest products in addition to providing many goods and services (IPGRI, 2001). The role of forests is particularly essential in the dry areas of CWANA with regard to their hydrological function in the ecosystem; under forest cover water infiltration is increased, runoff rates are reduced, and thus water availability may be improved. Furthermore, trees importantly protect land from degradation and desertification by preventing wind and water erosion.

Land degradation poses a major threat to sustainable forest development. Countries in the CWANA region, therefore, need to adopt more holistic approaches that are compatible with policy reform, technical guidelines and support systems to address rehabilitation and restoration through forest management and planted forest development. In many of these countries, the current rate of industrial development of planted forests barely keeps pace with losses from deforestation and the transfer of natural forests to protected status. It is possible not only to sustain but also to increase productivity in successive rotations of planted forests. This requires a clear definition of sound management of planted forests and their end use. It is necessary to integrate strategies for appropriate silviculture, tree improvement programs and nursery practices, matching species and provenance to sites, and to forest protection and harvesting practices.

AKST may address forestry priorities in the CWANA region that should include forest resources management, rehabilitation and restoration of degraded lands, development of forest lands and access to information technology and networks on forestry, and inclusion of forestry and forestry-related issues (IPGRI, 2001). Expanding forests and increasing their productivity may be achieved by using appropriate species, protecting forests from grazing, zoning, water harvesting and using unconventional water resources such as brackish water and treated wastewater for supplementary irrigation of forests in dry areas or seasons.

Agroforestry. Agroforestry systems are generally less widespread in dry areas than in humid environments in the tropics. This is mainly due to the role that competition between crops and trees plays, which is more important in dry than in humid areas. This competition requires cautious reflection on whether and how woody species may be combined with crops and livestock (see Kessler and Breman, 1991; Breman and Kessler, 1995). But agroforestry may provide extremely important products and services in dry areas (Table 5-1).

It is important to keep in mind that the drylands environment is quite different from environments where many well-known agroforestry practices such as alley cropping have developed. Innovative thinking is required to develop new approaches for using woody perennials together with crops and livestock, and certainly for adapting known technologies to the specific conditions. A participatory approach considering the needs and perceptions of all involved stakeholders is indispensable. Innovation, adaptation and participation will have to receive high priority in AKST with regard to agroforestry in CWANA.

5.2.1.5 Mechanization and labor organization in agricultural production

Adequate, locally adapted mechanization may substantially increase agricultural productivity, particularly in the case of field crops. Furthermore, a certain degree of mechanization may greatly support approaches and technologies to use scarce water resources more efficiently. Conservation agriculture practices such as conservation tillage, which help conserve soil productivity and biodiversity, often require some degree of mechanization. Conservation agriculture technologies will probably become more important in the future since they play an important role in adapting to climatic change and they substantially reduce greenhouse gas emissions. In addition, increasing labor constraints, particularly due to the increasing opportunity costs of labor, insufficient remuneration for agricultural work and the feminization of agriculture, further drive the need for mechanization.

However, mechanization has to be adapted to the specific agroecological and socioeconomic conditions of the farm enterprises. Since production systems and resource availability for farmers in CWANA differ greatly, strategies to improve mechanization must be specifically targeted. Whereas in certain areas, such as those that are resource poor or mountainous, improved crop–livestock integration may allow for simple mechanization through draft power, high-tech mechanization may be adequate in other parts of CWANA. The special case of Central Asia, where the degree of mechanization has considerably decreased since the collapse of the Soviet Union, clearly illustrates this need for specific approaches and strategies. Thus careful consideration of the specific conditions and cost-benefit analyses are necessary to choose adequate mechanization levels. Furthermore, changes in traditions or legal framework conditions may be required to introduce mechanization. For instance, land fragmentation limits increase in productivity through mechanization, and access to reasonably functioning credit and saving systems may be required for farmers to invest in the necessary equipment.

Future AKST relating to mechanization will have to consider changes in labor organization to a greater extent. Hitherto achievements in mechanization have generally unilaterally favored male workers; future efforts will have to focus much more on facilitating and easing the labor-intensive and tedious work of women and children. Awareness building and advocacy will be required to alter social preconceptions that associate machinery use with men and thus further limit women's use of technological improvements. This is particularly important in view of the increasing feminization of agriculture in many parts of the CWANA region, and it must allow for adequate and gender-balanced schooling and education.

5.2.1.6 Alternatives to conventional farming

Production of safe food remains a worldwide health concern. Agricultural chemicals like fertilizers, pesticides and fungicides are a prime food safety concern today. Jones et al. (2006) reports an increased risk of contamination with harmful microorganisms through irrigation water, spraying, cleaning, etc.

Such concerns have led to an increasing interest in developing and adopting alternative agricultural approaches to reduce the use of chemical pesticides and fertilizers. Integrated crop management—which includes integrated pest management—is an option to be considered in CWANA in the future, as is organic farming. Due to changes in consumer habits and perceptions, markets for sustainably produced and organically grown merchandise are growing, and they may represent an opportunity for quite a number of farmers in the CWANA region.

Integrated crop management (ICM) may be seen as a commonsense approach to farming. It combines the best of traditional methods with appropriate modern technology, balancing the economic production of crops with positive environmental management (BASIS/LEAF, 2004). Integrated

Table 5-1. Agroforestry options and related opportunities in dry areas of CWANA.

Agroforestry option	Opportunities
Production of livestock feed/fodder	Trees and shrubs (particularly leguminous species) may contribute to filling in the lack of nitrogen and protein in dry areas, supplying green fodder during the dry season. Trees and shrubs may be scattered in grazing land or be planted in "fodder banks" (fodder production during the dry season, "cut and carry"). Potential problem: lower palatability and digestibility due to tannins and other compounds.
Fruit production	Combining cropping and fruit production is traditional in the dry areas of CWANA; it is particularly attractive for areas where (supplemental) irrigation is possible. Nondemanding species such as *Ziziphus, Annona* and mango in the semiarid tropics, or olives, pomegranate and pistachios in the dry subtropics are intercropped with vegetables or cereals. Plastic tunnels and modern (drip) irrigation systems may render such systems quite productive and water efficient.
Wood production	Wood is used for fuel, construction, tools, art, fencing, etc. Production of firewood may substitute for livestock manure, which can thus be used to enhance and maintain soil fertility. Village woodlots are one approach.
Production of other goods	Trees can produce nontimber products such as traditional and modern medicines, gum arabic, fiber, dyes, cosmetics, oils, pesticides, silk (silkworm rearing). Diversifying production reduces risk and generates incomes, contributing to reducing poverty and improving livelihoods (Leakey et al., 2005).
Use of otherwise not usable water resources	Low-quality water may be used to increase productivity of agroforestry systems. Treated sewage effluent that cannot be used to produce food crops may still be suited to produce woody species, as certain trees and shrubs tolerate higher salt content than herbaceous plants (apply increasingly saline water to successively more salt-tolerant plants, Jorgensen et al., 1993).
Biodrainage	Deep roots of trees and shrubs may achieve good groundwater control, particularly at a local scale (see, e.g., Schofield et al., 1989 or Raper, 1998). This can considerably reduce the threat of salinization when groundwater tables come too close to the soil surface.
Soil fertility enhancement	Trees and shrubs may provide valuable additions to the soil by maintaining or increasing organic matter, by biological nitrogen fixation, through more efficient nutrient uptake (association vesicular arbuscular mycorrhiza) or by enhancing atmospheric nutrient input (dust and wet deposition). Trees may reduce loss of nutrients and organic matter, protecting the land against wind and water erosion by covering the soil surface with litter, reducing raindrop energy and wind velocity, providing physical barriers and protecting streambanks, or by retrieving nutrients from deep soil horizons.
Protection of soils against degradation, reclamation of degraded land	Protection measures using trees and shrubs include constructing contour vegetation strips, enforcing structural conservation measures such as stone bunds or terraces, protecting waterways, and stabilizing erosion gullies and sand dunes (Rocheleau et al., 1988; Baumer, 1990). Tree and shrub plantations on marginal lands prone to degradation may protect these resources from being cropped.
Protection of water resources	Contour buffer strips (reducing erosion, increasing sediment deposition, and facilitating water conveyance), riparian forest buffers (to reduce runoff and nonpoint source pollution from agricultural activities), and filter strips (vegetation for removing sediment, organic matter, and other pollutants from runoff and waste water) help in protecting water resources from pollution; use of excess and wastewater by trees; streambank protection; floodplain management ("waterbreaks").
Living hedges and fences	Protection of crops from freely roaming livestock through hedges and fences of indigenous thorny multipurpose trees and shrubs; use of large stem cuttings (e.g., of *Gliricidia sepium*) that quickly root and produce additional benefits (such as forage, green manure, or fruit) for fencing.
Reduced wind velocity (windbreaks)	Reduced wind velocity raises the relative humidity in the microclimate, which may reduce evapotranspiration. Windbreaks lessen wind erosion and sand shear, reduce loss of flowers in crops and deposition of dust (and thus nutrients) and snow (particularly important in the dry areas of Central Asia). The trees produce fuelwood, poles, fruit, fodder, mulch, etc. Potential problem: competition with crops for water and nutrients (Smith et al., 1997).
Provision of shade	Shade reduces energy expenditure of livestock for thermoregulation (cooling) in hot environments and thus increases productivity. It alters the microclimate and provides shelter.
Protection and conservation of biodiversity and wildlife	Trees provide a habitat for numerous forms of life and thus help in protecting and conserving biodiversity and wildlife. Trees on marginal lands may protect these areas from degradation, thereby helping conserve biodiversity.
Carbon sequestration	The introduction of agroforestry systems, together with grazing land management, is seen to represent a major opportunity to increase carbon sequestration in agriculture (IPCC, 2000).
Landscaping and beauty	Woody perennials may be used to order the landscape, by breaking monotony, furnishing landmarks (Baumer, 1990).

Source: Authors' elaboration.

production farming systems integrate natural resources and regulation mechanisms into farming activities to achieve maximum replacement of off-farm inputs and to secure sustainable production of high-quality food and other products through ecologically preferred and safe technologies. ICM focuses eliminating or reducing sources of present environmental pollution generated by agriculture and sustaining the multiple functions of agriculture while sustaining farm income (EU, 2002); therefore, research for and promotion of ICM systems may be of particular importance in CWANA countries where AKST is focusing on increasing productivity by intensifying production. Although yields currently tend to be a little lower in ICM as compared with conventional systems, a recent study (EU, 2002) suggests that, even in the highly productive agricultural systems in the European Union, it is possible to achieve similar levels of profitability using ICM techniques because lower revenues are balanced by reductions in production costs. Opportunities to achieve comparable or even increased profitability in less productive systems of CWANA are therefore considerable.

Integrated pest management (IPM) forms part of ICM and can be described as an effective and environmentally sensitive approach to pest management that relies on a combination of commonsense practices. Emphasis of IPM is on control, not eradication. The approach thus aims at keeping pests at acceptable levels (identified thresholds) by applying preventive cultural practices, identifying and monitoring pests, and applying mechanical, biological and (as a last resort) chemical controls when required. Since IPM programs use comprehensive information on the life cycles of pests and their interaction with the environment, research might be required to better adapt IPM strategies to CWANA conditions. This information, in combination with available pest control methods, is used to manage pest damage by the most economical means (which is particularly important for resource-poor CWANA farmers), and with the least possible hazard to people, property and the environment (USEPA, 2007). Although IPM's main focus is usually insect pests, IPM approaches may encompass diseases, weeds and any other naturally occurring biological crop threat.

Whereas ICM strategies still allow the use of agrochemicals, *organic agriculture* describes production systems that rely on ecosystem management rather than external agricultural inputs. Since organic agriculture renounces the use of synthetic inputs such as synthetic fertilizers and pesticides, veterinary drugs, genetically modified seeds and breeds, preservatives, additives and irradiation, it offers a production pathway that may be particularly attractive for resource-poor farmers in CWANA. In many developing countries, organic agriculture is adopted as a method to improve household food security or to achieve a reduction of input costs. Produce is not necessarily sold on the market or may be sold without a price distinction as it is not certified. Demand for organic products, however, is increasing not only in industrialized countries but—with increasing living standards—also in and around the CWANA region. Market opportunities for organic products are thus on the rise, both in view of exports and to satisfy inland demand; typically, organic exports are sold 20-25% higher than identical products produced on non-organic farms (Abou-Hadid et al., 2004). Vertical integration of market chains has proven to greatly benefit organic producers; small farmers may develop direct channels to deliver uncertified organic produce to consumers or organize themselves to have increased marketing strength in national and international markets.

Evidence is increasing that the transition to more ecological production practices does not compromise food security (IFOAM, 2007). Where external inputs have been high, yield reductions may occur during and after transition to organic farming, but organic agriculture may substantially increase yields in low-input areas (Pretty et al., 2006). In traditional rainfed systems, widespread in CWANA, organic agriculture has been demonstrated to outperform conventional agricultural systems under environmental stress conditions (FAO, 2007). Organic production additionally contributes to conserving biodiversity and natural resources, it may increase income or reduce costs, it produces safe and varied food, and it is sustainable in the long term. Therefore, organic agriculture should be an integral part of any agricultural policy aiming for food security and improved livelihoods (IFOAM, 2007). Organic farmers usually grow a variety of crops and rear livestock; this increases resilience of organic systems and may reduce production as well as market risks.

Whereas the use of genetically modified organisms (GMOs) within organic systems is not permitted during any stage of organic food production, processing or handling, modern *biotechnology*, if appropriately developed, could offer new and broad potential for contributing to food security. Biotechnologies developed over the last 30 to 40 years—such as tissue cultures, cell isolation, molecular diagnostics, marker-assisted selection and genetic engineering—are powerful instruments that can be used for different purposes. Most of these technologies are not controversial and can be used safely to increase food security. Since infrastructure requirements for certain technologies are not overly demanding, some of these technologies may well experience further development in the CWANA region.

An important subset of modern biotechnologies is *genetic engineering*, or the manipulation of an organism's genetic endowment by introducing, rearranging or eliminating specific genes through modern molecular biology techniques. Certain proponents of the GMO community are anticipating a second potential agricultural revolution, the Gene Revolution, in which modern biotechnology enables production of genetically modified crops that may be tailored to address ongoing agricultural problems in specific regions of the world, like developing high-yielding varieties tolerant to salinity and drought, pest and disease resistant. Although the genetically modified crop movement may have the potential to do enormous good, it also presents novel risks and has significant obstacles to overcome before it can truly be considered revolutionary. Evidence so far suggests that the technology has the potential to affect a wide range of plant and animal products and could have many consequences that extend beyond food production in agriculture (FAO, 2001).

Since biotechnology may facilitate increasing agricultural productivity, CWANA AKST should expand research and development efforts related to socially useful and environmentally friendly biotechnologies, including as appropriate the possible development of certain GMOs. However, consideration must be given to the potential benefits for food and nutrition security, and thereby for human health and well-being, on the one hand, and to the need to avoid risks to human health, social justice and the environment, on the other. Adequate safeguards must be put in place to ensure that all concerns are protected, including environmental concerns, while leaving options open for future generations.

Programs designed to bring the benefits of biotechnology to small-scale farmers in the CWANA region should be supported, while seeking to ensure that the aims and the effects of using such technologies serve to reduce hunger and malnutrition. Such programs may also be directed toward enhancing farmers' varieties or landraces that are already well adapted to local growing conditions, thereby adding specific value of interest to farmers.

The most urgent ethical task is to assess activities relating to food and agriculture in the light of their actual and potential impact on reducing poverty, hunger and malnutrition. There is a clear need to balance benefit to human health and the environment with risk. The risks are often unclear, speculative and impossible to test. The benefits of these new crops have not yet been fully demonstrated. People need to feel safe and assured that as far as possible their safety, their health and their beliefs have been taken into account before new forms of food products are introduced.

Although it is undoubtedly a useful exercise to observe the arguments and discussions other countries are having or have had when implementing agricultural biotechnology, it is in the end up to each country, whether industrialized or developing, to assess the benefits and risks as they may affect their own culture and environment, when deciding the best way forward (Kinderlerer and Adcock, 2005). As the potential impact of GMOs to both the environment and health is not entirely understood, many CWANA countries will probably take the precautionary approach and adopt the use of GMOs in farmers' fields only very cautiously. It is uncertain that biotechnology research, particularly related to GMOs, will gain ground in CWANA in the near future, because religious and other social factors may cause people in the region to be hesitant to accept GMO seeds and food crops. Since, however, the possibility exists that farmers start growing GMOs spontaneously, as observed in Pakistan, AKST systems and governments need to develop pertinent regulations and put them in place.

5.2.1.7 Adaptation to and mitigation of global climate change

Adaptation. Since the subregions in CWANA will be affected differently by global climate change, adaptation options will have to be site and situation specific. In many areas of CWANA water is projected to become even scarcer (IPCC, 2007a) and therefore improved *water resource management* and efficient water use will be crucially important. Particularly for rainfed agriculture, technologies such as water harvesting and supplemental irrigation will become particularly important (Pandey et al., 2003). Small-scale irrigation technologies (SIMINetwork, 2006) will gain importance, especially for poor farmers (e.g., for off-season production); access to functioning savings and credit systems will be a prerequisite for small-scale farmers to be able to make the necessary investments. Various water storage options (reservoirs of different sizes, groundwater storage and recharge) will have to be envisaged, particularly in the many areas in CWANA where summer precipitation will decrease (IPCC, 2007a). Investment in hydraulic infrastructure (rehabilitation, maintenance and new establishment) will be required to increase the reliability of water supply under increased water scarcity. Nevertheless, more frequent extreme rainfall events with high intensity will require increased focus on floodwater management, such as the design of dams and other infrastructure for flood protection, and soil surface management to reduce runoff and soil erosion through increased infiltration.

Various options exist to face the increasing threat of *land degradation* and desertification due to reduced vegetative cover as a result of changing climate, increased erosion by heavy rainfall, and climate-induced changes in land use, which leaves soils more vulnerable to degradation. Promoting vegetative soil cover and reducing soil disturbance should be the principle objectives in this regard. Cover crops and green manures, improved fallows and agroforestry practices, conservation tillage and adequate crop residue management will play important roles in protecting land against degradation induced by climate change. Soil and water conservation technologies in general (see Liniger and Critchley, 2007) will certainly gain importance in coping with the adverse effects of climatic changes in CWANA. Conservation agriculture has shown strong resilience to climatic abnormities in Central American highlands according to recent studies (Cherrett, 1999; Holt-Gimenez, 2002). Saltwater intrusion and increased salinity threats may require changes in production systems in certain areas, such as for flooded rice or aquaculture, besides adaptations with regard to species and varieties cultivated. Adaptations in rangeland management will become even more important in view of the predicted climatic changes in CWANA.

Cropping systems management will have to be adapted to changed climatic conditions. Changes may entail introducing new crops and varieties adapted for duration, tolerance and water demand, in crop rotations. Diversifying production portfolios as a strategy to cope with risk might become an important option. Currently underutilized crop species could play an important role in adapting cropping systems and varieties to changed climatic conditions. Introducing new cropping patterns adapted to site-specific conditions will require increased use of modern technologies such as crop simulation modeling and GIS for long-term planning to assess and reduce risks related to changed practices.

Crop breeding will have to focus particularly on improving tolerance to abiotic stress. For rainfed conditions, drought tolerance, early growth vigor for rapid establishment and phonological adaptation to changed climatic conditions are of particular importance. Improving heat tolerance will be a challenge for many crops, such as wheat and rice, to avoid significant yield reductions due to temperatures that will be generally or periodically higher. Since salinity problems will increase with saltwater intrusion and higher evapotranspiration, efforts to increase salinity tolerance of crop species will receive still more attention. Genomic tools might speed up conventional breeding efforts to achieve important breeding objectives. Maintaining biodiversity to exploit genetic diversity in semiarid ecosystems will facilitate adaptation to climate change.

Due to changing climatic conditions producers will have to cope with new and exacerbated *pest, disease and weed problems*. Warmer and more humid winters will lead to insect and pathogen overwintering ranges and their numbers expanding. Increased temperatures are also likely to facilitate expansion of highly damaging weeds such as striga. Integrated pest management will certainly gain importance, and the capacity of research, extension and farmers will have to be strengthened to be able to monitor thresholds, detect signs early, etc. Modeling efforts may need to be strengthened to understand pest–host dynamics under environmental change and to improve predicting how and where epidemics and other increased threats will occur.

Various *crop management practices* will facilitate adaptation to changed climatic conditions. Important aspects to consider include changes in soil preparation such as conservation tillage to improve nutrient and moisture retention and to prevent soil erosion; adaptation of planting and harvest dates or seeding densities; promotion of vigorous crop establishment through adequate soil fertility, addition of fertilizer to seed, seed priming, transplanting, supplemental irrigation; and other adapted management practices such as incremental fertilization.

Livestock production will probably be mainly affected by changes in feed availability because of rangeland and pasture productivity, and grain prices. Whereas intensively managed livestock systems have more potential for adaptation than crop systems, pastoral systems might need more attention since the rate of technology adoption is generally slower because changes in technology are viewed as risky (IPCC, 2007b).

Adaptation options in *coastal areas and marine fisheries* may have to include preventing development in coastal areas vulnerable to erosion, inundation and storm-surge flooding. "Hard" (dikes, levees, seawalls) or "soft" (beach nourishment, dune and wetland restoration, forestation) structures may be used to protect coasts. Storm warning systems and evacuation plans will have to be implemented, wetlands protected and restored, and estuaries and flooding plains maintained to preserve essential habitat for fisheries. Fishery management institutions will have to be modified and strengthened and policies to promote conservation of fisheries revised. Research and monitoring to better support integrated management of fisheries will be required.

However, all technological options to adapt to climatic changes will require an *enabling environment* that includes availability of financial resources, technology transfer, and cultural, educational, managerial, institutional, legal and regulatory practices. Affordability of such measures is, particularly for poor farmers, a prerequisite for their implementation. Access to functional savings and credit systems as well as to input and output markets is important. But also targeted support through adequate pricing policies or payments for ecosystem services will have to be considered to enhance the adaptive capacity of producers, especially in smallholder rainfed production systems and particularly in semiarid areas. Land and water access and use rights will have to be adapted to encourage both men and women farmers to invest in adapted technologies. Access to information, know-how and technology will have to be improved through better links between research, extension and farmers. Proactive risk management strategies will have to replace the currently prevailing reactive disaster management. Besides improved weather forecasting and access to reliable climate and weather information, early-warning networks and support agreements, such as the West African Comité permanent Inter-Etats de Lutte contre la Sécheresse dans le Sahel, may have to be developed. Insurance programs for farmers will encourage farmers to invest, improve the social resilience of poor rural populations, and mitigate the risks related to increased climatic variability. Capacity development will certainly be required to successfully face the challenges ahead and may also help in benefiting from the Flexible Mechanisms included in the Kyoto Protocol, such as the Clean Development Mechanism.

Mitigation. Beyond the use of biomass fuels to displace fossil fuels, the management of forests and rangelands and practices in agricultural production can play an important role in reducing current emissions of CO_2, CH_4 and N_2O and in enhancing carbon sinks (IPCC, 2007b). Woody perennials are particularly important in conserving and sequestering substantial amounts of carbon. Therefore, sustaining existing forest cover, slowing deforestation, regenerating natural forests and assisting the natural migration of tree species (e.g., through connecting protected areas and transplanting), establishing tree plantations and promoting agroforestry options are key land-use options for mitigating climate change. Such options as well as reducing conversion of grassland to cultivated land and setting aside ecological compensation areas may be all the more possible if agricultural production is intensified on surfaces less prone to degradation; to this end participatory land-use planning may be required.

Agricultural practices such as conservation tillage, rational management of crop residues (mulching, less burning of biomass) and proper fertilization (improved fertilizer-use efficiency, application matched with demand, incremental fertilization, use of legumes in rotations) may reduce greenhouse gas emissions and increase soil organic matter and thus carbon sequestration. Planting improved fallows and cover crops, improving rangeland management, avoiding soil erosion and particularly rehabilitating degraded lands further contribute to producing and maintaining soil organic matter (Duxbury, 2005; Lal, 2005; IPCC, 2007c).

Methane and N_2O emissions from rearing livestock may be substantially reduced in two ways. (1) Improving manure management, e.g., reducing anaerobic decomposition, may not only lessen emissions but allow for capitalizing on methane production by recovering CH_4 and using it to produce bioenergy. (2) Improving ruminant feeding is another way to reduce greenhouse gas emission from livestock husbandry; processing the feed for better digestibility, supplementing with nutrients and vitamins, or adding probiotics, yeasts and edible oils to animal feed may not only reduce CH_4 emissions but also increase productivity and thus result in a reduction of emissions per unit of product such as meat or milk. Recent experience, as in Australia, provides evidence that CH_4 production by rumen microorganisms may be controlled (Wright et al., 2004).

Methane emissions from paddy rice production may be significantly reduced through adapted water management, e.g., mid-season drainage combined with shallow flooding. Using ammonium sulfate fertilizer, which impedes CH_4 production, and considering new insights in rice CH_4 production in breeding programs may further decrease CH_4 emissions.

Although the strategy of soil carbon sequestration to mitigate climate change has been questioned, the likelihood that CO_2 will revert to the atmosphere because carbon sequestering practices might be abandoned seems rather small. Many of the practices that avoid greenhouse gas emissions and increase carbon sequestration can also improve agricultural efficiency and the economics of production. It is unlikely that farmers would abandon such win-win approaches unless competing demands for natural resources, mainly land, or some larger force compelled them to do so. The discussion however shows that practices to mitigate climate change have to be compatible with sustainable development; that is, they should also meet objectives unrelated to climate such as cost efficiency. AKST might therefore require capacity development in this regard and have to promote reforms of policies that currently encourage inefficient, unsustainable or risky farming, grazing and forestry practices.

5.2.1.8 Market orientation, diversification and risk management

Agrofood marketing, patterns and types of food consumption and food diversity are changing steadily as a result of development and globalization. Food markets are growing five times faster in the emerging market regions of Asia, Latin America, and central and eastern Europe than in the United States and western Europe (Reardon and Flores, 2006). National-level sales are growing at a country average of 7% in upper-middle-income countries like the Republic of Korea, the Czech Republic, Hungary, Poland, Argentina, Brazil, Chile, Mexico, South Africa, and 28% a year in lower-middle-income countries such as China, Bulgaria, Russia, Colombia, most of North Africa and West Asia, compared with about 2% a year in upper-income countries (Regmi and Gehlhar, 2005, citing data from www.euromonitor.com). With the expanding agrofood markets, food import bills in many CWANA countries are increasing, and agricultural imports constitute approximately one-fourth of CWANA's total merchandise imports (see Chapter 1 for geo-economics details).

Another added dimension to market reorientation is the spread of supermarkets and hypermarkets. In western Europe and the United States, the share of supermarkets went from none in 1920 to about 80% today—65% in Japan. The top four chains in the U.K. now have 50% of the market, chains in Germany 55%, France and Spain 60% (only 35% in the United States) (Cook, 2005). Besides this change in the structure of the food industry globally, there has been a rapid and recent transformation in how food industry firms source farm and processed products from producers. These changes are extremely important for local producers as well as for exporters. Because food industry firms are expanding the coverage of their procurement catchment areas, they are shifting their sourcing from traditional wholesale markets to the "new generation specialized wholesaler". Private standards of quality and safety are rapidly emerging (Reardon and Farina, 2001). This scenario provides an opportunity for AKST to be used to ensure that through enhanced production domestic agrofood products are substituted for imports. AKST can also help add postharvest value in storing, packaging, grading and labeling, by helping suppliers comply with standards and meet sanitary and phytosanitary requirements. It can build technology capital toward supplier upgrading, such as enabling suppliers to meet the tough new private standards like EurepGAP.

The volume of food trade increased 2.1 times from 1980 to 2003 (3.4% annually, faster than the annual rate of growth of GDP/capita in the world of 2.65%). Increases in trade were extremely uneven over product categories, with "nonstaples" the clear winners and grains the clear losers. Trade in fruits and vegetables grew by 330%; trade in meat, fish and seafood grew 300-400% from 1980 to 2003 (Reardon and Flores, 2006). Analysis of the production figures for cereals, fruits and vegetables in the CWANA region (see 2.1.2.1 for complete statistics on production) make it evident that in CWANA nonstaple crops are of secondary importance to cereals, which are and have long been one of the most important commodities in agricultural production. Due to the traditional and to a large extent nondiversified production trends in the region, CWANA has little exportable surplus of nonstaple crops and hence is not yet able to gain any significant benefit from new marketing opportunities that arise from trade of nontraditional crops.

In food economics, Bennett's Law states that as incomes rise, consumers switch into nonstaples and out of staples (Bennett, 1941). A strong middle class with higher spending power in CWANA is increasingly shifting from traditional to nonstaple food items, resulting in an increase in the food import bill. It is in this context that AKST can be used for agrofood diversification; market opportunities may be captured by producing products that are in high demand. Using AKST to produce nonstaple crops and off-season crops has great potential in CWANA, and it would not only contribute to increased income-earning opportunities for the growers (which in turn would reduce poverty and improve livelihoods) but also to a reduced import bill and rational use of foreign exchange at a macro level. Markets capitalizing on biodiversity as a source of food, herbal remedies and income are gradually emerging (Leaman et al., 1999). CWANA with its huge biodiversity hosts a large number of underutilized crops that might gain momentum in such markets (Giuliani, 2007). In many cases, the potential exists for more widespread use of these species. They include crops that could meet the needs of farmers wanting to increase yield from their land and consumers seeking a more natural and varied diet. They can offer opportunities for farmers to tap different markets and thus represent important new sources of income for rural people. Despite their local and potential importance, these species have been largely neglected by researchers. Information on their agronomic characteristics or nutritional value is often lacking, there is little genetic diversity available in gene banks for breeders to use and the seed industry largely ignores them. Therefore, improving the availability of information on underutilized crops demands more attention. Development of improved processing technologies and market analyses are required to capitalize on such "lost crops". New technologies such as molecular genetics and GIS will certainly play their part in developing conservation and use strategies. Participatory plant breeding approaches as well as marker-assisted breeding may allow obtaining improved plant material.

Diversifying production at various scales, from mixing seeds to integrating crops and livestock, will also substantially reduce production risks, particularly where higher income through sale of high-value products allows for pertinent investments such as small-scale irrigation. Diverse systems are generally more robust and resilient to shock and stresses, and thus better able to cope with risk (Haykazyan and Pretty, 2006; Werners et al., 2007). Crop diversification can considerably reduce the risks associated with pests and diseases, and the risk of crop failures due to such environmental conditions as climatic extremes and changes may be spread over a greater number of commodities. Diversification may also help in financial as well as market risk reduction and can thus contribute to stability, economic sustainability and improved livelihoods. It also allows for more flexibility and opportunities to adapt to changing framework conditions. Risk management strategies at the farm level may also include the choice of low-risk activities; although specialty crops such as tomato may offer the possibility of high gross returns they commonly have greater year-to-year production variability than the more common crops (Patrick, 1992).

Because of the multiple sources of risk, comprehensive strategies that integrate several responses to variability are often necessary for effective risk management. In addition to diversification these strategies may include choosing low-risk activities, dispersing production geographically, selecting and diversifying production practices, maintaining flexibility at production level; obtaining market information, spreading sales, practicing forward contracting, participating in government or other programs at marketing level; and insuring against losses, maintaining reserves, placing investments, acquiring assets, and limiting credit and leverage at the level of farm finances. The particular combination of risk-management responses an individual farmer uses will depend on the individual's circumstances, type of risks faced, and attitudes toward risk. Some risk responses act primarily to reduce the chance that an adverse event will occur, while others have the effect of providing protection against adverse consequences should the unfavorable event occur. Farmers find many different ways to implement these principal risk responses (Patrick, 1992). However, for farmers to choose certain framework conditions are prerequisite. These may include access to credit, insurance, markets and market information.

5.2.2 Rational management of natural resources

5.2.2.1 Water

Water management in and for agriculture has to be set in a broader perspective of integrated water resource management (IWRM) (GWP TAC, 2000; for links and resources regarding IWRM see InfoResources, 2003). IWRM aims at the coordinated development and management of water, land and related resources to maximize the resultant benefits in an equitable manner for all sectors and members of society without compromising the sustainability of ecosystems. Thus, IWRM pursues three major objectives: (1) efficiency by maximizing economic and social welfare derived from water resources and investments in providing water services; (2) equity in allocating water resources and services across different economic and social groups; and (3) environmental sustainability by not putting at risk the water system that we depend on for our survival.

IWRM considers water for people—drinking water and sanitation; water for food, i.e., for the farming sector including livestock and fisheries; water for nature—for preserving ecosystems; and water for other uses—industry, recreation, tourism, energy and transportation (SDC, 2005). IWRM therefore necessitates a holistic approach to management, considering the interdependencies within natural systems, but also the way that economic and social systems affect the demands placed on the resource base. It also requires a participatory approach, emphasizing the need for stakeholders to be involved in water development and management (including women as decision makers and water users). And IWRM requires understanding that the demands for water will inevitably outstrip the capacity of the resource base to deliver unless users become aware of the provision costs involved (including environmental costs). IWRM therefore represents a break with tradition, from sectoral to integrated management, from top-down to stakeholder and demand-responsive approaches, from supply to demand management, from command and control to more cooperative or distributive forms of governance, from closed expert-driven management organizations to more open, transparent and communicative bodies (GWP, 2006).

Since most countries in the Middle East and North Africa can be classified as having absolute water scarcity, and water demand from all sectors is expected to increase (Studer et al., 2002), IWRM is of particular importance in the region. Governments are currently the most active institutions in managing regional water resources; IWRM approaches, however, require participation of the different water users in managing water resources. Therefore, the role of governments will change from managing water resources to regulating the institutions involved in managing them. Regulations and laws will have to be adapted accordingly.

Efficiency of water use in agricultural production. Improving water-use efficiency in agriculture will have to include technical, economical, institutional and social options. Technical options include improving the infrastructural and organizational aspects of water conveyance and distribution systems; on the farm, they include improving the scheduling and practices of applying water in irrigated cropping—and making better use of precipitation in rainfed production. Since many farmers in CWANA may not be in the position to finance more efficient irrigation technology, they may have to be encouraged and assisted in moving toward more water-efficient systems. Furthermore, education and training will be needed to implement more water-efficient practices. However, water-demand management options such as water pricing developed with the participation of water users will have to complement technological options. Organizing users in water-user associations has proven effective in improving water management.

Water harvesting. Collecting, storing and concentrating precipitation at different scales (water harvesting) is an ancient technique dating back 4,000–5,000 years. It is currently under revival in response to the escalating water scarcity (Falkenmark et al., 2001). Harvested precipitation, i.e., collected runoff water, may be either diverted directly to the cropped area during the rainfall event ("runoff farming") or may be collected for irrigation or other purposes such as domestic use or livestock watering (Oweis et al., 1999). Runoff farming and using stored water for irrigation may be practiced at micro, meso and macro scale, and numerous technologies have been developed according to specific environmental and sociocultural conditions (see Critchley et al., 1991; Agarwal and Narain, 1997; Prinz et al., 2000; Prinz, 2002; Prinz and Malik, 2002; Mahnot et al., 2003; Oweis et al., 2004; CSE, 2006). Other water-harvesting techniques include floodwater harvesting, fog and dew harvesting, and groundwater harvesting by qanats, underground dams or special wells.

Water harvesting may not only tap unused water resources and thereby increase crop productivity and minimize the risk of crop failure in dry areas, it may also allow producing crops in environments where cropping is not feasible without such technologies. Furthermore, water harvesting may facilitate forestation or reforestation, fruit tree planting or agroforestry, and protect land from degradation and desertification (Prinz, 2002). Groundwater recharge for more sustainable use by different sectors represents another important benefit of water harvesting.

However, it is not clear if widespread use of water-harvesting technologies is achievable, since construction and maintenance costs, particularly the labor costs, are generally important. Furthermore, many water-harvesting projects require collective action at the community or watershed level, and land lost for catchment areas represents an opportunity cost that may deter small-scale farmers in land-scarce areas from adopting water-harvesting technologies (Rosegrant et al., 2002). Since certain technologies may require inputs that are too expensive for some farmers to supply, some intervention of state authorities may be needed (Prinz, 2002).

As rainwater harvesting should be an integral component of a farming system, a systems approach has to be followed and water-harvesting technologies should be combined with other improved management practices such as adequate fertilization, pest management, improved varieties, crop rotations, and efficient irrigation techniques. Applying remote sensing data and hydrological models at the watershed level may not only facilitate the identification of suitable water-harvesting sites and technologies but also help prevent problems between upstream and downstream water users and allow for supplying sufficient quantities of water for natural flora and fauna.

Since water harvesting operates at both the household or farm scale and the community or watershed scale, farming systems research must consider institutional and land-tenure issues, in which traditional and formal institutions may play a crucial role. Little research has been carried out in this respect, and thus AKST still faces important biophysical and socioeconomic knowledge gaps with regard to water harvesting. Extension and irrigation staff require more knowledge about water-harvesting techniques and the associated socioeconomic implications to achieve the potential gains in crop yields from water harvesting in combination with supplemental irrigation and improved farm management practices (Falkenmark et al., 2001). Since water-harvesting technologies originated in CWANA, a wealth of indigenous knowledge exists in the region that can be used to develop new practices and improve the efficiency of systems still in use today. For widespread adoption of such technologies, however, land-tenure systems will have to accommodate ownership or long-term use rights, so that farmers will be willing to invest in water-harvesting systems. Policies should encourage the required inputs for construction and maintenance.

Use of unconventional water resources. Rather than seeking pristine new water sources, a wide range of alternative water supplies will increasingly be used to meet demands. Reclaimed water, gray water, fog collection, recycled water, brackish water, saltwater, or desalinated water may all be considered usable for particular needs, and in fact may have environmental, economic or political advantages. Reclaimed water such as treated wastewater can be used to recharge groundwater aquifers, supply industry, irrigate certain crops, or augment potable supplies (Gleick, 2000).

However, using unconventional water resources may pose its own problems. Treated wastewater used in agriculture might entail health hazards and water-quality problems, requiring regulations regarding its treatment and reuse. Such regulations will particularly have to cover the responsibility of water polluters in treating their wastewater to make it safe to use (e.g., in agriculture) or to discharge into the environment. More training for farmers, water users and crop consumers will be needed to address issues related to health and water quality aspects.

Groundwater recharge. Groundwater resources are being overexploited in most CWANA countries (FAO, 1997; Aquastat, 2006). This is due to overpumping and also to reduced recharge related to diminishing infiltration rates caused by expansion of urban areas, inadequate land management, and climatic changes (Morris et al., 2003). Water-table elevations are dropping and seawater intrusion is becoming a common problem in many CWANA countries.

Maintaining and increasing aquifer recharge may counterbalance increased exploitation to a certain extent. Foremost, it is important to enhance natural recharge by adequate land management, i.e., by reducing runoff of precipitation. This may not only increase aquifer recharge but will also allow storing a greater part of the scarce precipitation in the soil for crops to use and will reduce erosion. The high evaporation rates in the CWANA region make groundwater storage particularly advantageous (UNEP/IETC, 2001).

Artificial recharge of groundwater aquifers is also a viable option. Artificial recharge can be achieved through surface spreading and preventing runoff or by direct well injection into the groundwater. Sources of recharge water may include precipitation and storm runoff, trapped from cultivated and uncultivated land, from urban areas, surface water, leaks in water supply systems, overirrigation, or treated wastewater (Morris et al., 2003; NEIWPCC, 2003). Artificial recharge may require temporary storage structures and water treatment in sedimentation tanks to improve water quality, particularly for treated wastewater. As a rule the quality of water recharged into an aquifer should be at least equal to or better than that of the groundwater, and water quality should be regularly monitored.

Water demand management. Options for managing water demand include technical, financial and economic measures, public awareness and active public participation in addition to institutional and legal measures (Gleick, 2000; Baroudy et al., 2005). Technical options for demand management include improving water infrastructure, rehabilitating existing water conveyance and distribution systems, lining earth ditches with concrete or other materials, replacing irrigation ditches with pipelines where affordable, rehabilitating old irrigation wells, rehabilitating hydraulic structure and irrigation networks, and installing water-measuring devices. At farm or field level, shifting to more efficient irrigation systems, improving irrigation practices, irrigating with proper scheduling, using unconventional water sources, cultivating more drought- and salinity-tolerant crops, and diversifying production systems with new crops and rotations that are more conservative in water use (Hillel, 2000) may reduce water demand considerably. Knowledge and understanding of farmers with regard to water-efficient technologies and practices will therefore have to be strengthened and increased.

However, besides technological options, demand for water will also have to be reduced through economic, administrative and social mechanisms (Baroudy et al., 2005). Whereas incentives may encourage investments into water-efficient technologies, water tariffs and water prices have a direct effect in controlling demand. In CWANA, farmers often have little reason to save water because irrigation water is easily accessible and farmers do not have to pay much for the water they use. Water-pricing policies may thus be efficient in general but remain highly controversial in the region. Therefore, water pricing should be used within a comprehensive framework to follow an IWRM approach in managing water resources. Usually water pricing includes the cost of water treatment where required, distribution and conveyance. It is important that water-pricing policies be developed with the participation of water users.

Financial and economic measures should be governed by two main principles: the user-pays principle and the polluter-pays principle. The water value should exceed the marginal cost of extracting and distributing the water. Water tariffs should be based on the full economic cost of the water; they should cover full operation and maintenance costs for the system. Flat tariff systems should be eliminated and farmers should be charged according to their actual consumption of water. Being charged by water consumption per unit area for irrigation gives farmers incentive to improve their water-use efficiency. Water cost from different sources and for different areas should be calculated according to the same principles.

Raising the degree of public awareness about rational water use and consumption patterns is important for effective demand management. A more promising approach than simple public campaigns is to strengthen the public's participation in controlling water demand and use, i.e., decentralization and participation in decision making regarding water management. Devising appropriate institutional frameworks adapted to the specific requirements and conditions of each country can establish and empower water-user associations, which have proven to increase water users' awareness and responsibility.

Virtual water trade. "Virtual water" is the water used to produce an agricultural product. Trade in agricultural products is thus also trade in virtual water. Trade of virtual water at national and particularly at international levels may reduce the pressure on scarce water resources and improve water-use efficiency globally (see World Water Council, 2004). By importing products requiring large amounts of water for production from areas with abundant water resources, water-scarce countries and areas may reduce the pressure on their own water resources and thus make water available for other purposes—the principle of comparative advantage. If the goods are imported from countries where less water is required to produce them, the global productivity of water may be increased. Since pressure on water resources may be eased through virtual water trade, investments in developing new water sources (such as dams or water transfer) may be reduced and negative side effects thereof diminished. Furthermore, the potential for conflict over water, which is particularly prominent in CWANA, may be reduced through virtual water trade (Allan, 2002).

However, virtual water trade also bears potential risks and drawbacks. These include ecological aspects (related transport, nutrient transfers, sustainability of production in exporting countries, alternative land use in importing countries) and economic concerns (how to afford imports; effects of imports on local agricultural production, rural development and consumer prices). Furthermore, opportunity costs of land and labor, as related to high unemployment rates, will have to be considered carefully in designing policies relating to virtual water trade. The greatest obstacle to the concept, however, lies in sociopolitical aspects, particularly in the geopolitical situation of most CWANA countries. People and countries are in general reluctant to become dependent on food imports; they may feel they will become restricted in autonomy and self-reliance. Thus implementing the virtual water concept requires consideration of important national goals such as food security (self-sufficiency or self-reliance), national security, economic growth and development (including poverty reduction, employment opportunities) and the quality of life in general. The virtual water concept may thus foster a more holistic approach to managing water resources, by linking water, food production, trade, consumption, food security, etc. It is a useful theory for developing policies targeting more productive water use and rational water-resource management.

However, research is still required to fill knowledge gaps on potentials and risks related to the concept. In addition, agreements on enabling framework conditions would have to be elaborated and implemented at the international (WTO) level. Such agreements would have to consider political, social and ecological aspects, assure the food security of importing countries and provide protection from abuse of dependencies through blackmail (Studer, 2005).

5.2.2.2 Soil and land

Since productive soil is the basis for agricultural production, soil and land degradation directly affects agricultural productivity and thus sustainability and development issues such as food security, nutritional quality, poverty reduction and overall development toward improved livelihoods (Steiner, 1996; Scherr, 1999; Penning de Vries et al., 2003). Furthermore, adequate soil management offers considerable opportunity for increased carbon sequestration, which may in turn mitigate climatic changes and their adverse effects on agricultural production and resource degradation.

Causes of soil and land degradation are closely related to numerous socioeconomic factors such as population growth and changing consumer habits; unprofitable farming due to low yields, high input cost, low farm-gate prices, and lack of access to markets; high risk, discouraging investment; and insecure land tenure and user rights (Studer et al., 2000). It has become apparent that AKST has to put more emphasis on tackling the reasons behind soil and land degradation, as well as developing and disseminating sustainable land-management practices that fit specific conditions.

Interventions to avoid, reduce and reverse soil and land degradation are required at different levels, and have to be coordinated and synchronized (Steiner, 1996). Most obvious are interventions at the *plot or field level.* Numerous practices and technological options fostering sustainable land management are available through both traditional and modern knowledge. However, options such as cover cropping, terracing, green manuring, conservation tillage and rotations with leguminous crops have to be adapted to the specific agroecological and socioeconomic conditions of the farm enterprises. This requires trial and evaluation methods that can be extrapolated to other locations. Modeling approaches supply information more rapidly than field trials and are considerably less costly. Further, various case scenarios may be simulated and explored, and results may be upscaled or transferred to other environments, particularly if modeling is combined with GIS or remote sensing. However, modeling approaches require a solid database, which in many countries of the region is not yet established.

A major problem with regard to sustainable land management is that the profitability of particular measures is often low or not directly obvious, especially in view of such continuously rising opportunity costs as labor. Technology development has therefore to focus on satisfying the short-term economic interest of *farmers and households.* Focus on promoting sustainable land use will thus have to shift to increasing productivity by maintaining, improving and stabilizing yields rather than on conserving soils (Scherr, 1999). Furthermore, technologies introduced will have to be compatible with the farmer's farming system and risk-avoidance strategy. Risk reduction is a particular primary concern for smallholder farming in the adverse environments of CWANA dry areas. Thus farmers may find quite acceptable recommendations that do not necessarily improve profitability but promise greater yield security and stability (Steiner, 1996).

For many interventions focusing on soil and land management a *community or watershed* approach will be necessary. Land-use planning, possibly coupled with changes in land tenure, zoning rules, control of agricultural land conversion and management of common lands, involves issues that are typically dealt with at this level (Scherr, 1999). Particularly in marginal lands, spatial concentration and intensification of production should be encouraged to achieve more profitable production and simultaneously protect fragile land from degradation (Scherr, 1999). Improving present land use and identifying alternatives to inappropriate land use have to be negotiated in a multilevel stakeholder approach, in which it is important to integrate local authorities and national administrations (Hurni et al., 1996). Participatory approaches and the formation of farmer or land-user associations, village committees or cooperatives may facilitate mutual understanding and collective action (Steiner, 1996; Penning de Vries et al., 2003).

At the *national level,* policies and legislation relating to socioeconomic issues and institutional aspects regarding agricultural research and extension will play a major role for achieving more sustainable soil and land management. Flanking measures to reduce the demographic pressure of high population growth rates may relieve pressure on land (Steiner, 1996). Generating nonagricultural employment opportunities, agricultural opportunities in other areas and opportunities in forest management (Scherr, 1999) may contribute to increased income in rural areas. This change for the better should in turn favor investments in agriculture, often facilitating measures and practices for more sustainable land use. Developing credit and savings schemes may help farmers organize and finance investments in land improvements and conservation.

The effects of pricing policies regarding inputs and outputs as influenced by market liberalization and protective measures, tariffs and taxes, or other incentives and charges may vary in different countries. Higher producer prices often stimulate investment in agriculture and may thus lead to increased productivity and more sustainable land management. Ensuring market access for farmers is a prerequisite for pricing policies to produce impact, often requiring infrastructure development. Internalizing such external effects of land degradation as off-site costs of erosion (Steiner, 1996; Penning de Vries et al., 2003) and recognizing the multifunctionality of agriculture may assist in appropriate pricing policies.

Land tenure legislation must guarantee long-term land-use rights to owners and leaseholders if land users are expected to invest in long-term soil conservation measures. However, often such rules need to be tailored to such local conditions as traditional land rights and the interests of different stakeholders. Local institutions are generally in a better position to adapt and enforce such regulations than national entities. The value of land is an essential component of land law because it greatly determines the commitment of land users to use this production factor in a sustainable way (Steiner, 1996).

Another important domain of national policy relates to agricultural *research and extension.* To assume their task of solving the problems of land users, researchers and extension agents require a clear definition of target groups and recommendation domains, and precise information on the decision-making criteria of land users, who often apply decision-making measures other than those specialists use. Participatory technology development may foster the adoption of interventions (Steiner, 1996); improve integration of research, technology development and extension; and facilitate learning from and disseminating results of successful land management practices and approaches (Steiner, 1996; Penning de Vries et al., 2003). Initiatives such as WOCAT (World Overview of Conservation Approaches and Technologies) or IASUS (International Actions for the Sustainable Use of Soils) and UNCCD may further facilitate such exchange and allow for enhanced cooperation (Hurni et al., 1996, 2006). Drawing on local farmers' know-how and traditional indigenous techniques may facilitate development of appropriate adapted technologies (Steiner, 1996). Remote sensing, GIS and time-series data may allow for exploration of relationships between soil-quality changes and farm management, local economic and social conditions, and the policy environment. Thus soil-related information may be incorporated in economic and policy modeling for more holistic and integrated analysis of problems and solutions, and allow for the evaluation of different scenarios and identification of priority areas for action (Scherr, 1999; Penning de Vries et al., 2003).

In summary, approaches to avoid and reverse soil and land degradation should generally consider

- Following a participatory, multidisciplinary systems approach (sectorwide thinking; Scherr, 1999)
- Following the principle of subsidiarity in decision making (decisions should be delegated to the lowest possible level) (Hurni et al., 1996)
- Fitting and targeting the specific environment (regarding development pathways, farming systems, soil types, degree of degradation, etc.; Scherr, 1999)
- Combining indigenous traditional wisdom with modern knowledge and technologies (such as remote sensing, GIS, and simulation modeling)

With regard to the rational management of soil resources, AKST in CWANA will have to target the following major focus areas:

- *Increasing or maintaining soil fertility and quality.* To counteract negative nutrient balances in many CWANA countries, AKST will have to focus on more efficient use of nutrients, e.g., by developing nutrient management systems for specific soils (Scherr, 1999) or by splitting fertilizer applications. Nutrient inputs will have to be increased, requiring access to and affordability of mineral fertilizers; the complementary use of organic fertilizers from crop residues, manure, compost and green manure will have to be encouraged; and the benefits of biological nitrogen fixation through legumes in rotations, green manure or cover crops will have to be better exploited. Increasing problems of micronutrient deficiencies and depletion will have to be explored and solved. Loss of nutrients will have to be avoided wherever possible by rapidly incorporating manure, combating erosion, etc. Ways will have to be explored to better close nutrient cycles by recapturing nutrients currently discarded in water bodies or dumped as waste elsewhere.
- Adequate *organic matter management* is particularly essential in CWANA since organic matter decomposes rapidly in high temperatures. Organic matter increases nutrient availability through direct addition and may enhance nutrient use efficiency by improving cation exchange capacity (CEC). Furthermore, increasing the organic matter content in soils improves water-holding capacity, which is extremely important in the dry areas of CWANA, and enhances soil structure, which reduces susceptibility to wind and water erosion and promotes soil fauna and flora. Increasing organic matter in soils also presents a big opportunity to act as a sink for carbon sequestration, thereby offering potential to mitigate global warming.
- *Combating wind and water erosion* remains a major challenge in CWANA. Cropping systems, rotations and cropping practices aiming at year-round soil cover should be envisaged wherever possible, although this is not always possible in the dry areas of CWANA. Using harvest residues in mulching, strip cropping (possibly with perennial vegetation), bunds, ridges, terraces, etc., will have to be promoted and profitability of suggested measures assured. Conservation tillage and economically productive cover crops or perennials integrated in crop rotations may be as economical as conventional cropping. Developing low-cost soil conservation and rehabilitation techniques such as control of water flow over land will have to receive major attention (Scherr, 1999).
- *Protecting and conserving vegetative cover and quality.* Since vegetative cover is key to soil protection, maintaining and—where required—restoring flora and fauna are fundamental for sustainable land use (Scherr, 1999). Appropriate grazing management and protection of land susceptible to degradation (e.g., against inappropriate cropping) will have to receive particular attention in the future.
- Practices to *avoid salinization* of highly productive irrigated land are well known and consist of improving system- and farm-level water management regimes and the necessary investments in proper drainage systems. AKST will have to investigate diversification options into higher-value crops to justify the required investments. Methods to use saline lands and low-cost options to control or reverse salinization will also have to receive major attention (Scherr, 1999).
- *Reducing pollution* of soils (as well as of water and air) is particularly important in the more intensive production systems that will probably develop in many areas of CWANA. In this regard, regulating the use of agrochemicals and disposing of agrochemical and livestock waste will have to play a major role in protecting agricultural soils from pollution (Scherr, 1999). Raising awareness and understanding about pollution problems will have to accompany such regulations; lessons learned in other, e.g., industrialized parts of the world will be of particular value.
- Particularly challenging with regard to sustainable land management is the *development of new lands*—reclamation of land never cultivated before, such as practiced in Egypt. Whereas such new lands hold considerable potential, because their low disease, pest and weed pressure raise the opportunity for organic production, their development may bear great difficulties, in particular with regard to building up and maintaining soil fertility and the high susceptibility of marginal lands to degradation.
- Stopping *sand encroachment* will represent a major task with regard to protecting productive soils in many

countries of CWANA (Dupuy et al., 2002; ACSAD et al., 2004). Many different methods of combating this phenomenon are known (for a review see Ramadan and Al Sudairawi, 2005), which allows AKST in CWANA to capitalize on lessons learned not only in the region but also in Sahelian countries and in China and Mongolia.

5.2.2.3 Biodiversity

Biodiversity is undoubtedly being lost in many parts of the globe, often at a rapid pace. This loss poses a serious threat to agriculture and the livelihoods of millions of people. Conserving biodiversity and using it wisely is a global imperative. Biodiversity provides the foundation for our agricultural systems. It provides the sources of traits to improve yield, quality, resistance to pests and diseases, and traits than can adapt to changing environmental conditions such as global warming.

Loss of biodiversity requires countermeasures such as increased efforts toward conservation by different means (see Table 5-2). Conservation may be in situ or ex situ, in either natural or seminatural habitat, or in some purpose-built environment (Braun and Ammann, 2002). The choice of one technique or the other, or a combination of both, will depend on the particular case. In situ conservation will involve maintaining and protecting natural habitats, while botanical gardens and seed banks are used for ex situ conservation. Both of the latter require precise knowledge of taxonomy. Today, conservation also embraces various components of agrobiodiversity like crop varieties, landraces, semidomesticates and crop relatives. The methods of biotechnology can be applied to the study of virtually any biological phenomenon and will in some cases have practical applications for maintaining biodiversity. Conversely, threats to biodiversity by biotechnology also need to be considered.

Different approaches to conserving biodiversity and different ways of using genetic resources are described here in more detail. One approach is on-farm management that involves maintaining crop species on farms or in home gardens; ICM may play an important role in this approach. Wild populations regenerate naturally and are dispersed naturally by wild animals and winds and in water courses. A second approach is the in situ conservation of forests and other wild plant species, often carried out through, but not limited to, designated protected areas such as national parks and nature reserves. In addition and depending on the type of species to be conserved, different ex situ conservation methods may be used. A complementary conservation strategy can be defined as "the combination of different conservation actions, which together lead to an optimum sustainable use of genetic diversity existing in a target gene pool, in the present and future" (Bioversity, 2007). It should not be forgotten that the main objective in any plant genetic resource conservation program is to maintain the highest possible level of genetic variability present across the gene pool of a given species or crop, both in its natural range and in a germplasm collection.

Plant genetic resource conservation and use may greatly benefit from applying modern developments in molecular genetics. CWANA countries could benefit from the program the International Plant Genetic Resources Institute (IPGRI) has identified, which includes the following components (IPGRI, 2001):

- Capacity building, with an ultimate goal of providing the genetic resources community with tools in the molecular area, emphasizing development in the context of research, gene bank management and germplasm use.
- Research that includes information on genetic diversity and location of useful genes and alleles in germplasm collections.

Table 5-2. **Approaches to conserve biodiversity.**

Intervention	Description
Move toward sustainable agriculture	Develop means to support sustainable agricultural production that minimizes negative effects on natural biodiversity
Keep bioregional perspective	Adopt a broader perspective of agriculture to capitalize on shared regional opportunities offered
Share inventories and information	Survey, inventory and disseminate information for enhancing agricultural development
Conserve genetic resources	Develop national ex situ storage facilities
Manage genetic resources on farm	Enhance on-farm management of genetic diversity (landraces and traditional breeds of livestock) and explore ways to manage this diversity in a more efficient and durable way.
Broaden the genetic basis of crops and livestock	Efforts are needed to widen the genetic base of modern cultivars and breeds, usually very narrow.
Conserve biodiversity in natural ecosystems	Natural and seminatural ecosystems contain wild species, races and populations of great importance for food and agriculture; efforts need to be developed to strengthen maintenance of these ecosystems across the region.
Broaden cultivated crop portfolio	Efforts need to be directed toward widening diversity of farm crops and breeds to promote agricultural sustainability

Source: Street and Khalikulov, 2004.

- Storage, management and analysis of molecular marker information obtained from screening germplasm collections and linking this information to existing traditional data.
- Policy and biosafety where IPGRI closely monitors developments and helps partners in national programs to define their stance for issues related to the conservation of diversity, proprietary concerns and protection of the environment.

Modern information technologies such as GIS used to characterize the geographical distribution of wild plants, or new electronic technologies for monitoring the environment such as Planetor, a computer program for analyzing environmental problems (Hawkins and Nordquist, 1991) may greatly contribute to conserving the environment and biodiversity. But new strategies and policies to conserve the biodiversity and improve research on biodiversity are additionally required.

Efforts on biodiversity conservation can learn from context-specific local knowledge and institutional mechanisms such as cooperation and collective action; intergenerational transmission of knowledge, skills and strategies; concern for well-being of future generations; reliance on local resources; restraint in resource exploitation; an attitude of gratitude and respect for nature; management, conservation and sustainable use of biodiversity outside formal protected areas; and transfer of useful species among households, villages and the larger landscape (Pandey, 2003, 2004).

Traditions are reflected in a variety of practices regarding the use and management of trees, forests and water:

- Collection and management of wood and non-wood forest products
- Traditional ethics, norms and practices for restrained use of forests, water and other natural resources
- Traditional practices to protect, control production and regenerate forests
- Cultivation of useful trees in cultural landscapes and agroforestry systems
- Creation and maintenance of traditional water-harvesting systems such as tanks along with planting tree groves close by

These systems support biodiversity, although not necessarily natural ecosystems, and help reduce harvest pressure (Pandey, 2004).

Traditional knowledge associated with biological resources is an intangible component of the resource itself. Traditional knowledge has the potential of being translated into commercial benefits by providing leads for developing useful products and processes. These valuable leads save time, money and investment of the modern biotech industry into any research and product development. Hence, a share of benefits must accrue to creators and holders of traditional knowledge.

Options for protecting traditional knowledge, innovations and practices include (1) documentation of traditional knowledge, (2) a patent system for registering innovations, and (3) development of a sui generis (only example of its kind) system (WTO, 2000).

5.2.2.4 Livestock and fish

The threat of extinction for many species, both known and as yet undiscovered, grows ever greater as whole ecosystems vanish, human populations proliferate, and human-mediated interference increases (Ryder et al., 2000). Whereas a laudable effort is being made to organize seed banks for plants, no such organized attempts to store genetic material exist for many species of either vertebrate or invertebrate animals. There are worldwide attempts to coordinate and store samples of DNA for every endangered animal species in DNA libraries or to freeze cells or tissues that could readily yield DNA for captive breeding programs.

Captive breeding provides an insurance strategy against extinction and for some species may be the only hope of survival. It requires input from population genetics to preserve high levels of genetic diversity, and from reproductive physiologists to promote the establishment of pregnancies, for example by artificial insemination. Cryopreservation of gametes and embryos has a role to play, while in the future, nuclear replacement cloning from established cell lines might prove of value. Such strategies may succeed in saving a small fraction of endangered species, at least for a time (Ryder et al., 2000). These tools will be particularly powerful when used in conjunction with efforts to conserve the habitats in which populations restored by DNA techniques can live.

West Asia and Mediterranean North Africa are endowed with considerable genetic diversity in small ruminants—various breeds of sheep and goats that are adapted to a range of arid and semiarid environmental conditions. But these local breeds may be endangered through intensified production systems and uncontrolled crossbreeding with exotic breeds. Therefore, it is important to think of possible ways to conserve the genetic diversity of these local breeds, which may be valuable in the future.

One way of preserving genetic diversity is ex situ conservation by storing frozen semen in gene banks. Another way is in situ conservation. The best way forward would be a combination of both conservation approaches, but the costs of ex situ conservation might be high. Storage facilities could be shared by different countries, thus reducing costs for each country.

In aquaculture, broodstock is either obtained from the wild or domesticated in the hatchery. Depending on the wild is not enough for optimum aquaculture production. In the hatchery, broodstock must be managed to ensure genetic resources are conserved, to maintain the desirable characters of the farmed species and to avoid problems of inbreeding (Bartley, 1998).

Genetic processes such as hybridization, chromosome set manipulation and sex reversal are used in aquaculture to improve breeds. Genetic technologies can also be used to reduce the environmental risks of exotic species escaping from the aquaculture facilities. To reduce the effects of changing genetic resources of organisms produced in hatcheries, several protocols have been prepared that demonstrate the best methods for choosing the origin and number of parents from specific fish species.

5.2.2.5 Institutional considerations

CWANA member countries are encouraged to become party to the International Treaty on Plant Genetic Resources for Food and Agriculture if they have not already done so. Its objectives are the conservation and sustainable use of plant genetic resources for food and agriculture and the fair and equitable sharing of benefits derived from their use, in harmony with the Convention on Biological Diversity, for sustainable agriculture and food security.

CWANA countries are to benefit from IPGRI efforts that support the conservation and use of neglected and underutilized crop species. IPGRI assesses the diversity and conservation status of a wide range of neglected crops through participatory regional programs, and implements activities to enhance both these varieties and their marketing. Neglected and underused crop species—also known as orphan crops—have been overlooked by scientific research and by development workers, despite the fact that they play a crucial role in food security, income generation and food culture for the rural poor. This lack of attention means that the potential value of these crops goes unrealized. It also places them in danger of continued genetic erosion and ultimate disappearance, further restricting development options for the poor. IPGRI is attempting to safeguard the genetic resources and associated knowledge through ex situ and in situ conservation across the CWANA region in areas where their genetic diversity is highest, like in Turkmenistan, Syria and Tunisia. In addition, IPGRI is improving its understanding of the agromorphological and market-driven traits and exchange of materials and experiences across countries, which will strengthen country capacity in commercializing and promoting the multiple uses of such crops.

The goal of ICARDA's Genetic Resource Unit is to conserve and use the biodiversity of ICARDA's mandate crops: wheat, barley, lentil, kabuli chickpea, fava bean, and pasture and forage species and their associated rhizobia. Its gene bank serves as a repository for a world collection of these crops and their wild relatives; crops that are of vital importance, not only to the CWANA region, but to the world at large.

With regard to conserving the diversity of threatened and wild fish species, different international organizations working with fisheries such as FAO have made efforts to change the criteria for adding new marine species to the list of endangered species. Fishery authorities in the countries concerned were also encouraged to participate in the convention in related subjects. Marine protected areas are well placed to conserve fish biodiversity as they can protect critical habitats. The Convention on Biological Diversity was ratified in, 1995 with the main objectives of conservation of biological diversity and sustainable use of its components. This convention plays an important role in conserving aquatic biodiversity. The FAO Code of Conduct for Responsible Fisheries (CCRF, also ratified in 1995) is another important tool for conserving aquatic biodiversity. Both CBD and CCRF have similar articles regarding the introduction of alien species. Both treaties encourage countries to notify their neighbors about any introduction and to establish a database or information system regarding introduction of aquatic organisms. They both also encourage the countries to monitor the aquatic environment and conserve genetic diversity. Countries should develop a code of best practices for responsible introduction of alien species.

Other actions that could be envisaged at CWANA country level to foster the conservation of biodiversity:

- Developing national genetic resources legislation
- Establishing an IPR system
- Placing NBSAPs in the mainstream of the national development plans of the country
- Synergizing implementation of CBD action plans and other conventions such as UNCCD and the UN Convention on Climate Change
- Becoming party to the Cartagena Biosafety Protocol to safeguard against GMO release through transboundary movement

5.2.3 Capacity development and knowledge management

Insofar as scientific and technical progress in the region is concerned, a number of trends and opportunities have occurred: the adoption of new technologies, particularly biotechnology and ICT, privatization of state-owned enterprises and trade liberalization, a greater role for development agencies in agricultural and rural economies, and increased international collaboration through the ecoregional approach and South-to-South programs (IPGRI, 2001).

Advances in scientific knowledge across a broad range of disciplines will be required to develop more and better food and fiber products with improved nutritional quality, to reduce food and commodity yield losses due to pests and diseases; ensure healthy livestock, sustainable fisheries, aquaculture and forestry sectors; manage water more efficiently; prevent and reverse land degradation; and conserve and manage genetic diversity (El-Beltagy, 2005).

A focused and appropriate research agenda is required to meet these challenges that are supported by public investment. Unfortunately, public investment in agricultural research and development is declining, while private sector investment is increasing in the OECD (Organization for Economic Co-operation and Development). Private sector investments tend to focus on commodities produced for OECD markets and often neglect the needs of the poor. Thus increased investment by the private sector will not meet the demand for diversified agricultural products and improved rural livelihoods via the required multisectoral approach that covers economic, environmental, ethical and social considerations.

Given the decline in public sector investment in developing countries at a time when the challenges to apply science and technology are urgent, there is a need to consider carefully the agenda for future agricultural research and development efforts. This agenda must also include public debates on controversial issues such as the development and deployment of genetically modified organisms and other aspects of modern biotechnology (Thomas et al., 2003).

In a study conducted by the World Bank and FAO in ten developing countries (Rivera et al., 2005) including three CWANA countries: Egypt, Morocco, and Pakistan, it was concluded that these countries do not yet appear to possess a totally integrated and operative agriculture knowledge and information system (AKIS), although all appear to want to move in that direction and be making significant progress. Agricultural education, research and extension still tend to operate as three separate systems (or subsystems).

In the following section we address capacity development options to support sharing, exchanging and disseminating knowledge generated through AKST systems in its subsystems of education, research and extension, and to integrate these subsystems in the CWANA region. The emergence of ICT in the last decade has opened new avenues in knowledge management that could play important roles in meeting the prevailing challenges relating to ICT and knowledge management.

5.2.3.1 Information and knowledge produced by AKST institutions

The ultimate objectives of AKST activities are to come up with results that can advance research more in certain areas, and engender technologies that AKIS stakeholders can use to increase production, conserve the environment, etc. The following subsections describe the options proposed to meet challenges related to sharing, exchanging and disseminating knowledge and technologies generated from AKST activities and that are most needed by growers, extension workers, researchers and decision makers.

Mechanisms and infrastructure for *sharing and exchanging* agricultural knowledge generated from research at national and regional levels should be enhanced. Many research activities are repeated due to the lack of such mechanisms and infrastructure at the national level. Researchers can find research papers published in international journals and conferences more easily than finding research papers published nationally in local journals, conferences, theses and technical reports.

Mechanisms and infrastructure for *transferring technologies* produced as the result of research to growers either directly or through intermediaries (extension subsystem) should be strengthened. Knowledge and technologies fostering agricultural production and environment conservation are examples. Although many extension documents exist in the region, produced by national agricultural research and extension systems to inform growers about the latest recommendations concerning different agricultural practices, these documents are not disseminated, updated or managed to respond to the needs of extension workers, advisers and farmers. This is also true for technical reports, books and research papers related to production.

Indigenous knowledge must be kept as a heritage for new generations. It is available through experienced growers and specialists in different commodities. These inherited agricultural practices are rarely documented, but they embody a wealth of knowledge that researchers need to examine thoroughly.

Economic and social knowledge must also be made easily accessible to different stakeholders at operational, management and decision-making levels, so that those responsible will be able to make appropriate decisions regarding the profit making of certain technologies and their effect on resource-poor farmers.

All these types of knowledge must be made available to the *education subsystem* to keep students up to date with the latest developments.

5.2.3.2 Integration of education, research and extension subsystems

In a case study conducted by ICARDA (Belaid et al., 2003), recommendations are made to strengthen stakeholders links in national agricultural systems in CWANA. To achieve strong and reliable links among all agricultural stakeholders, the different AKST institutions must be strengthened. The following options are proposed to strengthen these institutions in the CWANA region.

Option 1: Develop institutional capacity. Throughout the priority-setting process that ICARDA provided in the CWANA region in 2003 (Belaid et al., 2003), it became clear that these institutions were not well equipped in resources, organization and representation to adequately address the priority needs of the region. It is therefore essential to strengthen these institutions to enable them to fully play their role in implementing, disseminating and diffusing information that can be used in practice. The acute lack of capacity in other key disciplines such as social sciences, combined with the shift in research focus towards relatively "new" issues such as alleviating poverty and managing natural resources requires capacity development to meet this gap of AKST to adequately implement the subregional research agenda.

Option 2: Develop agricultural extension. Agricultural extension is needed in the CWANA region to educate professional agriculturists (including farmers) who may further enlarge or refine this body of knowledge.

Option 3: Improving agricultural education. Agricultural universities and institutes need to adopt and reorient curricula for new requirements with special training programs in extension development and education and new technologies. Education and training should embrace new scientific achievements and innovations. A full comprehensive training cycle, integrated with science, can ensure that production systems adopt the outputs. New methods of delivering services and new schemes of organization of training that result from the revolutionary changes in information and telecommunication areas (distance and correspondence learning) will be able to cover all levels of rural society. Using a complex approach in education as a unified system should take into account the inputs at all levels in the education hierarchy, including higher (universities and institutes) and secondary vocational education (colleges and academic lyceums) that contribute to development of agricultural education and human capacity as well as humanitarian and social capacity, and their role in renovating, reorganizing and reorienting agricultural production systems. Technical renovation of laboratory and experimental equipment, facilities and materials could be achieved by creating conditions suitable for research and experiments in the classroom and in the field. This could be achieved through providing specialized machinery, equipment and tools necessary for experimental activities. Private investment and funding affect the focus of the education; research and extension result in new subjects and specializations being added to the curriculum. Therefore, integrating such subjects as "international trade" and "agricultural products marketing" means finding staff with the necessary qualifications, knowledge and skills to teach these subjects or retraining staff to do so.

Higher education institutions must have their own production farms to give the students and farmers practical training and research sites for researchers. Creating this kind of training centers that are linked simultaneously to higher education and retraining of specialists is extremely important for strengthening and developing farmers' movements. AKST systems of different countries have their own development priorities and programs. Collaborative scientific programs may stipulate that the national system must act jointly on a specific crop or aspect of scientific research. Such program requirements will affect the agricultural education system and the entire process of developing innovations. Methods should be developed and shared among CWANA countries, possibly in well-known universities and research centers in the region.

Integrating education, research and extension is a principal task for CWANA to accomplish to achieve its sustainable development goals (Figure 5-2).

Options for strengthening the integration of AKST stakeholders include:

Option 1: Involvement of AKST stakeholders. The gap analysis has identified the insufficient involvement of many AKST stakeholders (Belaid et al., 2003). Considering the mandates of these institutions and the important contribution they could make to improving agricultural production and developing capacity of the region, their involvement in the region should be significantly enhanced (Box 5-1).

Option 2: Regional cooperation. To reorient regional cooperation and facilitate implementation of the identified regional AKST institutions, some key approaches were identified, such as networks, coordination meetings and traveling workshops. Existing networks, such as among universities and research institutes, need to be reviewed and consolidated. International centers in CWANA like ICARDA may facilitate such reviews and consolidation. Joint projects (interuniversity, between research centers) can enhance regional collaboration and cooperation, through various types of projects that include education, research and extension. The key innovation of the CWANA regional priority-setting exercise was to set the right conditions for a dialogue where "nontraditional stakeholders", i.e., farmers, NGOs, the private sector and grassroots organizations, would play a central role. CWANA countries have the opportunity to use cooperational links independently, create a special regional association of agricultural education, research and extension institutions, and interconnect them with a unified network. In view of the complexity of challenges facing the region it is unlikely that AKST regional institutions will be able to satisfactorily address them on their own. This in turn highlights the urgency of establishing strategic partnerships to tackle the problems of developing agricultural capacity in the region. The strategic partnership should seek to link the education, research and extension initiatives to the development goals of alleviating poverty and improving food security.

Option 3: Applying a participatory approach. Participatory bottom-up approaches can be done at three levels: (1) regional, (2) CWANA subregional, and (3) North Africa, West Asia, Central Asia and Caucasus. The valuable lessons learned from previous case studies (Belaid et al., 2003; Thomas et al., 2003) have triggered the need to develop mechanisms that will expand collaboration and dialogue through sustainable links and strategic partnerships with "nontraditional" stakeholders, especially NGOs, farmers, grassroots organizations and the private sector—the ultimate clients of agricultural innovation system products and innovations. In national agricultural information systems, the collaborative relationships with universities, NGOs, the private sector, farmers, and farmers' organizations are, by and large, at an embryonic stage and need therefore to be significantly consolidated (Belaid et al., 2003).

5.2.3.3 Agricultural knowledge management using information and communication technology

The central purpose of knowledge management is to transform information and intellectual assets into enduring value

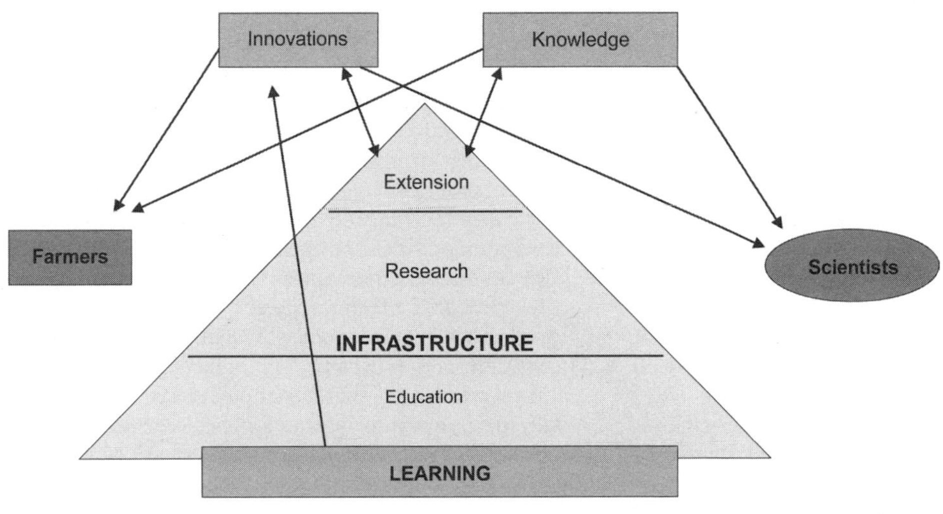

Figure 5-2. *Integration of education, research and extension.* Source: Authors' elaboration.

Box 5-1. Changes in the innovation process in Central Asia (Source: Babu, 2006)

The innovation process presents itself as a process of creating and spreading innovations and consists of three components: (1) innovation as a new idea, knowledge, a result of research; (2) introduction of innovations in practical activities; (3) diffusion of innovation, through products, services or technology in new places and conditions.

The analysis of the situation in the Central Asia and Caucasus (CAC) region as a whole has shown that countries follow principally new policy reforms in the process of transition to market economy. The innovation development in the region is through creating national innovation systems in all sectors including agriculture. The national agricultural innovation system (NAIS) is a collection of legal and economic aspects involving research innovation, their introduction, dissemination, use and realization into national policy. The system (NAIS) is based on close-fitting intercoupling data. These processes reflected the development of agriculture as a whole and the breakup of the old centrally planned economic relationships. They have brought about a sharp decline in production of agricultural products. When structural reforms are completed, the economic situation will be normalized.

What are the reasons behind economic dependency on agricultural development? The economic dependency of the CAC countries on agriculture is primarily for food security as well as inputs for processing industries and employment for rural population. The well-being of nearly half of the population depends on agriculture in the countries of CAC. Climate and environmental conditions favor agricultural development. Efficient agricultural innovation systems are a guarantee for the economic growth and stability of agricultural development. This was also experienced by the developed countries. Similar processes exist in the countries of eastern Europe, where these processes started early. The process of reform and restructuring were reflected in the innovative sphere of agriculture first of all.

The previous command system affected all areas of innovation systems. In education, the study and introduction of production scientific research achievements was introduced through planned command. The enormous facilities financed scientific institutions according to the plan made for developing the science. However, due to a lack of working mechanisms for introducing these actions appropriately and adequately with respect to the desire and interest of the producers, many achievements did not find any use—even though many of them received patents. The coefficients of success of their introduction and returns were occasionally reduced to zero. The chief thing was to report.

Today the situation and incumbent relations have changed, but problems have not decreased. For example, system competitive grants are present in nearly every country. Their principles are competition, priority setting, urgency, usefulness, cost performance, etc. Specific approaches are employed: (1) selecting the best innovation projects for introducing farming facilities; (2) building on the readiness of today's farmers to introduce innovations by undergoing the necessary process of education and consultation; and (3) introducing the proposed innovation project to confident farmers and other agricultural commodity producers in case the project realizes.

(Metcalfe, 2005). The basic idea is to strengthen, improve and propel the organization by using the wealth of information and knowledge that the organization and its members collectively possess (Milton, 2003). It has been pointed out that a large part of knowledge is not explicit but tacit (Schreiber et al., 1999). This is true for knowledge in agricultural science and technology where a lot of good practices are transferred without being well documented in books, papers or extension documents.

To manage the knowledge properly, ICT is needed. Studies on using information systems for rural development can be found in FAO (2000) and ICARDA (2006). In CWANA, existing efforts in collecting appropriate knowledge need to be coordinated and made available through ICT to the end users: researchers, extension workers, students and growers. Making this knowledge available electronically on the Web will make it sharable, exchangeable, accessible, and available all the time to these users (Figure 5-3).

A *database management system* is the core of information and knowledge management. This technology can be used in different applications:

- Building a national agriculture research information system (NARIS) needs to include research outcomes, projects, institutions and researchers in every country, and a regional research information system that works as a portal for all the NARIS. An example NARIS has been developed at the agriculture research center in Egypt (ARC, 2007).
- Managing global market information, analyzing this information, making local market information available on the Internet, assuring product quality control and providing product traceability will help any country gearing toward export-led growth economy.
- Developing an information system of indigenous agricultural practices can enable researchers to examine this knowledge and decide on its usefulness for sustainable development. Such a system will also keep this knowledge for future generations before it disappears as a result of advanced technologies.
- Developing an information system recording matured technologies that on a trial basis have proven successful and success stories that have achieved economic growth will strengthen the interaction between inventors and innovators. This will lead to an innovation-driven economic growth paradigm.

Multimedia information systems are needed to store and retrieve images confirming the occurrence of certain disorders,

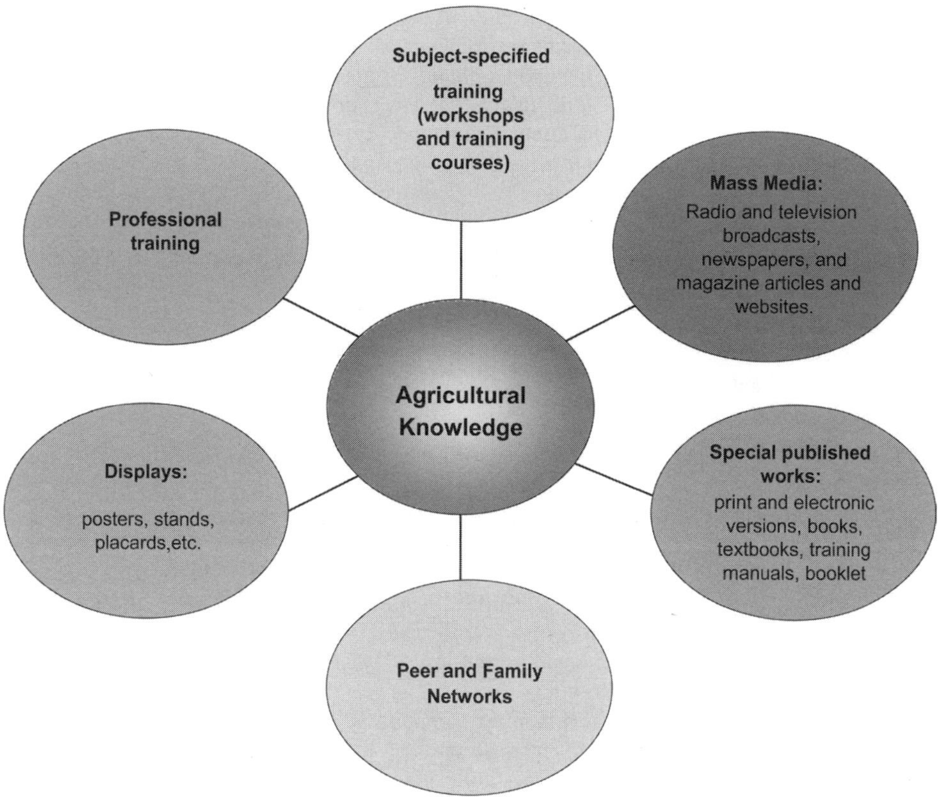

Figure 5-3. *Channels for dissemination of agricultural knowledge and information.* Source: Authors' elaboration.

and videotapes and audiotapes describing how to perform agricultural operations.

Geographic information systems (GIS) are needed to store databases about natural resources with a graphical user interface that enables users to access these data easily using geographical maps.

Decision support system techniques are needed in many applications:

- Simulating and modeling methods can be used to build computer systems that can model and simulate the effect of different agricultural production policies on the economy and the environment to help top management make decisions.
- Using expert systems technology to improve crop management and track its effect on conserving natural resources is elaborated in Rafea (1999). This technology may also be appropriate for keeping indigenous knowledge (Rafea, 1995, 1998, 2000).

Modern ICT—Internet and Web technology—is needed to make these systems available regionally and globally. Accessing the Internet will bring a wealth of information to all agriculture stakeholders in rural and urban areas and will help in overcoming the digital divide.

As most farmers in CWANA have no hands-on experience or access to digital networks, leaders of national agricultural research and extension systems should be encouraged to consider the ICT option. Training farmers and extension workers, including women, in ICT will help them access a lot of useful information if each country tries to develop contents in the language people are using (Box 5-2).

5.2.4 Policy adaptations required to realize options

Many of the options related to technological advances, capacity development and knowledge management, which may facilitate overcoming the challenges AKST faces in CWANA, can only be realized and yield impact if policies are adapted accordingly. Policy adaptations may be required at different levels and in various domains.

5.2.4.1 Land-use and land-tenure rights

Land-tenure legislation must guarantee long-term land-use rights to owners and leaseholders if land users are expected to invest in enhancing the productivity and long-term conservation of land. Land-use planning, zoning rules, and management of common lands require participatory approaches to consider the often-conflicting interests of different stakeholders. Since land tenure and use regulations need to be tailored to local conditions, decentralization may foster community or watershed level approaches that integrate the roles of local authorities and national administrations. Specific adaptations to land-use and land-tenure policies in CWANA might include the following:

- Reduce local government interference in land privatization and land use.

Box 5-2. Using ICT in the CWANA Region

In 1987, officials at the Egyptian Ministry of Agriculture and land reclamation recognized expert systems as an appropriate technology for speeding development in the agricultural sector. To realize this technology, in 1989, the ministry initiated the Expert Systems for Improved Crop Management Project (ESICM) in conjunction with the Food and Agriculture Organization of the United Nations (FAO) and the United Nations Development Programme (UNDP). The project began in mid-1989 and the Central Laboratory for Agricultural Expert Systems (CLAES) joined the Agricultural Research Center (ARC) in 1991. Through the development, implementation and evaluation of knowledge-based decision support systems, CLAES is helping farmers throughout Egypt optimize the use of resources and maximize food production. A dozen expert systems have been developed for horticulture and field crop management.

In 2000, the Virtual Extension and Research Communication Network (VERCON) project was funded by the FAO Technical Cooperation Program (TCP) to develop a Web-based information system to strengthen the link between research and extension (CLAES, 2002; FAO, 2003). Several expert systems have been made available on this network in addition to other modules. Two expert systems on the diagnosis of the sheep and goat (CLAES, 2006a) and bovines (CLAES, 2006b) are available on the CLAES Web site.

In collaboration with ICARDA, CLAES has developed three regional expert systems for wheat (CLAES, 2006c), faba (CLAES, 2006d) and barley (ICARDA, 2006).

CLAES also developed the National Agricultural Research Management Information System (NARIMS) through a project funded by FAO/TCP. This system has five modules: Institutes Information System, Researchers Information Systems, Projects Information Systems, Publication Information System, and National Research Program Information System (CLAES, 2007). The Association for Agricultural Research Institutes in the Near East and North Africa (AARINENA) plans to implement this system in all member countries (AARINENA, 2004).

- Review and amend land legislation to ensure that it unambiguously defines suitable land-ownership, use and inheritance rights, and the conditions under which land can be expropriated.
- Strengthen the judicial agencies responsible for land ownership, to ensure that they are independent, transparent and accessible and that they provide adequate protection for land users.
- Develop a legal and institutional framework that ensures that land users and owners have clear, secure rights to use, own and transfer property and that defines and supports the state's role as ultimate land custodian. Future procedures and administrative structures should be low cost, accessible by all, transparent and conducive to the efficient operation of land markets and secured credit transactions.
- Develop a system rights to land ownership, land use and land transfer that ensures that producers have full incentives to increase agricultural production and to use their land in a sustainable manner.
- Ensure that these rights are fully transferred to all producers through the issue of land-use titles, and that producers have the right to choose whether they operate as individual farmers or as collectives, formed according to their preferred means of association (family, village, etc.).
- Develop an active market for selling and leasing land and land-use rights.

5.2.4.2 Integrated water resources management

The coordinated development and management of water and related resources (IWRM) depends to a great extent on developing and implementing appropriate and coherent policies. Pertinent policies should be elaborated with the participation of all stakeholders in IWRM to allow consideration of water demands in sectors other than agriculture. The basic principles of IWRM could be applied through the following:

- Promoting transparent decision making, decentralized governance and a participatory approach to water operations under the principle of subsidiarity (at lowest competent authority level).
- Promoting managerial, financial and institutional innovations at all levels including new models of cooperation among the various stakeholders and the introduction of water pricing and water rights to encourage rational and efficient allocation of water, discourage waste, enhance water quality and ensure adequate water services.
- Reconciling the competing objectives of countries and sectors (power operation, flood control, irrigation, industrial and domestic supply, and environment), decreasing conflicts in water use, and supporting regional cooperation and information exchange.
- Fostering demand-responsive versus supply-oriented approaches. This requires that water users and consumers be engaged in selecting, financing, implementing and managing water services that meet their demands and willingness to pay.
- Promulgating policies regarding water-resource management that embrace water-demand management, development of currently untapped water sources, water quality conservation and transboundary collaboration.
- Managing water demand, which may include policies to improve the efficiency of water use in agriculture (e.g., regulations regarding the use of efficient irrigation systems); financial and economic measures such as rational water-pricing options (possibly considering special arrangements for the poor) or the use of incentives and disincentives; and virtual water trade, bearing in mind food security and sociopolitical aspects. Raising public awareness about rational water use and consumption patterns is prerequisite for the implementation and success of such policies.
- Developing currently untapped water sources, concentrating on improving sustainable delivery of surface wa-

ter by adequate investments and projects, and on using unconventional water resources. Measures for managing economic demand may help finance investments and incentives required to promote innovation and reduce risks related to the development of pertinent technologies and projects.

- Developing or adapting and then enforcing policies related to water quality. This might require investments in monitoring infrastructure and capacity development.
- Collaborating across boundaries to address common problems and appropriate strategies to reduce water shortages through improving the management of water resources. This might require amendments to judicial systems to deal with water disputes and conflicts.

5.2.4.3 Management of genetic resources and biodiversity

Principal policy instruments for conserving and managing biodiversity are land-use planning and zoning, such as establishing protected areas. Modern technologies such as GIS may greatly facilitate adequate land-use planning. Considering international incentives such as the Kyoto Flex-Mex mechanisms (e.g., the Clean Development Mechanism) may encourage land-use planning that conserves biodiversity. Protection of biodiversity may be linked with adapted land-use opportunities to find solutions acceptable to various stakeholders. Practices in ecoagriculture such as agroforestry, compensation areas and biodiversity-enhancing landscape elements may considerably contribute to biodiversity conservation. Direct payments for functions to conserve and maintain the ecosystem such as biodiversity conservation may encourage farmers to adopt such practices; in some industrialized countries (e.g., Switzerland) direct payments are linked to ecologically friendly land use and management. However, adoption of practices conserving biodiversity is generally dependent on secure land tenure or use rights (e.g., for improved rangeland management. Implementing NBSAPs developed through CBD may facilitate biodiversity conservation as well as make use of this treasure in CWANA. With regard to aquatic resources, the FAO Code of Conduct for Responsible Fisheries serves as an important tool for conserving aquatic biodiversity.

Invasive alien species are a considerable threat to biodiversity and can disturb both agricultural and natural systems devastatingly. Besides policies regulating the importation of living plants and plant material, legislation has to cover aspects such as responsible aquaculture, trafficking of unprocessed wood and use of ballast water.

Furthermore, appropriate biosafety regulations need to be included in country-led sustainable development strategies to face the potentials and challenges related to biotechnology. Policy analysis and development should consider risk assessment, capacity building in research and regulatory systems, and communication and public outreach. Policies should guide research for the poor (e.g., by protecting their intellectual property rights), protect against potential health risks, address possible ecological risks, and regulate the private sector (Pinstrup-Andersen, 1999).

5.2.4.4 Markets and trade

As the markets in which agricultural products compete are changing rapidly, measures to increase output must be accompanied by measures that improve the ability to compete in these markets. The objective should be to add more value rather than produce more, by providing appropriate framework conditions for reorienting and improving production and processing. Only if market organization in CWANA countries is improved will stakeholders in agricultural value chains fully capitalize on increased agricultural production.

- Adequate *input and output pricing* policies are key for enhancing agricultural production while conserving the natural environment. Price stability is extremely important so that farmers can invest and innovate rather than be defensively risk averse. Using targeted subsidies and direct payments (e.g., providing environmental services through agriculture) in the framework of coherent market policies may still be envisaged to promote innovation and more market-oriented production.
- Producers, processors and traders need access to *credits, markets* (to close the gap between rural areas and urban centers) and reliable *market information*, particularly in view of more diversified and market-oriented production. This may enable them to identify and introduce a portfolio of agricultural products that corresponds to consumer demand in major domestic and export markets. Developed public market information services can strengthen the position of various stakeholders in the market chain by providing regular information daily or weekly by newsletter, radio, television or mobile phone on product prices in major regional markets. Adaptation to information policies may be required to let value-chain stakeholders capitalize on relevant information available.
- Appropriate *technologies and infrastructure* are required for well-functioning value chains. Processing facilities at different levels may substantially reduce postharvest losses, and together with the development of agribusiness provide additional income along the value chain, particularly if diversified production (with more focus on nonstaples) targets newly emerging market opportunities (organic products, supermarkets, etc.). Abandoning state interference and policies encouraging investments by the private sector will encourage a shift toward market-oriented agriculture. Strengthened links between research, extension and farmers, possibly by including the private sector, may help implement the required technologies and infrastructure.
- Introducing modern, low-cost farm management systems to improve yields and product quality requires adequate policies. Regulations and procedures associated with seed testing and certification may have to be modernized, and restrictions on the import and use of high-performing seed varieties from other countries relaxed. Vertical integration and professional value-chain management facilitate quality and safety management at the relevant levels and allow complying with newly emerging standards. Investments in infrastructure and pertinent legislation (e.g., appropriate food safety and biosafety procedures and regulation, revision and modernization of product standards) may be required to improve postharvest management and assure quality control to comply with international standards.
- Import and export policies and trade arrangements have to provide an enabling framework for well-functioning

domestic markets. Coherent policies require an integration of AKST and agricultural production in national development strategies and plans to reconcile conflicting views and ambitions with regard to national goals such as national security, food sovereignty (virtual water trade, etc.), economic growth and development, and quality of life. To strengthen the position of CWANA producers and agrofood businesses in international, globalized markets the negotiating capacity in trade talks may have to be further developed since proactive engagement in trade negotiations and active participation in international programs and initiatives will be required. Reducing the costs and delays associated with border transit procedures and intensifying current efforts to create a low-cost, green corridor that gives improved access to neighboring markets (trade agreements with regional trading partners) may represent further options for making CWANA markets more efficient. Principles of good governance such as representation, transparency, accountability and civil society participation may ensure that social and environmental concerns will be better represented in negotiation processes and resulting agreements.

5.2.4.5 Risk management and property rights

Besides risk-reducing approaches at production level, such as investments in supplemental irrigation facilities or diversification of production, policies have to provide a framework that promotes innovation by reducing associated risks. Well-functioning savings and credit schemes and the development of insurance programs for farmers will encourage farmers to make necessary investments and implement innovative technologies and approaches. Proactive risk management strategies and policies will have to replace the currently prevailing reactive disaster management. This is particularly important in view of the increasing threat of more frequent extreme events caused by climatic change such as droughts, storms and floods, and possible abrupt changes in globalized markets. Improved social safety nets and compensatory policies may also help safeguard the disadvantaged against likely negative effects of structural adjustments and reforms.

Although intellectual property rights are intended to stimulate innovation, enhancing investment in research and access to potentially useful technologies from abroad, they are based on a paradigm of market-led development that contrasts with the traditional approach in agricultural research, which focused on sharing ideas and producing public goods. Currently the issue of IPR is particularly discussed in the domain of plant breeding and biotechnology. The IPR regulatory environment needs to be reshaped to facilitate the generation, dissemination, access to and use of AKST. IPR regulations will have to balance private and communal rights while considering national interests and benefits for local communities. Benefits based on local and traditional knowledge will particularly have to be protected and shared in an equitable manner.

5.2.4.6 Institutional reform and role of government

In CWANA countries where private sector institutions are weak and reform still has far to go, the state retains major responsibilities. Important domains on which government should focus its resources and activities include policy formulation, guidance on legislation and regulation, and provision of essential public services in the areas of seed and plant protection, animal health, border control, food safety, and product standards and certification. Preparing a coherent medium-term sector strategy can form the basis not only for policy formulation but also for ministry input into budget preparation, public investment planning, and specific policies and legislation relating to land use and land reform, trade, taxation, market activity and competition, rural finance, research and extension.

Production targets should be discontinued as a policy instrument and replaced with growth in value addition, household income and export revenue. The efficiency with which essential public services are provided leaves room for improvement in many CWANA countries, and comprehensive modernization of current regulatory practices (including product standards) may be required. There are great opportunities to transfer activities such as seed and livestock breeding to the private sector, and it seems advantageous to allocate more resources to providing information and support to producers and agribusinesses on land privatization and market activity.

The ability of public institutions to reorient their activities is currently constrained by limited awareness of what is appropriate in a market economy, a reluctance to change old approaches, and a limited allocation of human and financial resources for formulating policy. Weak budgetary resources and a significant need to retrain staff and boost output are further constraints.

5.3 Implications of Various Options and Possible Mitigation Measures

Although options to achieve SDGs presented in the preceding sections will contribute to reducing hunger and improving socioeconomic conditions in CWANA, certain options to increase agricultural productivity may have negative environmental effects or be associated with social or economic drawbacks.

The increasing use of pesticides and the related pollution of the ecosystems is a big concern. The trend toward genetically uniform crops increases the potential for serious disaster by eliminating the many different strains of a given crop that farmers previously used. But government policies perpetuate conventional agriculture and discourage farming practices that could make agriculture more sustainable (Chrispeels and Sadava, 2003).

Many modern agricultural practices in the CWANA region are not environmentally sustainable as they have negative aspects (e.g., habitat conversion due to agriculture, soil erosion, and pollution from chemical pesticides and fertilizers). Modern agricultural chemicals have largely contributed to increased crop production, but they also have negative side effects such as groundwater pollution, interference in terrestrial and marine biodiversity, and health hazards to producers and consumers. Therefore, other technical options such as biological control, integrated pest management, integrated crop management, good agricultural practices and organic farming provide great opportunities for making agricultural production more sustainable (Clay, 2004).

New technologies need to be developed and implemented for using and managing limited water resources. These technologies will have to focus, on one side, on quantitative aspects, like increasing water supply and decreasing water demand; on the other hand, conserving water quality will have to receive more attention in the future. AKST in this regard will not only have to concentrate on protecting water resources against pollution from agricultural activities as well as from other sources; it will also have to explore ways to use water of lower quality in agriculture, and to better match the quality of water supplied to its specific use, considering that water of varied quality will be allocated among sectors. Progress in AKST is especially important. Agriculture is by far the biggest consumer of freshwater resources in the region, and progress in AKST will free up water for other sectors, including the environment, which will progressively need more good-quality water.

Measures to balance the effects on natural resources of options fostering agricultural production will have to include public awareness, public education and sufficient regulation. Public awareness and education in this regard should include training farmers in integrated pest management and organic farming practices to reduce the use of chemical pesticides and insecticides, and to improve their knowledge about what fertilizers and nutrients different crops require. Regulations are also required to protect public health and protect natural resources, including soils and water, from degradation.

In the following section, negative externalities related to changes in agricultural production (particularly relating to intensification of production) will be discussed and possible mitigation options presented.

5.3.1 Intensification of crop production

5.3.1.1 Use of agrochemicals

As water is the most restricting factor for agricultural development in the dry CWANA region, emphasis will be placed on using water more efficiently and increasing production per unit of water applied. This will result in more intensive agriculture and will increase the use of agricultural fertilizers and pesticides.

Extending the use of *chemical and organic fertilizers* will result in increasing concentrations of different ions and cations in the soil. This might result in increasing soil salinity, particularly where irrigation water additionally adds minerals to soils, if leaching of salts is insufficient as observed in greenhouse production in the Middle East. Leaching nutrients, on the other hand, may negatively affect water quality with possible effects on human and animal health and eutrophication of water sources. Fertilization according to soil fertility and crop requirements, based on regular soil fertility assessment as well as incremental fertilization and the use of slow-release fertilizers may mitigate such problems. Precision agriculture using both modern tools (such as GIS) and simple techniques (such as fertilizing according to leaf color) may greatly support adequate fertilization (Bahu and Gulati, 2005).

The extensive use of *crop protection products* such as pesticides and herbicides may result in increasing the content of such substances and their nonbiodegradable derivates in soils and water and ultimately in agricultural products, which may compromise food quality and safety. Therefore, the use of crop protection products should be limited wherever possible. Adequate crop rotations and ecological compensation areas may reduce pest, disease and weed pressure. Pesticides and other crop-protection agents should be used according to monitoring and thresholds; IPM and ICM strategies and technologies are available that allow for a minimal, targeted, efficient and still-effective use of crop protection products, although they may have to be adapted to specific local conditions. Organic agriculture, avoiding the use of chemical crop protection agents, not only reduces such risks to a minimum but also has the potential to target growing markets in CWANA and other, mainly industrialized, regions.

Both excessive fertilization and use of crop protection agents may negatively affect biodiversity. Whereas high nutrient loads mainly affect species diversity, organometallic compounds and other chemicals may also affect genetic diversity (Vogt et al., 2007). Since the loss of genetic variation is more difficult to notice than that of species variation, it is important to understand the effects that different pollutants have on ecosystems and on species and their genes.

Monitoring environmental indicators and parameters relating to possible pollution of natural resources by organic fertilizers and agrochemicals will be important for maintaining a healthy resource base. Food quality and safety monitoring and control in accordance with pertinent legislation and regulations that may need to be developed will be necessary to prevent health problems and to comply with international standards.

5.3.1.2 Mechanization

Extensive use of mechanization may result in losing soil organic matter and thus soil fertility. It can also reduce macropore spaces in soils and thus decrease soil aeration. Mechanization may compact the soil, and tillage hardpan may form, which will reduce the permeability of subsoils for water and roots. Furthermore, frequent and thorough working of the soil may negatively affect biodiversity. Conservation tillage options and adequate crop rotations may overcome such negative effects of mechanization.

Promoting mechanization may also have socioeconomic consequences. Depending on machinery and maintaining it with fuel and spare parts may increase the risk of debt, particularly if framework conditions promote overmechanization. Extension services therefore will need to support farmers in assessing the cost efficiency of investment in mechanization and relate it to their capacities, and savings and credit systems need to allow for pertinent investments. Since machinery is mostly designed for use by men, women might be left with an increased burden of tedious nonmechanized work. To counteract such increased gender imbalances, mechanization will have to consider the feminization of agriculture and design implements that are suitable for women.

5.3.1.3 Reduction in diversity

Intensification of agricultural production in recent decades has generally been associated with simplification of agricultural systems (Haykazyan and Pretty, 2006). In addition to the negative effects of mechanization, biodiversity loss and a declining diversity in agriculture itself (Haykazyan and Pretty, 2006) have resulted from pesticide and fertilizer use, expansion of cropped surfaces, land degradation and the use of a narrow range of high-yielding crop varieties based on conventional breeding or biotechnology. To counteract these trends, appropriate legislation is required that fosters ICM practices and promotes the provision of ecosystem services through agriculture.

However, awareness about the importance of biodiversity is generally low in CWANA, particularly among farmers. Initiatives to conserve biodiversity have so far mainly been successful only if tangible benefits were perceived. It is therefore necessary to raise awareness about the benefits of conserving biodiversity such as reduced pest and disease pressure in diversified systems and resilience of diverse agricultural systems and landscapes to shocks and stresses. Emerging markets for biodiversity-based products may offer economic incentives to land users. Participatory approaches and shared management, possibly including payments or other incentives for ecosystem services, may represent acceptable approaches to encourage the adoption of conservation practices. In situ and ex situ conservation of biodiversity, participatory decentralized breeding approaches, and integration of local, traditional knowledge and experiences in AKST may further contribute to conserving biodiversity in intensified production systems.

5.3.2 Intensification of livestock production

Interaction between livestock and the environment has been a subject of global debate in recent years, focusing on negative aspects such as global warming with little attention being paid to the positive attributes. It is not disputed that livestock contribute to greenhouse gas and other atmospheric emissions, which contribute to climate change; it has been reported that livestock account for 35-40% of all anthropogenic emissions of methane (Steinfeld et al., 2006). Livestock also contribute to carbon dioxide emissions through basic metabolism and respiration. These negative factors of livestock rearing may be reduced considerably, and there are also many positive elements of livestock–environment interactions that should not be overlooked.

5.3.2.1 Zoonoses and other diseases

The perpetuation and transmission of a group of diseases shared between humans and other vertebrates that act as reservoirs of infection is an important hazard for human health that arises from livestock rearing. These diseases are popularly known as zoonoses. Some like trichinosis and salmonellosis are exclusively or partially food-borne through meat, milk or eggs (Majok and Schwabe, 1996; Payne and Wilson, 1999). A second group that is emerging is thought to be perpetuated through human–animal interactions, such as avian influenza (bird flu). The current challenge is that there are no known technologies or management practices acceptable to farming communities to deal with these groups of infections. Zoonoses are among the most complicated human diseases from an epidemiological standpoint and therefore controlling them requires an extremely good understanding. This knowledge first has to be developed in most of the CWANA region. Important methods for dealing with this group of infections are to promote surveillance and reporting by collaborating with medical and public health experts, and to conduct education and awareness campaigns about how people are exposed through animals and their products and the risks of such infections. Good hygiene practices in processing foods of animal origin are further safeguards for insuring food safety for human consumption. These practices will have to be improved as technologies become more advanced with the acquisition of new scientific knowledge.

5.3.2.2 Residues

Health risks also occur as a result of livestock management practices. Currently, human hazards from chemical residues in foods of animal origin may not be an issue of major proportion in CWANA because the costs of such products are too expensive for most resource-poor small-scale farmers. Mineral supplements such as sulfur and phosphorus, which ruminants must have to utilize nitrogen, as yet have little if any market in CWANA. The use of anabolic steroids and other compounds used as feed additives for cattle to promote faster growth, as well as elevation of natural levels of somatotropins in cattle, pigs, poultry and sheep are not yet widespread. However, such hazards may become of major concern in the near future as chemicals, medicines and agricultural by-products for feeding livestock become less expensive and more widely used (Smith et al., 2005), and as awareness rises about health risks with foods of animal origin. The challenge, therefore, is to design and operate organized multilevel systems for detecting and assessing the environmental hazards and monitoring environmental quality (Schwabe, 1984). Possible technological approaches for consideration involve the following:

- Evolve appropriate mechanisms for recognizing and detecting new hazards through their effects upon animals.
- Develop protocols for animal testing of potential hazardous substances.
- Use specific health indicators to monitor the environment.

Assays for toxicity and safety of manufactured chemical substances of diverse nature—drugs, food additives, pesticides, and other agricultural and industrial chemicals—that are currently used to assess human risk will probably have to be complemented by a range of tests with appropriate animal species (Newberne, 1980; Squire, 1981) to detect not only acute toxicities but also mutagenic, carcinogenic, teratogenic or other chronic effects.

Dealing with issues of residues of growth hormones, antibiotics, feed additives, heavy metals, etc., in livestock products is, however, often problematic because farmers want their animals to grow fast to reach market weight quickly. Nevertheless, health risks due to residues are a concern and have therefore to be considered.

5.3.2.3 Manure and waste management

Besides the close interaction between livestock and humans and the consumption of animal products, direct human health effects of pollution from rearing livestock pose an additional threat to human health and the environment, particularly in intensified production systems. High levels of nitrates in water may lead to disease such as the blue baby syndrome (Pretty and Conway, 1988). More importantly, manure, through contaminated water or fresh produce, can carry a range of serious human pathogens, with high incidence of morbidity and mortality, particularly in babies and young children. These pathogens, often asymptomatic in livestock, vary from bacterial pathogens such as *E. coli, Campylobacter, Salmonella,* and *Leptospira* spp., and protozoan agents (*Cryptosporidium*) to viruses such as the hepatitis A virus (World Bank, 2005).

The potential threats of pollution to the environment and human health will have to be watched cautiously. Removing policy distortions that promote artificial economies of scale in livestock production, developing approaches to let resource-poor producers capitalize from the benefits of increased livestock production, and regulating environmental and public health concerns will represent important challenges for CWANA decision makers. Technological options for improved handling and storage of animal waste exist and may be promoted by adding value to these wastes by using them in biogas digesters for energy production and as fertilizers. Areawide integration of specialized crop and livestock production may help to reinstall the link between livestock and crop production, not only on the farm but on a regional scale. By fostering collaboration between specialized livestock operations and crop farmers, animal waste may be recycled in an environmentally and economically beneficial way (Menzi et al., 2003).

5.3.3 Fisheries and aquaculture

About 200 types of diseases are known that can be transferred from foods to humans (FAO, 2000). Fish products can be a source of disease due to general food habits, rate of consumption, type of product and species of fish. Therefore, fish producers have established and applied a system called Hazard Analysis & Critical Control Points to eliminate or reduce the adverse health effects of fish products.

Several organic substances such as dioxin and inorganic substances such as cadmium and mercury can affect fish quality. Their deleterious effects increase if they exceed the maximum allowable limits and if they occur in closed seas and rivers (FAO, 2004). Even if these substances occur in food fishes in low quantity, their incidence may be of concern for people who eat fish daily and for pregnant women, infants and children who eat large quantities of fatty fishes.

The safety of aquaculture products is important as production from aquaculture has increased and become available to more people. Antibiotics are frequently used in aquaculture to prevent or treat diseases. Therefore, responsible limiting of the use of antibiotics is important for sustainability aquaculture and for safety of fish products. Responsible use implies determining the maximum residue level of these antibiotics and ensuring that these levels are not exceeded.

Food safety in fisheries is important and has been endorsed by international agreements. The FAO Code of Conduct for Responsible Fisheries clearly requests countries to develop their fisheries in a way that does not result in environmental degradation or health problems for people.

5.3.4 Water management

Changing water use for agriculture will entail implications beyond the agricultural sector. Consideration of externalities through application of IWRM may avoid negative effects on the environment and society.

5.3.4.1 Conflict over water resources

As a result of increasing water demands for agriculture, depletion of water sources and conflicts over them are expected to increase in CWANA. These conflicts may arise at various levels such as among sectors using water or among different user groups. Regional conflicts between countries over shared water resources will probably intensify; conflicts over surface and groundwater sources may accelerate and add to existing tensions and conflicts in the region. Possible conflict areas might include the Tigris, Euphrates, Jordan, Indus and Nile river basins. Conflicts over groundwater sources might include the countries of North Africa and the Middle East. Mitigation of such conflict potential may include demand management in different facets, bi- and multilateral negotiations and agreements, or an increase in virtual water trade—e.g., by producing and exporting high-value crops with low water requirements and importing water-intensive crops from a subsidized world market (Allan, 2002). Negotiations should respect the Helsinki rules and guidelines (International Law Association, 1967) and foster their implementation to avoid conflict; regional parties need to cooperate to formulate regional solutions for water shortages.

Nearly all water resources in the region are being used. Therefore, water shortages are expected to result in more pressure on the agricultural sector to divert water from agriculture to other uses such as industrial and domestic sectors. This will result in conflict among sectors and internal socioeconomic and political tensions. Each country of the region will need to address these conflicts specifically, but participation of all concerned stakeholders in pertinent discussions and negotiations will greatly facilitate solutions that allow for optimized economic and social welfare derived from water resources, their equitable allocation and their environmental sustainability. The systems of water rights and water allocation will have to be adapted in all countries of the region, addressing water allocation among the sectors and respecting historical water rights of the different users. The promotion of water-user groups (or water-user associations) who jointly manage and organize water distribution may improve the efficiency of water use and the distribution and management of water resources while at the same time empowering local populations and relieving public institutions. If all concerned stakeholders can participate in such associations internal conflicts over water distribution and water allocation may be mitigated.

Improving on-farm water management and the efficiency of water distribution can reduce return flows and possibly reduce recharge of certain groundwater aquifers. Since existing resources are fully used in most countries of the region, reducing return flows and improving the efficiency of water use at the upstream end of any river basin might result in reducing water availability for downstream users, thereby increasing conflicts over water resources. For example, when surface water systems are replaced by pressurized irrigation systems or if surface irrigation efficiency is improved (e.g., using surge irrigation), tail water runoff will reduce. Upstream users will increase their irrigated areas as a result of water savings, but downstream users might be adversely affected because return flows to them are reduced. Again, following the IWRM principles, having all concerned stakeholders participate in planning and implementing significant changes in water management may mitigate such conflict potential.

5.3.4.2 Depletion and development of water resources

Another effect of improved irrigation efficiency, particularly on highly permeable soils, is reduced seepage to unconfined aquifers, which may reduce the safe yield of such aquifers and possibly decrease the amount of water available from them. Together with increasing agricultural and domestic demands this situation may additionally deplete renewable and nonrenewable water sources. Many countries in the region—Saudi Arabia, Jordan and Libya—have been using nonrenewable sources. It is expected that these sources will be depleted in the future, and new water sources will have to be found. In other areas, the renewable water sources have been depleted beyond their safe yield capacities and thus their water quality has been deteriorating. An example is the Gaza Strip, where groundwater resources have been used beyond their natural recharge capacity. Seawater intrusion and intrusion from brackish groundwater aquifers have now deteriorated these resources. Besides enhancing natural recharge, such as through appropriate land management, artificial recharge of groundwater aquifers may reduce problems associated with decreased groundwater availability.

The development of new water resources, however, may entail deleterious side effects. Creating new (particularly large) reservoirs may not only flood fertile valley bottoms but dislocate the local population and destroy property, habitats and cultural heritage. Having local communities participate in decision making, establishing smaller-size structures or reducing demand may avoid the necessity for large dams and reservoirs.

5.3.4.3 Use of unconventional water

As water resources are limited in the region, the use of marginal water such as brackish and treated wastewater will increase. However, the use of unconventional water resources may be associated with certain problems. Using treated wastewater in agriculture might entail health hazards and create water-quality problems that will have to be addressed. Contaminating crops with harmful microorganisms such as *Salmonella* in lettuce and onion or *E. coli* in sprouted seed are potential risks associated with using wastewater for irrigation (Jones et al., 2006), and nematodes and pathogens in soils occur more frequently. Using marginal water such as drainage, saline or brackish water, and wastewater may also affect soil and water quality negatively. Accumulation of salts, heavy metals and other substances in soils and water will have to be prevented by establishing and enforcing pertinent legislation and control. Regulations regarding wastewater treatment and reuse will particularly have to cover the responsibility of water polluters in treating their wastewater to a standard acceptable for safe use, as in agriculture, or for disposal in the environment. Increased awareness among farmers, water users and crop consumers will be required to address issues related to health and water-quality aspects.

5.4 Uncertainties

The preceding sections of this chapter have demonstrated that there is a whole range of technological, institutional and policy options through which AKST can contribute toward achieving SDGs. If appropriate countermeasures and precautions are considered, even possibly associated negative implications of these options may be dealt with and mitigated.

The future, however, bears uncertainties related to environmental framework conditions. Important changes and developments that are difficult or even impossible to foresee may affect agriculture and the role and effect of AKST considerably. Uncertainties are arising in various domains such as the geopolitical situation, global markets and trade (international trade regimes for agricultural inputs and products), supply and demand for agricultural products (e.g., biofuel vs. food and related effects on prices and the environment), price developments for inputs (e.g., energy prices) and outputs, climatic changes and unstable weather patterns (with their effect on resource quality and availability), the ability to tackle human, animal and plant diseases, and acceptance of genetically modified foods.

5.4.1 Global markets and trade

Weather-related production shocks, energy price trends, investment in biofuel capacity, economic growth prospects and future agricultural policy developments are among the main uncertainties affecting the prospects for world agricultural markets (OECD-FAO, 2006). A major uncertainty is the outcome of the Doha Development Agenda of multilateral trade negotiations. If trade barriers and support for agricultural production are substantially lowered, world prices for a number of agricultural commodities as well as trade may rise considerably. Outside the Doha negotiations, however, bilateral or regional free trade agreements may increase trade in agricultural products between members.

Increased trade opportunities coupled with higher product prices may change the focus of agricultural production and related AKST toward more export-oriented strategies. Whereas producers might benefit from such developments, poor consumers in urban areas particularly might suffer from higher food prices. Emergence of new markets for biofuels, carbon trading and biodiversity preservation also open new opportunities yet to be tapped (World Bank, 2007).

Domestic policy changes in important producer and export countries such as the United States represent further uncertainties. The prospects for world agricultural markets are highly dependent on economic developments in Brazil, China and India, three of the world's agricultural giants. Outbreaks of animal diseases such as BSE or avian influenza may greatly influence demand and have significant consequences for producers. Shifts in demand from an affected commodity to another may occur briskly, and markets of affected countries may close up. Animal diseases may thus cause major disruptions in the meat sector, which will be further transmitted to feed markets (OECD-FAO, 2006).

5.4.2 Energy prices

Higher energy prices, as for crude oil, directly impinge on agricultural production costs. Energy is used directly to operate machinery, and indirectly through such inputs as fertilizers and pesticides, the production of which is particularly energy demanding (World Bank, 2007). Increasing energy prices would thus raise production costs, which would be translated into higher commodity prices both regionally and internationally. As the share of energy in production costs is substantially higher for crops than for livestock, crop production is particularly affected by changing energy prices (OECD-FAO, 2006). However, since intensive livestock production is strongly based on cereals and oilseed meal, livestock products will be affected as well, although to a lesser extent. Higher energy prices are therefore expected to reduce trade volumes of most commodities, particularly crops, all the more so because transport cost will also increase. On the other hand, a further increase in crude oil prices may promote a shift towards bioenergy production.

5.4.3 Bioenergy

Developments in bioenergy production represent a major uncertainty for agricultural production and markets, and for achieving SDGs in general. High energy prices combined with increased biofuel production from food crops could lead to large increases in food crop prices by affecting both supply and demand (World Bank, 2007). Commodity prices for crops such as maize, wheat, oilseed and sugar may rise drastically (by 30-75%) (World Bank, 2007) and competition between food and feed uses and nonfood uses for particular crop sectors may result in major production and market changes (OECD-FAO, 2006). Not only would bioenergy crops be affected; through cross-commodity influence, production and availability of traditional foods and feeds might decrease. Furthermore, increased bioenergy production might accelerate land conversion from forest to agricultural use or from extensive to intensive production, which may—together with the escalating demand for livestock products—considerably affect the environment negatively through deforestation and degradation of land and water resources.

It is important to note that the currently observed boom in producing bioenergy is mostly based on public support and encouragement (OECD-FAO, 2006). Such support may create market distortions that need to be better understood before pertinent policies are put in place. However, the economics of bioenergy, and particularly its positive and negative externalities, are not yet well understood and depend critically on local circumstances (Avato, 2007). These knowledge gaps related to increased bioenergy production call for investment in AKST research and development to produce more sustainable technologies that are adapted to smallholder farming systems. Research needs to develop second-generation biofuels that rely on agricultural and timber wastes instead of food crops, thus reducing the pressure on food crop prices and possibly contributing to the supply of more environmentally friendly supplies of biofuels (World Bank, 2007).

5.4.4 Climate change

Global warming is one of the areas of greatest uncertainty for agriculture (World Bank, 2007). So far, not all effects of climate change on agricultural production and yields have been included in crop–climate models. Nonlinearity of yield response to temperature above threshold levels can result in high losses with moderate temperature increases that are not yet considered. The combined effect of higher average temperatures plus variability of temperature and precipitation, more frequent and intense droughts and floods, and reduced availability of water for irrigation is likely to affect yields negatively, even globally, and can be devastating for agriculture in many tropical regions. Assumptions about the magnitude of the effect of carbon fertilization are still debated although they are critical for predicting whether crop yields will increase under elevated CO_2 concentrations. Climate change is also increasing production risks in many farming systems, reducing the ability of farmers and rural societies to manage risks on their own.

Uncertainty regarding what climatic changes to expect is even higher in view of increasing evidence that these changes are happening at a pace faster than that until recently foreseen (World Bank, 2007). Proactive strategies and research are therefore crucial to face these uncertainties.

5.4.5 Genetically modified organisms (GMOs)

Worldwide, many people are eating genetically modified foods with no adverse affects on human health having been reported in peer-reviewed scientific literature. However, there could still be long-term effects on human health that have not yet been detected (genetically modified foods have been available for fewer than ten years). Although many field trials have been held, and in some parts of the world there has been large-scale commercial planting of genetically modified crops, work done has been insufficient to fully assess environmental effects, especially in the biodiversity-rich tropics (OECD, 2000). Modern biotechnology has opened up new avenues and opportunities in a wide range of sectors, from agriculture to pharmaceutical production. Nevertheless, the scale of the global debate on GMOs is unprecedented. This debate, which is intensive and at times emotionally charged, has polarized scientists, food producers, consumers and public interest groups as well as governments and policy makers (FAO, 2001). Today, it is not clear to what extent the incredibly rapid expansion of genetically modified crop production and use in animal (particularly fish) production will continue, particularly in the developing world. Due to the intensity of the debate over GMOs, new discoveries may have massive effects, particularly on the demand side. In addition, neglected investment of GMO developers in traits and crops that will benefit the poor and weak regulatory capacity and systems fuel public distrust and ignite opposition of various interest groups to widespread GMO use (World Bank, 2007).

People in general are directly interested in technological developments, yet obstacles to their participating in making decisions must be acknowledged and overcome. The public has not been adequately informed about applying gene technology to food production and the potential consequences on consumer health or the environment. With the confusing array of claims, counterclaims, scientific disagreement and misrepresentation of research that is present in the media, the public is losing faith in scientists and government. Widely communicated, accurate and objective assessments of the benefits and risks associated with genetic technologies should involve all stakeholders. Experts have the ethical obligation to be proactive and to communicate in terms that the layperson can understand. More opportunities are needed that enable the exchange of information among scientists, corporate representatives, policy makers and the public at large. Including members of the public on advisory committees set up to formulate laws, regulations and policies would help ensure that their perspectives are fairly represented (FAO, 2001).

5.4.6 Investment in AKST

Investments in AKST have hugely accelerated growth and reduced poverty in much of the developing world. However, although agricultural productivity improvements have been closely linked to investments in AKST, market failures have led to serious underinvestment. Trade subsidies and national policies that reduce incentives to farmers in developing countries are a disincentive to public and private investment in AKST (World Bank, 2007).

Increasing public and private investment in AKST and strengthening institutions and partnerships with the private sector, farmers and civil society are now essential to bridge the knowledge divides, strengthen user demand for AKST, increase competitiveness, and ensure that the poor participate and benefit. These investments will be even more important in the future, with rapidly changing markets, growing resource scarcity, and greater uncertainty from multiple threats. Ways to increase investments in AKST exist, such as by forming coalitions of producer interests around particular commodities or value chains, to lobby for more public funding and for producers to cofinance AKST. In addition, institutional reforms will be needed to make investing in public AKST more attractive and to make funding more transparent and open to a wider range of research providers in universities, civil society and the private sector (World Bank, 2007).

5.5 Ways Forward

A whole range of technological, institutional and policy options exist to overcome the major challenges for attaining sustainable development goals in the CWANA region. Although the options presented can positively affect achieving sustainable development goals, some might at first sight be associated with negative externalities, particularly with regard to the environment and natural resources. However, provided that appropriate precautions and countermeasures are implemented, most of these implications can be mitigated.

Technological options alone generally cannot bring about the hoped-for changes. Framework conditions have to be favorable for technological achievements to be successfully implemented. Economic aspects, institutional arrangements, and political decisions and regulations have to form a coherent framework in which AKST and its achievements can flourish. Using natural resources, employing research, training and extension methods, and educating the public, making it aware and getting its participation are all required and must be balanced to achieve optimal results.

Many of the options presented are valid for most countries in the region. However, these options will have to be adapted to the specific environments targeted. Furthermore, the options will receive different priorities in the various CWANA countries. Each country will need to develop strategic plans to select and prioritize policies according to its local circumstances and needs.

References

AARINENA. 2004. Minutes of AARINENA ICT Steering Committee Meeting, Tehran, Iran. Available at http://www.aarinena.org/rais/monthly/2ndSCMhome.htm.

Abou-Hadid, A., K.H. Batanouny, A.S. Jabarine, and A.A. Kader. 2004. Proposal for expanding the crop mandate of ICARDA to include horticultural crops. ICARDA, Aleppo.

ACSAD/CAMRE/UNEP 2004. State of desertification in the Arab world. Updated study. ACSAD, Damascus.

Agarwal, A., and S. Narain (ed). 1997. Dying wisdom: Rise, fall, and potential of India's traditional water harvesting systems. Centre for Science and Environment, New Delhi.

ARC (Agriculture Research Center). 2007. National Agriculture Research Information Management System (NARIMS). Available at http://www.arc.sci.eg.

Aillery, M., and N. Gollehon. 2003. Irrigation water management. *In* R. Heimlich (ed) Agricultural resources and environmental indicators. 2003. Agric. Handbook No. AH722. USDA, Washington, DC.

ALAWUC. 2002. Report of the NGO/CSO Preparatory Meeting for the Second Session of the Agriculture, Land and Water Use Commission for the Near East (ALAWUC). ALAWUC/NE/02/REP. March 2002. FAO Regional Office for the Near East, Cairo.

Allan, J.A. 2002. Hydro-peace in the Middle East: Why no water wars? A case study of the Jordan River Basin. SAIS Rev. 22(2):255-272. Available at http://www.soas.ac.uk/waterissues/occasionalpapers/OCC45.pdf.

Allen, R.G., L.S. Pereira, D. Raes, and M. Smith. 1998. Crop evapotranspiration: Guidelines for computing crop water requirements. Irrig. Drain. Pap. 56. FAO, Rome.

Aquastat. 2006. AQUASTAT [Online]. Available at http://www.fao.org/ag/agl/aglw/aquastat/main/index.stm. FAO, Rome.

Avato, P.A. 2007. Bioenergy: An assessment. World Bank, Washington, DC.

Babu, S. 2006. Policy reforms and agricultural development in central Asia. Springer, New York.

Bahu, S., and A. Gulati. 2005. Economic reforms and food security: The impact of trade and technology in South Asia. Haworth Press, NY.

Baroudy, E., A.A. Lahlou, and B. Attia. 2005. Managing water demand: Policies, practices, and lessons from the Middle East and North Africa forums. IWA Publ., London and IDRC, Ottawa.

Bartley, D.M. 1998. Genetics and breeding in aquaculture: current status and trends. Pp. 13-30. *In* D.M. Bartley and B. Basurco (eds) Genetics and breeding of Mediterranean aquaculture species. Cah. Options Mediterr. 34:1022-1379.

BASIS/LEAF. 2004. The BASIS/LEAF integrated crop management course. Available at http://www.basis-reg.com/downloads/Syllabus%20ICM.PDF.

Baumer, M. 1990. Agroforestry and desertification. CTA, Wageningen.

Belaid, A., M. Solh, and A. Mazid. 2003. Setting agricultural research priorities for the Central and West Asia and North Africa region. Available at http://www.icarda.org/ARP_CWANA/ARPS_FINAL_REPORT1.htm. ICARDA, Syria.

Bennett, M.K. 1941. Wheat in national diets. Wheat Studies 18(2):37-76.

Bioversity. 2007. Conservation and use Biodiversity themes. Available at http://www.bioversity international.org/Themes/Conservation_and_Use/index.asp. Bioversity Int., Rome.

Blench, R.M. 1998. Rangeland degradation and socio-economic changes among the Bedu of Jordan: Results of the 1995 IFAD survey. Pp. 397-423. *In* V.R. Squires and A.E. Sidahmed (eds) Drylands: Sustainable use of rangelands in the twenty-first century. IFAD, Rome.

Blench, R.M. 2000. Extensive pastoral livestock systems: Issues and options for the future. FAO–Japan Cooperative Project GCP/JPN/005/JPN. Collection of information on animal production and health. FAO, Rome.

Blench, R.M. 2001. Pastoralism in the new millennium. FAO Anim.Health Product. Ser. 150. FAO, Rome.

Bonnis, G., and R. Steenblik. 1998. Water, agriculture and the environment. The OECD Observer No. 212. OECD, Paris.

Braun, R., and K. Ammann. 2002. Biodiversity: The impact of biotechnology. Encycl. Life Support Systems. EOLSS Publ., Oxford.

Breman, H., and J.J. Kessler. 1995. Woody plants in agro-ecosystems of semi-arid regions—with an emphasis on the Sahelian countries. Adv. Ser. Agric. Sci. 23. Springer, Berlin.

Brouwer, C., K. Prins, and M. Heibloem. 1989. Irrigation water management: Irrigation scheduling. Training Man. 4. FAO, Rome.

CA. 2007. Water for food, water for life: A comprehensive assessment of water management in agriculture. Earthscan, London and IWMI, Colombo.

Carney, D. 2002. Sustainable livelihoods approaches: Progress and possibilities for change. DFID, London.

CGIAR. 2006. Who we are. Available at http://www.cgiar.org/who/index.html. CGIAR, Washington, DC.

Chapagain, A.K., and A.Y. Hoekstra. 2003. Virtual water flows between nations in relation to trade in livestock and livestock products. Value of Water Res. Rep. Ser. 13. UNESCO-IHE (Institute for Water Education), Netherlands.

Cherrett, I. 1999. Soil conservation by stealth. Int. Agric. Dev. 19(3):13-15.

Chrispeels, M.J., and D.E. Sadava (ed). 2003. Plants, genes and crop biotechnology. 2nd ed. ISBN 0-7637-1586-7. Jones and Bartlett Publ., Boston.

CLAES. 2002. Egyptian implementation of the virtual extension and research communication network. Available at http://www.vercon.sci.eg. CLAES, Egypt.

CLAES. 2006a. Sheep and goat expert system. Available at http://www.vet.claes.sci.eg/sheep. CLAES, Egypt.

CLAES. 2006b. Bovine expert system. Available at http://www.vet.claes.sci.eg/bovine. CLAES, Egypt.

CLAES. 2006c. Wheat expert systems. Available at http://www.claes.sci.eg/wheates. CLAES, Egypt.

CLAES. 2006d. Vicia faba expert system. Available at http://fabaes.claes.sci.eg. CLAES, Egypt.

CLAES. 2007. National agricultural research management information system. Available at http://www.arc.sci.eg. CLAES, Egypt.

Clay, J. 2004. Review of world agriculture and the environment. Crop Prot. Res. Inst., Washington, DC.

Cook, R. 2005. Efficiency or market dominance: Evidence from the fresh produce industry. Presentation Ann. Meeting Food Distribution Res. Soc., October. Washington, DC.

Critchley, W., K. Siegert, C. Chapman, and M. Finkel. 1991. Water harvesting: A manual for the design and construction of water harvesting schemes for plant production. FAO, Rome.

CSE (Centre for Science and Environment). 2006. Available at http://www.rainwaterharvesting.org.

Delgado, C., M. Rosegrant, H. Steinfeld, S. Ehui, and C. Courbois. 1999. Livestock to 2020: The next food revolution. Food Agric. Environ. Disc. Pap. 28. IFPRI, Washington, DC, FAO, Rome, and ILRI, Nairobi.

Delgado, C.L., N. Wada, M.W. Rosegrant, S. Meijer, and M. Ahmed. 2003. Outlook for fish to 2020: Meeting global demand. Available at http://www.ifpri.org/pubs/fpr/pr15.pdf). IFPRI, Washington, DC, and WorldFish Center, Penang.

DFID. 2001. Sustainable livelihoods guidance sheets. Sec. 8, Glossary. Available at http://www.livelihoods.org/info/info_guidancesheets.html#1. DFID, London.

DFID. 2007. Livelihoods connect: Creating sustainable livelihoods to eliminate poverty. Available at http://www.livelihoods.org/. DFID, London.

Dupuy, A., C. Lee, and T. Schaaf (eds). 2002. Combating desertification: freshwater resources and the rehabilitation of degraded areas in the drylands. Proc. Int. Sem. N'Djamena, Chad, 30 Oct-4 Nov 2000. UNESCO-MAB Drylands Ser. 1. UNESCO, Paris.

Duxbury, J.M. 2005. Reducing greenhouse warming potential by carbon sequestration in soils: opportunities, limits, and tradeoffs. Pp. 435-450. *In* R. Lal et al. (eds) Climate change and global food security. Taylor and Francis, NY.

El-Beltagy, A. 2002. Central and West Asia and North Africa: Challenges and opportunities. ICARDA, Aleppo.

El-Beltagy, A. 2005. Central and West Asia and North Africa: Challenges and opportunities. Available at http://www.cgiar.org/meetings/agm05/stakeholders_docs/agm05_stake_cwana.pdf. Keynote at the CGIAR Ann. Gen. Meeting, 2005 (AGM05), Marrakech.

EU (European Union). 2002. Integrated crop management systems in the EU. Amended final report for European Commission DG Environment. Agra CEAS Consulting. 1882/BDB/May 2002.

Falkenmark, M., P. Fox, G. Persson, and J. Rockstrom. 2001. Water harvesting for upgrading of rainfed agriculture: Problem analysis and research needs. SIWI Report 11. Stockholm Int. Water Inst., Sweden.

FAO. 1997. Irrigation in the Near East region in figures. Water Rep. No. 9. FAO, Rome.

FAO. 2000. Agricultural knowledge and information systems for rural development (AKIS/RD). Available at http://www.fao.org/sd/EXdirect/EXre0027.htm. FAO, Rome.

FAO. 2001. Genetically modified organisms, consumers, food safety and the environment. Ethics Ser. 2. FAO, Rome.

FAO. 2003. Virtual extension and research communication network. Available at http://www.fao.org/sd/2003/KN10053_en.htm. FAO, Rome.

FAO. 2004. The state of world fisheries and aquaculture, 2004. Available at ftp://ftp.fao.org/docrep/fao/007/y5600e/y5600e00.pdf. FAO, Rome.

FAO. 2007. Frequently asked questions on organic agriculture. Available at http://www.fao.org/organicag/fram11-e.htm. FAO, Rome.

Gate, R., and M. Lee. 2005. Definition of social sustainability. City of Vancouver Policy Report—Social Development. Available at http://www.city.vancouver.bc.ca/ctyclerk/cclerk/20050524/documents/p1.pdf.

Gilbert, R., D. Stevenson, H. Girardet, and R. Stren. 1996. Making cities work. Earthscan, London.

Giuliani, A. 2007. Developing markets for agrobiodiversity. Bioversity Int. and Earthscan, London.

Gleick, P.H. 2000. The changing water paradigm: a look at twenty-first century water resources development. Water Int. 25(1):127-138.

Guendel, S. 2002. Peri-urban and urban livestock keeping in East Africa: A coping strategy for the poor? Scoping study commissioned by the Livestock Production Prog. (LPP). Nat. Resources Inst., UK.

GWP. 2006. GWP toolbox for integrated water resources management. Available at http://www.gwptoolbox.org/en/index.html. Global Water Partnership, Stockholm.

GWP TAC. 2000. Integrated water resources management. Global Water Partnership TAC Background Pap. 4. SE-105 25. Global Water Partnership, Stockholm.

Hawkins, R., and W. Nordquist. 1991. Planetor: A computer program for analyzing environmental problems. *In* Agriculture and the environment. U.S. Gov. Printing Off., Washington, DC.

Haykazyan, V., and J. Pretty. 2006. Sustainability in Armenia: New challenges for the agricultural sector. Centre Environ. Society Occas. Pap. 2006-1. Univ. Essex, Colchester.

Hillel, D. 2000. Salinity management for sustainable irrigation: Integrating science, environment, and economics. World Bank, Washington, DC.

Holt-Gimenez, E. 2002. Measuring farmers' agroecological resistance after Hurricane Mitch in Nicaragua: A case study in participatory sustainable land management impact. Agric. Ecosyst. Environ. 93(1-3):87-105.

Hurni, H. 1996. Precious earth: From soil and water conservation to sustainable land management. ISCO and Centre Dev. Environ., Berne.

Hurni, H., M. Giger, and K. Meyer (ed) 2006. Soils on the global agenda: Developing international mechanisms for sustainable land management. IUSS and Centre Dev. Environ., Berne.

ICARDA. 2006. Utilization of intelligent systems for plant protection. Available at http://uispp.icarda.cgiar.org/uispp/. ICARDA, Aleppo.

IFAD. 2002. IFAD strategy for rural poverty reduction in the Near East and North Africa. IFAD, Rome.

IFOAM. 2007. Organic agriculture and food security. Available at http://www.ifoam.org/organic_facts/index.html. IFOAM, Bonn.

InfoResources. 2003. Integrated Water Resources Management (IWRM): A way to sustainability. Available at www.inforesources.ch/pdf/focus1_e.pdf.

ILA (International Law Association) (ed). 1967. The Helsinki rules on the uses of the waters of international rivers. Pp. 484-532. *In* Report of the Fifty-Second Conference, Helsinki, 14-20 Aug 1966. ILA, London.

IPCC. 2000. Land use, land use change and forestry. Special Report of the IPCC. Cambridge Univ. Press, Cambridge.

IPCC. 2007a. Climate change, 2007: The physical science basis. Contribution of Working Group I to the Fourth Assessment Report. Cambridge Univ. Press, Cambridge.

IPCC. 2007b. Climate change, 2007: Impacts, adaptation and vulnerability. Working Group II Contribution Fourth Assessment Report. Summary for Policymakers. Available at http://www.ipcc.ch. Cambridge Univ. Press, Cambridge.

IPCC. 2007c. Climate change, 2007: Mitigation of climate change. Contribution of Working Group III to the Fourth Assessment Report. Available at http://www.ipcc.ch. Cambridge Univ. Press, Cambridge.

IPGRI. 2001. Regional Report CWANA 1999-2000. Available at http://www.bioversityinternational.org/Publications/Pdf/821.pdf. IPGRI, Rome.

Irz, X., L. Lin, C. Thirtle, and S. Wiggins. 2001. Agricultural growth and poverty alleviation. Dev. Policy Rev 19.

Jones, K., J. Heaton, M. Adams, D. Kennedy, and T. Brocklehurst. 2006. Microbiological hazards and prevention in primary production. *In* Food safety in crop production. 16-20 Jan 2006. Universities of Lancaster and Salford. Cumbria.

Jorgensen, G.S., K.H. Solomon, and V. Cervinka. 1993. Agroforestry systems for on-farm drain water management. Center Irrig. Tech. Project Rep. California State Univ., Fresno.

Kessler, J.J., and H. Breman. 1991. The potential of agroforestry to increase primary production in the Sahelian and Sudanian zones of West Africa. Agrofor. Syst. 13:41-62.

Kinderlerer, J., and M. Adcock. 2005. Agricultural biotechnology, politics, ethics, and policy. *In* S.W. Omamo and K. Von Grebmer (eds) Biotechnology, agriculture, and food security in southern Africa. IFPRI, Washington, DC, and FANRPAN, Harare.

Lal, R. 2005. Climate change, soil carbon dynamics, and global food security. Pp. 11-143. *In* R. Lal et al. (eds) Climate change and global food security. Taylor and Francis, New York.

Leakey, R.R.B., Z. Tchoundjeu, K. Schreckenberg, S.E. Shackleton, and C.M. Shackleton. 2005. Agroforestry tree products (AFTPs): targeting poverty reduction and enhanced livelihoods. Int. J. Agric. Sustain. 3(1):1-23.

Leaman, D.J., H. Fassil, and I. Thormann. 1999. Conserving medicinal and aromatic plant species: Identifying the contribution of the International Plant Genetic Resources Institute. Study commissioned by IDRC. IPGRI, Rome.

Lien, G., J.B. Hardaker, and O. Flaten. 2007. Risk and economic sustainability of crop farming systems. Agric. Syst. 94(2):541-552.

Liniger, H., and W. Critchley (eds). 2007. Where the land is greener: Case studies and analysis of soil and water conservation initiatives worldwide. WOCAT, CTA, Wageningen, FAO, Rome, UNEP, Nairobi and CDE, Berne.

Lipton, M., J. Litchfield, R. Blackman, D. De Zoysa, L. Qureshy, and H. Waddington. 2003. Preliminary review of the impact of irrigation on poverty with special emphasis on Asia. FAO, Rome.

Mahnot, S.C., D.C. Sharma, A. Mishra, P.K. Singh, and K.K. Roy. 2003. Water harvesting and management. Practical Guide Ser. 6. SDC/Intercooperation Çoord. Unit, Jaipur.

Majok, A.A., and C.W. Schwabe. 1996. Development among Africa's migratory pastoralists. Bergin and Garvey, NY.

Menzi, H., P. Gerber, and H. Steinfeld. 2003. Area-wide integration of livestock and crop activities as a solution to allow sustainable growth of intensive livestock production in South East Asia. Workshop presentation at the ZIL/SFIAR Conf. Rio + 11: Towards a new paradigm in agricultural research—The challenge of sustainable. 6 Jun 2003. ETH Zurich.

Metcalfe, A.S. (ed). 2005. Knowledge management and higher education: A critical analysis. [Online]. Available at http://site.ebrary.com/lib/aucairo/Doc?id=10084481&ppg=14. Inform. Sci. Publ.

Milton, N. 2003. Knowledge management. Available at http://www.epistemics.co.uk/Notes/40-0-0.htm.

Morris, B.L., A.R.L. Lawrence, P.J.C. Chilton, B. Adams, R.C. Calow, and B.A. Klinck. 2003. Groundwater and its susceptibility to degradation: A global assessment of the problem and options for management. Early Warning and Assessment Rep. Ser. RS.03-3. UNEP, Nairobi.

Morton, J., and R. Matthewman. 1996. Improving livestock production through extension: Information needs, institutions and opportunities. ODI Natural Perspect. 12, London.

Nasr, M. 1999. Assessing desertification and water harvesting in the Middle East and North Africa: Policy implications. ZEF Disc. Pap. Dev. Policy No. 10. Zentrum für Entwicklungsforschung, Bonn.

NEIWPCC (New England Interstate Water Pollution Control Commission). 2003. Artificial groundwater recharge. White Paper, NEIWPCC. Lowell, MA.

Newberne, P.M. 1980. Naturally occurring food borne toxicants. Pp. 463-496. *In* R.S. Goodhart and M.E. Shils (eds) Modern nutrition in health and disease. Lea and Febiger, Philadelphia.

Ngaido, T., M. McCarthy, and M. Di Gregorio. 2002. International Conference on Policy and Institutional Options for the Management of Rangelands in Dry Areas. Workshop summary paper. IFPRI, CAPRi Working Pap. 23. IFPRI, Washington, DC.

OECD. 2000. Genetically modified foods—widening the debate on health and safety. OECD Edinburgh Conf. Sci. Health Aspects Genetically Modified Foods. OECD consultation with nongovernmental organizations on biotechnology and other aspects of food safety. OECD, Paris.

OECD-FAO. 2006. OECD-FAO agricultural outlook, 2006-2015. OECD, Paris and FAO, Rome.

Oweis, T., and A. Hachum. 2001. Reducing peak supplemental irrigation demand by extending sowing dates. Agric. Water Manage. 50:109-123.

Oweis, T., A. Hachum, and A. Bruggeman (eds). 2004. Indigenous water-harvesting systems in West Asia and North Africa. ICARDA, Aleppo.

Oweis, T., A. Hachum, and J. Kijne. 1999. Water harvesting and supplementary irrigation for improved water use efficiency in dry areas. SWIM Pap. 7. IWMI, Colombo.

Pandey, D.N. 2003. Cultural resources for conservation science. Conserv. Biol. 17(2):633-635.

Pandey, D.N. 2004. Traditional knowledge systems for biodiversity conservation [Online]. Available at http://www.infinityfoundation.com/mandala/t_es/t_es_pande_conserve.htm.

Pandey, D.N., D. Narayan, A.K. Gupta, and D.M. Anderson. 2003. Rainwater harvesting as an adaptation to climate change. Curr. Sci. 85(1):46-59.

Patrick, G.F. 1992. Managing risk in agriculture. Coop. Ext. Serv., Purdue Univ., West Lafayette.

Payne, W.J.A., and R.T. Wilson. 1999. An introduction to animal husbandry in the tropics. 5th ed. Cambridge Univ. Press, Cambridge.

Penning de Vries, F.W.T., H. Acquay, D. Molden, S.J. Scherr, C. Valentin, and O. Cofie. 2003. Integrated land and water management for food and environmental security. CA Res. Rep. 1. Comp. Assessment Sec., IWMI, Colombo.

Perry, B., A. Nin Pratt, K. Sones, and C. Stevens. 2005. An appropriate level of risk: balancing the need for safe livestock products with fair market access for the poor. Pro-Livestock Policy Initiative Working Pap. 23. FAO, Rome.

Pinstrup-Andersen, P. 1999. Biotechnology for developing-country agriculture: Problems and opportunities. Developing appropriate policies. Focus 2, Brief 9. IFPRI, Washington, DC.

Pretty, J.N., A.D. Noble, D. Bossio, J. Dixon, R.E. Hine, F.W.T. Penning de Vries, et al. 2006. Resource-conserving agriculture increases yields in developing countries. Environ. Sci. Tech. 40(4):114-119.

Pretty, J.N., and G.R. Conway. 1988. The blue-baby syndrome and nitrogen fertilisers: A high risk in the tropics? IIED, London.

Prinz, D. 2002. The role of water harvesting in alleviating water scarcity in arid areas. P. 627. *In* M. Al-Rashed et al. (eds) Proc. Int. Conf. Water Resources Manage. Arid Regions. A.A. Balkema, Netherlands.

Prinz, D., and A. Malik. 2002. Runoff farming. Dept. of Rural Engineering. Univ. Karlsruhe, Germany.

Prinz, D., S. Wolfer, and K. Siegert. 2000. Water harvesting for crop production. FAO training course. Available at http://www.fao.org/ag/agL/aglw/wharv.htm. FAO, Rome.

Rafea, A. 1995. Expert systems as a tool for information technology in agriculture. International informatics access 1995. Singapore.

Rafea, A. 1998. Agriculture. *In* J. Liebwitz (ed) Handbook of applied expert systems. CRC Press, Boca Raton.

Rafea, A. 1999. Natural resource conservation and crop management expert systems. *In* G. Kersten et al. (eds) Decision support systems for sustainable development. IDRC and Kluwer Acad. Publ., Boston.

Rafea, A. 2000. Information strategy for the West Asia and North Africa (WANA) region [Online]. Available at http://www.egfar.org. GFAR, Rome.

Ramadan, A., and M. Al Sudairawi. 2005. Soil stabilisation methods in Kuwait: History and future. Eur. J. Sci. Res. 11(1):57-75.

Raper, G.P. 1998. Agroforestry water use in Mediterranean regions of Australia. Water and Salinity Issues Agrofor. No. 2, RIRDC Publ. No. 98/62, RIRDC Project No. DAW-35A. Rur. Ind. Res. Dev. Corp., Kingston, Australia.

Reardon, T., and E. Farina. 2001. The rise of private standards of food quality and safety: illustrations from Brazil. Int. Food Agric. Manage. Rev. 4:413-421.

Reardon, T., and L. Flores. 2006. Customized competitiveness—strategies for horticultural exporters: Central America focus with lessons from and for other regions. Food Policy 31(6):483-503.

Regmi, A., and M. Gehlhar (ed). 2005. New directions in global food markets. ERS Agric. Inform. Bull. 794. Available at www.ers.usda.gov. USDA, Washington, DC.

Rivera, W.M., M.K. Qamaer, and H.K. Mwandemere. 2005. Enhancing coordination among AKIS/RD actors: an analytical and comparative review of country studies on agricultural knowledge and information systems for rural development (AKIS/RD). FAO, Rome.

Rocheleau, D., F. Weber, and A. Field-Juma. 1988. Agroforestry in dryland Africa. ICRAF, Nairobi.

Rosegrant, M.W., and X. Cai. 2000. Water scarcity and food security: alternative futures for the 21st century. J. Water Sci. Tech. 43:61-70.

Rosegrant, M., X. Cai, S. Cline, and N. Nakagawa. 2002. The role of rainfed agriculture in the future of Global food production. EPTD Disc. Pap. 90. IFPRI, Washington, DC.

Rosegrant, M.W., C. Ringler, T. Benson, X. Diao, D. Resnick, J. Thurlow et al. 2006. Agriculture and achieving the Millennium Development Goals. Available at http://www.ifpri.org/pubs/cp/agmdg.asp. IFPRI, Washington, DC.

Ryder, O.A., A. McLaren, S. Brenner, Y.-P. Zhang, and K. Benirschke. 2000. DNA banks for endangered animal species. Science 288:275-277.

Scherr, S. 1999. Soil degradation—a threat to developing-country food security by, 2020? Food Agric. Environ. Disc. Pap. 27. Available at http://www.ifpri.cgiar.org/pubs/catalog.htm. IFPRI, Washington, DC.

Schofield, N.J., I.C. Loh, P.R. Scott, J.R. Bartle, P. Riston, R.W. Bell, et al. 1989. Vegetation strategies to reduce stream salinities of water resource catchments in south-west Western Australia. P. 81. Water Resource Dir Rep. WS33. Water Authority of Western Australia, Leederville, WA.

Schreiber, G., H. Akkermans, A. Anjewierden, R. De Hoog, N. Shadbolt, W. Van De Velde, et al. 1999. Knowledge engineering and management: The common KADS methodology. MIT Press, Cambridge, MA.

Schwabe, C.W. 1984. Veterinary medicine and human health. 3rd ed. Williams and Wilkins, Baltimore.

SDC (Swiss Agency Dev. Coop.). 2005. Water, 2015—policy principles and strategic guidelines for integrated water resource management. Available at http://www.deza.ch/ressources/deza_product_en_1750.pdf. SDC, Berne.

Seckler, D., and U. Amarasinghe. 2000. Water supply and demand, 1995 to 2025: Water scarcity and major issues. *In* World water supply and demand. IWMI, Colombo.

Seré, C., and H. Steinfeld. 1996. World livestock production systems: current status, issues and trends. Animal Prod. Health Pap. 127. FAO, Rome.

Sidahmed, A.E. 1996. The rangelands of the arid/semi-arid areas: challenges and hopes for the 2000s. Keynote address to Symp. D: Range management. Int. Conf. Desert Dev. Arab Gulf Countries, 23-26 Mar 1996. KISR, Kuwait.

SIMINetwork. 2006. Smallholder irrigation market initiative website [Online]. Available at http://www.siminet.org.

Smith, D.M., P.G. Jarvis, and J.C.W. Odongo. 1997. Sources of water used by trees and millet in Sahelian windbreak systems. J. Hydrol. 198:140-153.

Smith, T., S.H. Godfrey, E. Ssewannyana, P.J. Buttery, and E. Owen (ed). 2005. Small stock in development. Proc. workshop on enhancing the contribution of small livestock to the livelihoods of resource-poor communities. Masaka, Uganda, 15-19 Nov 2004. NRI, Aylesford, Kent.

Solomon, K.H. 1988. Irrigation system selection. Irrigation notes. California State Univ., Fresno.

Squire, R.A. 1981. Ranking animal carcinogens: a proposed regulatory approach. Science 214:877-880.

Steiner, K.G. 1996. Causes of soil degradation and development approaches to sustainable soil management. GTZ, GmbH, Eschborn. Margraf Verlag, Weikersheim.

Steinfeld, H., P. Gerber, T. Wassenaar, V. Castel, M. Rosales and C. de Haan. 2006. Livestock's long shadow: Environmental issues and options. Livestock, Environ. Dev. Initiative. FAO, Rome.

Street, K., and Z. Khalikulov. 2004. Agro-biodiversity conservation in Central Asia and the Caucasus: A focus on ICARDA's recent collaborative work. Pp. 259-260. *In* J. Ryan et al. (eds) Agriculture in Central Asia: Research for development. ICARDA, Aleppo.

Studer, C. 2005. Basics important to understand the concept of virtual water and policy relevance of the concept. Presentation at the Egyptian-Swiss Workshop on Virtual Water Strategy in Integrated Water Resource Manage., 5-8 Jun 2005. El Kanater, Egypt.

Studer, C., J. Jamaleddin, and A. Thust. 2002. Framework paper for the UNCCD Thematic Programme Network 4 (TPN4) on Water Resources Manage. Agric. Arid, Semi-Arid and Sub-Humid Areas of the Asia Reg. Action Prog. Damascus and Bonn.

Studer, C., M.A. Zöbisch, and R. Tutwiler. 2000. Land degradation—nature and consequences, reasons behind, and strategies to address it. Reg. Workshop on Degradation and Rehabilitation of Marginal Lands in the Arab Region, 2-4 Jul 2000. Cairo.

Thomas, R., A. Aw-Hassan, A.B. Salkini, and R. LaRovere, 2003. Minutes of consultation for the CWANA region on the Proposed IAASTD, Egypt. 25-26 Feb 2003 [Online].

Available at http://www.agassessment.org/docs/cairoreport.pdf. ICARDA, World Bank and the Dep. Agric. Land Reclamation, Egypt.

Turner, N.C. 1986. Crop water deficits: A decade of progress. Adv. Agron. 39:1-51.

UNEP/IETC. 2001. Sourcebook of alternative technologies for freshwater augmentation in West Asia. IETC Tech. Publ. Ser. 8. Available at http://www.unep.or.jp/ietc/Publications/TechPublications/TechPub-8f/B/Groundwater1.asp. UNEP/IETC, Shiga, Japan.

University of Arizona. 1999. Arizona master gardener manual: an essential reference to gardening in the desert Southwest. Available at http://ag.arizona.edu/pubs/garden/mg/irrigation/. Univ. Arizona, Tucson.

University of Reading. 2005. Environmental challenges in farm management. Course documentation. Available at http://www.ecifm.rdg.ac.uk/.

USEPA. 2007. Integrated pest management (IPM) and food production. Fact sheet. Available at http://www.epa.gov/pesticides/factsheets/ipm.htm. USEPA, Washington, DC.

Van Duivenbooden, N., M. Pala, C. Studer, and C.L. Bielders. 2000. Cropping systems and crop complementarity in dryland agriculture to increase soil water use efficiency: A review. Netherlands J. Agric. Sci. 48:213-236.

Vogt C., C. Nowak, J.B. Diogo, M. Oetken, K. Schwenk, J. Oehlmann. 2007. Multi-generation studies with *Chironomus riparius* — Effects of low tributyltin concentrations on life history parameters and genetic diversity. Chemosphere 67(11):2192-200.

Werners, S.E., F. Incerti, M. Bindi, M. Moriondo, and F. Cots. 2007. Diversification of agricultural crops to adapt to climate change in the Guadiana River Basin. Int. Conf. Climate Change, Hong Kong. 29-31 May 2007. Available at http://www.hkie.org.hk/iccc2007/docs/PDF/Oral%20Papers/W01.PDF.

World Bank. 2005. Managing the livestock revolution: Policy and technology to address the negative impacts of a fast-growing sector. ARD Rep. 32725-GLB. p. 63 [Online]. Available at http://siteresources.worldbank.org/INTARD/Resources/Livestock_final+no+maps.pdf. World Bank, Washington, DC.

World Bank. 2007. World development report, 2008: Agriculture for development. World Bank, Washington, DC.

World Water Council. 2004. E-conference synthesis: Virtual water trade—conscious choices. Mar 2004. WWC Publ. Vol. 2, Marseille, France.

Wright, A.D.G., P. Kennedy, C.J. O'Neill, A.F. Toovey, S. Popovski, S.M. Rea et al. 2004. Reducing methane emissions in sheep by immunization against rumen methanogens. Vaccine 22:3976-3985.

WTO. 2000. Protection of biodiversity and traditional knowledge—the Indian experience. WT/CTE/W/156, IP/C/W/198. Available at http://www.twnside.org.sg/title/cteindia.htm. WTO, Geneva.

Annex A
CWANA Authors and Review Editors

Armenia
Ashot Hovhannisian • Ministry of Agriculture

Austria
Maria Wurzinger • University of Natural Resources & Applied Life Sciences

Egypt
Sonia Ali • Zagarid University
Ahmed Abd Alwahed Rafea • American University of Cairo
Mohamed Abo El wafa Mahmoud Gad • GTZ
Azza Emara • Agricultural Research Institute, Agricultural Research Center

France
Jean Albergel • IRD Department of Earth and Environment
Celine Dutilly-Diane • CIRAD
Alia Gana • National Center for Scientific Research CNRS/ LADYSS-
Selma Tozanli • Mediterranean Agronomic Institute of Montpellier

Iran
Hamid Siadat • Independent

Jordan
Saad M. Alayyash • Jordan Badia Research and Development Center
Ruba Al-Zubi • Ministry of Environment
Mahmud Duwayri • University of Jordan
Muna Yacoub Hindiyeh • Jordan University of Science and Technology
Lubna Qaryouti • Ministry of Agriculture/Rangeland Directorate
Rania Suleiman Shatnawi • Ministry of Environment

Kyrgyzstan
Rafael Litvak • Research Institute of Irrigation

Lebanon
Roy Antoine Abijaoude • Holy Spirit University

Morocco
Mohamed Moussaoui • Independent

Oman
Younis Al Akhzami • Ministry of Agriculture and Fisheries
Dawood Al-Yahyai • Marine Sciences and Fisheries Centre
Abdallah Mohamed Omezzine • University of Nizwa, Oman

Pakistan
Iftikhar Ahmad • National Agricultural Research Centre
Mukhtar Ahmad Ali • Centre for Peace and Development Initiatives
Yameen Memon • Government Employees Cooperative Housing Society
Farzana Panhwar • SINDTH Rural Women's Uplift Group
Syed Wajid Pirzada • Pakistan Agricultural Research Center
Abid Suleri • Sustainable Development Policy Institute (SDPI)
Ahsan Wagha • Damaan Development Organization/GEF/SGP

Palestine
Jamal Abo Omar • AnNajah National University
Jad E Isaac • Applied Research Institute – Jerusalem
Thameen Hijawi • Palestinian Agricultural Relief Committees (PARC)
Numan Mizyed • An-Najah National University
Azzam Saleh • Al-Quds University

Sudan
Ali Taha Ayoub • Ahfal University for Women
Asha El Karib • ACORD
Aggrey Majok • Independent
Ahmed S.M. El Wakeel • NBSAP (a UNDP-IUCN-HCENR Project)

Switzerland
Christoph Studer • Swiss College of Agriculture

Syria
Nour Chachaty • Independent
Alessandra Galie • ICARDA
Stefania Grando • ICARDA
Theib Yousef Oweis • ICARDA
Manzoor Qadir • ICARDA
Kamil H. Shideed • ICARDA

Tajikistan
Sanginov Sanginboy Rajabovich • Soil Science Research Institute of Agrarian Academy of Sciences

Tunisia
Mohamed Annabi • Institut National de la Recherche Agronomique de Tunisie
Rym Ben Zid • Independent
Mustapha Guellouz • IAASTD CWANA, DSIPS – Diversification Program, ICARDA
Kawther Latiri • Institut National de la Recherche Agronomique de Tunisie

Lokman Zaibet • Ecole Supérieure d'Agriculture de Mograne, Zaghouan

Turkey

Gulcan Eraktan • University of Ankara
Yalcin Kaya • Trakya Agricultural Research Institute
Ayfer Tan • Aegean Agricultural Research Institute
Ahu Uncuoglu • Tubitak Research Institute for Genetic Engineering and Biotechnology (RIGEB)
Fahri Yavuz • Ataturk University

United Kingdom

Sajid Kazmi • Middlesex University Business School

United Arab Emirates

Abdin Zein El-Abdin • Lootah Educational Foundation

Uzbekistan

Sandjar Djalalov • Independent
Alisher A. Tashmatov • Ministry of Finance

Annex B

Peer Reviewers

Afghanistan
Samuel Kugbei • FAO

Armenia
Samvel Avetisyan • Ministry of Agriculture
Davit Loqyan • Ministry of Agriculture

Egypt
Abou Hadid Ayman • Agricultural Research Center
Mahmoud Mohamed Fahim • Soil, Water and Environment Research Institute
Rashad Ahmad Abo Elenein • Agriculture Research Center
Mustapha Malki • KariaNet

France
Louis Aime Aumaitre • EAAP / INRA
Claude Torre • AFD

Iran
M. Reza Mehrnejad • Pistachio Research Institute
Mohammad Hassan Roozitalab • Agricultural Research and Education Organization

Jordan
Ruba Al Zubi • Ministry of Environment
Abdel Nabi Fardous • National Center for Agricultural Research and Technology Transfer
Nasri Haddad • ICARDA WARP
Rida A. Shibli • Jordan University of Science and Technology

Kenya
Marcus Lee • UNEP

Morocco
Abdelkader Allali • Ministry of Agriculture, Rural Development and Fishing
Kamal Mohamed • Independent
Rachid Mrabet • INRA

Oman
Khalfan Al-Naabi • Ministry of Agriculture & Fisheries
Saleem Nadaf • Ministry of Agriculture & Fisheries
Abdallah Mohamed Omezzine • University of Nizwa Oman

Poland
Ursula Soltysiak • AgroBio Test

South Africa
Raymond Auerbach • Rainman Landcare Foundation

Sudan
Ali Taha Ayoub • Ahfad University for Women
Asha Elkarib • ACORD Organization

Syria
Ahmed Amri • ICARDA
Abdallah Bari • ICARDA
Celine Dutilly-Diane • CIRAD/ICARDA
Alessandra Galie • ICARDA
Kenneth Street • ICARDA
Richard Thomas • ICARDA

Tunisia
Rym Ben Zid • Independent
Mohamed Aziz Darghouth • Ecole Nationale de Médecine Vétérinaire
Riadh Zghal • STARC

Turkey
Yalcin Kaya • Trakya Agricultural Research Institute
Yuksel Tuzel • Ege University

UAE
Ahmed Almasoum • United Arab Emirates University

UK
Philip Bubb • UNEP
Emyr Thomas • Independent

USA
Philip Bereano • University of Washington
Ralph Crawford • US Forest Service
Doug Gurian-Sherman • Union of Concerned Scientists
Michael Hansen • Consumers Union of US
Lisa Kitinoja • Extension Systems International
Susan J. Owens • USDA

Uruguay
Ariel Julio Castro Tabo • Universidad de la República

Annex C
Glossary

Agriculture A linked, dynamic social-ecological system based on the extraction of biological products and services from an ecosystem, innovated and managed by people. It thus includes cropping, animal husbandry, fishing, forestry, biofuel and bioproducts industries, and the production of pharmaceuticals or tissue for transplant in crops and livestock through genetic engineering. It encompasses all stages of production, processing, distribution, marketing, retail, consumption and waste disposal.

Agricultural biodiversity Encompasses the variety and variability of animals, plants and microorganisms necessary to sustain key functions of the agroecosystem, its structure and processes for, and in support of, food production and food security.

Agricultural extension Agricultural extension deals with the creation, transmission and application of knowledge and skills designed to bring desirable behavioral changes among people so that they improve their agricultural vocations and enterprises and, therefore, realize higher incomes and better standards of living.

Agricultural innovation Agricultural innovation is a socially constructed process. Innovation is the result of the interaction of a multitude of actors, agents and stakeholders within particular institutional contexts. If agricultural research and extension are important to agricultural innovation, so are markets, systems of government, relations along entire value chains, social norms, and, in general, a host of factors that create the incentives for a farmer to decide to change the way in which he or she works, and that reward or frustrate his or her decision.

Agricultural population The agricultural population is defined as all persons depending for their livelihood on agriculture, hunting, fishing or forestry. This estimate comprises all persons actively engaged in agriculture and their non-working dependants.

Agricultural subsidies Agricultural subsidies can take many forms, but a common feature is an economic transfer, often in direct cash form, from government to farmers. These transfers may aim to reduce the costs of production in the form of an input subsidy, e.g., for inorganic fertilizers or pesticides, or to make up the difference between the actual market price for farm output and a higher guaranteed price. Subsidies shield sectors or products from international competition.

Agricultural waste Farming wastes, including runoff and leaching of pesticides and fertilizers, erosion and dust from plowing, improper disposal of animal manure and carcasses, crop residues and debris.

Agroecological Zone A geographically delimited area with similar climatic and ecological characteristics suitable for specific agricultural uses.

Agroecology The science of applying ecological concepts and principles to the design and management of sustainable agroecosystems. It includes the study of the ecological processes in farming systems and processes such as: nutrient cycling, carbon cycling/sequestration, water cycling, food chains within and between trophic groups (microbes to top predators), life cycles, herbivore/predator/prey/host interactions, pollination, etc. Agroecological functions are generally maximized when there is high species diversity/perennial forest-like habitats.

Agroecosystem A biological and biophysical natural resource system managed by humans for the primary purpose of producing food as well as other socially valuable nonfood goods and environmental services. Agroecosystem function can be enhanced by increasing the planned biodiversity (mixed species and mosaics), which creates niches for unplanned biodiversity.

Agroforestry A dynamic, ecologically based, natural resources management system that through the integration of trees in farms and in the landscape diversifies and sustains production for increased social, economic and environmental benefits for land users at all levels. Agroforestry focuses on the wide range of work with trees grown on farms and in rural landscapes. Among these are fertilizer trees for land regeneration, soil health and food security; fruit trees for nutrition; fodder trees that improve smallholder livestock production; timber and fuelwood trees for shelter and energy; medicinal trees to combat disease; and trees that produce gums, resins or latex products. Many of these trees are multipurpose, providing a range of social, economic and environmental benefits.

AKST Agricultural knowledge, science and technology (AKST) is a term encompassing the ways and means used to practice the different types of agricultural activities, and including both formal and informal knowledge and technology.

Alien Species A species occurring in an area outside of its historically known natural range as a result of intentional or accidental dispersal by human activities. Also referred to as introduced species or exotic species.

Aquaculture The farming of aquatic organisms in inland and coastal areas, involving intervention in the rearing process to enhance production and the individual or corporate ownership of the stock being cultivated. Aquaculture practiced in a marine environment is called mariculture.

Average Rate of Return Average rate of return takes the whole expenditure as given and calculates the rate of return to the global set of expenditures. It indicates whether or not the entire investment package was successful, but it does not indicate whether the allocation of resources between investment components was optimal.

Biodiversity The variability among living organisms from all sources, including inter alia, terrestrial, marine and other aquatic ecosystems and the ecological complexes of which they are part; including diversity within species and gene diversity among species, between species and of ecosystems.

Bioelectricity Electricity derived from the combustion of biomass, either directly or co-fired with fossil fuels such as coal and natural gas. Higher levels of conversion efficiency can be attained when biomass is gasified before combustion.

Bioenergy (biomass energy) Bioenergy is comprised of bioelectricity, bioheat and biofuels. Such energy carriers can be produced from energy crops (e.g., sugar cane, maize, oil palm), natural vegetation (e.g., woods, grasses) and organic wastes and residues (e.g., from forestry and agriculture). Bioenergy refers also to the direct combustion of biomass, mostly for heating and cooking purposes.

Biofuel Liquid fuels derived from biomass and predominantly used in transportation. The dominant biofuels are ethanol and biodiesel. Ethanol is produced by fermenting starch contained in plants such as sugar cane, sugar beet, maize, cassava, sweet sorghum or beetroot. Biodiesel is typically produced through a chemical process called transesterification, whereby oily biomass such as rapeseed, soybeans, palm oil, jatropha seeds, waste cooking oils or vegetable oils is combined with methanol to form methyl esters (sometimes called "fatty acid methyl ester" or FAME).

Bioheat Heat produced from the combustion of biomass, mostly as industrial process heat and heating for buildings.

Biological Control The use of living organisms as control agents for pests (arthropods, nematodes mammals, weeds and pathogens) in agriiculture. There are three types of biological control:

Conservation biocontrol: The protection and encouragement of local natural enemy populations by crop and habitat management measures that enhance their survival, efficiency and growth.

Augmentative biocontrol: The release of natural enemies into crops to suppress specific populations of pests over one or a few generations, often involving the mass production and regular release of natural enemies.

Classical biocontrol: The local introduction of new species of natural enemies with the intention that they establish and build populations that suppress particular pests, often introduced alien pests to which they are specific.

Biological Resources Include genetic resources, organisms or parts thereof, populations, or any other biotic component of ecosystems with actual or potential use or value for humanity.

Biotechnology The IAASTD definition of biotechnology is based on that in the Convention on Biological Diversity and the Cartagena Protocol on Biosafety. It is a broad term embracing the manipulation of living organisms and spans the large range of activities from conventional techniques for fermentation and plant and animal breeding to recent innovations in tissue culture, irradiation, genomics and marker-assisted breeding (MAB) or marker assisted selection (MAS) to augment natural breeding. Some of the latest biotechnologies, called "modern biotechnology", include the use of *in vitro* modified DNA or RNA and the fusion of cells from different taxonomic families, techniques that overcome natural physiological reproductive or recombination barriers.

Biosafety Referring to the avoidance of risk to human health and safety, and to the conservation of the environment, as a result of the use for research and commerce of infectious or genetically modified organisms.

Blue Water The water in rivers, lakes, reservoirs, ponds and aquifers. Dryland production only uses green water, while irrigated production uses blue water in addition to green water.

BLCAs Brokered long-term contractual arrangements (BLCAs) are institutional arrangements often involving a farmer cooperative, or a private commercial, parastatal or a state trading enterprise and a package (inputs, services, credit, knowledge) that allows small-scale farmers to engage in the production of a marketable commodity, such as cocoa or other product that farmers cannot easily sell elsewhere.

Catchment An area that collects and drains rainwater.

Capacity Development Any action or process which assists individuals, groups, organizations and communities in strengthening or developing their resources.

Capture Fisheries The sum (or range) of all activities to harvest a given fish resource from the "wild". It may refer to the location (e.g., Morocco, Georges Bank), the target resource (e.g., hake), the technology used (e.g., trawl or beach seine), the social characteristics (e.g., artisanal, industrial), the purpose (e.g., commercial, subsistence, or recreational) as well as the season (e.g., winter).

Carbon Sequestration The process that removes carbon dioxide from the atmosphere.

Cellulosic Ethanol Next-generation biofuel that allows converting not only glucose but also cellulose and hemicellulose—the main building blocks of most biomass—into ethanol, usually using acid-based catalysis or enzyme-based reactions to break down plant fibers into sugar, which is then fermented into ethanol.

Climate Change Refers to a statistically significant variation in either the mean state of the climate or in its variability, persisting for an extended period (typically decades or longer). Climate change may be due to natural internal processes or external forcing, or to persistent anthropogenic changes in the composition of the atmosphere or in land use.

Clone A group of genetically identical cells or individuals that are all derived from one selected individual by vegetative propagation or by asexual reproduction, breeding of completely inbred organisms, or forming genetically identical organisms by nuclear transplantation.

Commercialization The process of increasing the share of income that is earned in cash (e.g., wage income, surplus production for marketing) and reducing the share that is earned in kind (e.g., growing food for consumption by the same household).

Cultivar A cultivated variety, a population of plants within a species of plant. Each cultivar or variety is genetically different.

Deforestation The action or process of changing forest land to non-forested land uses.

Degradation The result of processes that alter the ecological characteristics of terrestrial or aquatic (agro)ecosystems so that the net services that they provide are reduced. Continued degradation leads to zero or negative economic agricultural productivity.

For loss of *land* in quantitative or qualitative ways, the term *degradation* is used. For water resources rendered unavailable for agricultural and non-agricultural uses, we employ the terms *depletion* and *pollution*. *Soil degradation* refers to the processes that reduce the capacity of the soil to support agriculture.

Desertification Land degradation in drylands resulting from various factors, including climatic variations and human activities.

Domesticated or Cultivated Species Species in which the evolutionary process has been influenced by humans to meet their needs.

Domestication The process to accustom animals to live with people as well as to selectively cultivate plants or raise animals in order to increase their suitability and compatibility to human requirements.

Driver Any natural or human-induced factor that directly or indirectly causes a change in a system.

Driver, direct A driver that unequivocally influences ecosystem processes and can therefore be identified and measured to different degrees of accuracy.

Driver, endogenous A driver whose magnitude can be influenced by the decision-maker. The endogenous or exogenous characteristic of a driver depends on the organizational scale. Some drivers (e.g., prices) are exogenous to a decision-maker at one level (a farmer) but endogenous at other levels (the nation-state).

Driver, exogenous A driver that cannot be altered by the decision-maker.

Driver, indirect A driver that operates by altering the level or rate of change of one or more direct drivers.

Ecoagriculture A management approach that provides fair balance between production of food, feed, fuel, fiber, and biodiversity conservation or protection of the ecosystem.

Ecological Pest Management (EPM) A strategy to manage pests that focuses on strengthening the health and resilience of the entire agro-ecosystem. EPM relies on scientific advances in the ecological and entomological fields of population dynamics, community and landscape ecology, multi-trophic interactions, and plant and habitat diversity.

Economic Rate of Return The net benefits to all members of society as a percentage of cost, taking into account externalities and other market imperfections.

Ecosystem A dynamic complex of plant, animal, and micro-organism communities and their nonliving environment interacting as a functional unit.

Ecosystem Approach A strategy for the integrated management of land, water, and living resources that promotes conservation and sustainable use in an equitable way.

An ecosystem approach is based on the application of appropriate scientific methodologies focused on levels of biological organization, which encompass the essential structure, processes, functions, and interactions among organisms and their environment. It recognizes that humans, with their cultural diversity, are an integral component and managers of many ecosystems.

Ecosystem Function An intrinsic ecosystem characteristic related to the set of conditions and processes whereby an ecosystem maintains its integrity (such as primary productivity, food chain biogeochemical cycles). Ecosystem functions include such processes as decomposition, production, pollination, predation, parasitism, nutrient cycling, and fluxes of nutrients and energy.

Ecosystem Management An approach to maintaining or restoring the composition, structure, function, and delivery of services of natural and modified ecosystems for the goal of achieving sustainability. It is based on an adaptive, collaboratively developed vision of desired future conditions that integrates ecological, socioeconomic, and institutional perspectives, applied within a geographic framework, and defined primarily by natural ecological boundaries.

Ecosystem Properties The size, biodiversity, stability, degree of organization, internal exchanges of material and energy among different pools, and other properties that characterize an ecosystem.

Ecosystem Services The benefits people obtain from ecosystems. These include provisioning services such as food and water; regulating services such as flood and disease control; cultural services such as spiritual, recreational, and cultural benefits; and supporting services such as nutrient cycling that maintain the conditions for life on Earth. The concept "ecosystem goods and services" is synonymous with ecosystem services.

Ecosystem Stability A description of the dynamic properties of an ecosystem. An ecosystem is considered stable if it returns to its original state shortly after a perturbation (resilience), exhibits low temporal variability (constancy), or does not change dramatically in the face of a perturbation (resistance).

Eutrophication Excessive enrichment of waters with nutrients, and the associated adverse biological effects.

Ex-ante The analysis of the effects of a policy or a project based only on information available before the policy or project is undertaken.

Ex-post The analysis of the effects of a policy or project based on information available after the policy or project has been implemented and its performance is observed.

Ex situ Conservation The conservation of components of biological diversity outside their natural habitats.

Externalities Effects of a person's or firm's activities on others which are not compensated. Externalities can either hurt or benefit others—they can be negative or positive. One negative externality arises when a company pollutes the local environment to produce its goods and does not compensate the negatively affected local residents. Positive externalities can be produced through primary education, which benefits not only primary school students but also society at large. Governments can reduce negative externalities by regulating and taxing goods with negative externalities. Governments can increase positive externalities by subsidizing goods with positive externalities or by directly providing those goods.

Fallow Cropland left idle from harvest to planting or during the growing season.

Farmer-led Participatory Plant Breeding Researchers and/or development workers interact with farmer-controlled, managed and executed PPB activities, and build on farmers' own varietal development and seed systems.

Feminization The increase in the share of women in an activity, sector or process.

Fishery Generally, a fishery is an activity leading to harvesting of fish. It may involve capture of wild fish or the raising of fish through aquaculture.

Food Security Food security exists when all people of a given spatial unit, at all times, have physical and economic access to sufficient, safe and nutritious food to meet their dietary needs and food preferences for an active and healthy life, and that is obtained in a socially acceptable and ecologically sustainable manner.

Food Sovereignty The right of peoples and sovereign states to democratically determine their own agricultural and food policies.

Food System A food system encompasses the whole range of food production and consumption activities. The food system includes farm input supply, farm production, food processing, wholesale and retail distribution, marketing, and consumption.

Forestry The human utilization of a piece of forest for a certain purpose, such as timber or recreation.

Forest Systems Forest systems are lands dominated by trees; they are often used for timber, fuelwood, and non-wood forest products.

Gender Refers to the socially constructed roles and behaviors of, and relations between, men and women, as opposed to sex, which refers to biological differences. Societies assign specific entitlements, responsibilities and values to men and women of different social strata and sub-groups.

Worldwide, systems of relation between men and women tend to disadvantage women, within the family as well as in public life. Like the hierarchical framework of a society, gender roles and relations vary according to context and are constantly subject to changes.

Genetic Engineering Modifying genotype, and hence phenotype, by transgenesis.

Genetic Material Any material of plant, animal, microbial or other origin containing functional units of heredity.

Genomics The research strategy that uses molecular characterization and cloning of whole genomes to understand the structure, function and evolution of genes and to answer fundamental biological questions.

Globalization Increasing interlinking of political, economic, institutional, social, cultural, technical, and ecological issues at the global level.

GMO (Genetically Modified Organism) An organism in which the genetic material has been altered anthropogenically by means of gene or cell technologies.

Governance The framework of social and economic systems and legal and political structures through which humanity manages itself. In general, governance comprises the traditions, institutions and processes that determine how power is exercised, how citizens are given a voice, and how decisions are made on issues of public concern.

Global Environmental Governance The global biosphere behaves as a single system, where the environmental impacts of each nation ultimately affect the whole. That makes a coordinated response from the community of nations a necessity for reversing today's environmental decline.

Global Warming Refers to an increase in the globally averaged surface temperature in response to the increase of well-mixed greenhouse gases, particularly CO_2.

Global Warming Potential An index, describing the radiative characteristics of well-mixed greenhouse gases, that represents the combined effect of the differing times these gases remain in the atmosphere and their relative effectiveness in absorbing outgoing infrared radiation. This index approximates the time-integrated warming effect of a unit mass of a given greenhouse gas in today's atmosphere, relative to that of carbon dioxide.

Green Revolution An aggressive effort since 1950 in which agricultural researchers applied scientific principles of genetics and breeding to improve crops grown primarily in less-developed countries. The effort typically was accompanied by collateral investments to develop or strengthen the delivery of extension services, production inputs and markets and develop physical infrastructures such as roads and irrigation.

Green Water Green water refers to the water that comes from precipitation and is stored in unsaturated soil. Green water is typically taken up by plants as evapotranspiration.

Ground Water Water stored underground in rock crevices and in the pores of geologic materials that make up Earth's crust. The upper surface of the saturate zone is called the water table.

Growth Rate The change (increase, decrease, or no change) in an indicator over a period of time, expressed as a percentage of the indicator at the start of the period. Growth rates contain several sets of information. The first is whether there is any change at all; the second is what direction the change is going in (increasing or decreasing); and the third is how rapidly that change is occurring.

Habitat Area occupied by and supporting living organisms. It is also used to mean the environmental attributes required by a particular species or its ecological niche.

Hazard A potentially damaging physical event, phenomenon and/or human activity, which my cause injury, property damage, social and economic disruption or environmental degradation.

Hazards can include latent conditions that may represent future threats and can have different origins.

Household All the persons, kin and non-kin, who live in the same or in a series of related dwellings and who share income, expenses and daily subsistence tasks. A basic unit for socio-cultural and economic analysis, a household may consist of persons (sometimes one but generally two or more) living together and jointly making provision for food or other essential elements of the livelihood.

Industrial Agriculture Form of agriculture that is capital-intensive, substituting machinery and purchased inputs for human and animal labor.

Infrastructure The facilities, structures, and associated equipment and services that facilitate the flows of goods and services between individuals, firms, and governments. It includes public utilities (electric power, telecommunications, water supply, sanitation and sewerage, and waste disposal); public works (irrigation systems, schools, housing, and hospitals); transport services (roads, railways, ports, waterways, and airports); and R&D facilities.

Innovation The use of a new idea, social process or institutional arrangement, material, or technology to change an activity, development, good, or service or the way goods and services are produced, distributed, or disposed of.

Innovation system Institutions, enterprises, and individuals that together demand and supply information and technology, and the rules and mechanisms by which these different agents interact.

In recent development discourse agricultural innovation is conceptualized as part and parcel of social and ecological organization, drawing on disciplinary evidence and understanding of how knowledge is generated and innovations occur.

In situ Conservation The conservation of ecosystems and natural habitats and the maintenance and recovery of viable populations of species in their natural habitats and surroundings and, in the case of domesticated or cultivated species, in the surroundings where they have developed their distinctive properties and were managed by local groups of farmers, fishers or foresters.

Institutions The rules, norms and procedures that guide how people within societies live, work, and interact with each other. Formal institutions are written or codified rules, norms and procedures. Examples of formal institutions are the Constitution, the judiciary laws, the organized market, and property rights. Informal institutions are rules governed by social and behavioral norms of the society, family, or community. Cf. Organization.

Integrated Approaches Approaches that search for the best use of the functional relations among living organisms in relation to the environment without excluding the use of external inputs. Integrated approaches aim at the achievement of multiple goals (productivity increase, environmental sustainability and social welfare) using a variety of methods.

Integrated Assessment A method of analysis that combines results and models from the physical, biological, economic, and social sciences, and the interactions between these components in a consistent framework to evaluate the status and the consequences of environmental change and the policy responses to it.

Integrated Natural Resources Management (INRM) An approach that integrates research of different types of natural resources into stakeholder-driven processes of adaptive management and innovation to improve livelihoods, agroecosystem resilience, agricultural productivity and environmental services at community, eco-regional and global scales of intervention and impact. INRM thus aims to help to solve complex real-world problems affecting natural resources in agroecosystems.

Integrated Pest Management The procedure of integrating and applying practical management methods to manage insect populations so as to keep pest species from reaching damaging levels while avoiding or minimizing the potentially harmful effects of pest management measures on humans, non-target species, and the environment. IPM tends to incorporate assessment methods to guide management decisions.

Intellectual Property Rights (IPRs) Legal rights granted by governmental authorities to control and reward certain products of human intellectual effort and ingenuity.

Internal Rate of Return The discount rate that sets the net present value of the stream of the net benefits equal to zero. The internal rate of return may have multiple values when the stream of net benefits alternates from negative to positive more than once.

International Dollars Agricultural R&D investments in local currency units have been converted into international dollars by deflating the local currency amounts with each country's inflation ration (GDP deflator) of base year 2000. Next, they were converted to US dollars with a 2000 purchasing power parity (PPP) index. PPPs are synthetic exchange rates used to reflect the purchasing power of currencies.

Knowledge The way people understand the world, the way in which they interpret and apply meaning to their experiences. Knowledge is not about the discovery of some final objective "truth" but about the grasping of subjective culturally conditioned products emerging from complex and ongoing processes involving selection, rejection, creation, development and transformation of information. These processes, and hence knowledge, are inextricably linked to the social, environmental and institutional context within which they are found.

Scientific knowledge: Knowledge that has been legitimized and validated by a formalized process of data gathering, analysis and documentation.

Explicit knowledge: Information about knowledge that has been or can be articulated, codified, and stored and exchanged. The most common forms of explicit knowledge are manuals, documents, procedures, cultural artifacts and stories. The information about explicit knowledge also can be audiovisual. Works of art and product design can be seen as other forms of explicit knowledge where human skills, motives and knowledge are externalized.

Empirical knowledge: Knowledge derived from and constituted in interaction with a person's environment. Modern communication and information technologies, and scientific instrumentation, can extend the "empirical environment" in which empirical knowledge is generated.

Local knowledge: The knowledge that is constituted in a given culture or society.

Traditional (ecological) knowledge: The cumulative body of knowledge, practices, and beliefs evolved by adaptive processes and handed down through generations. It may not be indigenous or local, but it is distinguished by the way in which it is acquired and used, through the social process of learning and sharing knowledge.

Knowledge Management A systematic discipline of policies, processes, and activities for the management of all processes of knowledge generation, codification, application and sharing of information about knowledge.

Knowledge Society A society in which the production and dissemination of scientific information and knowledge function well, and in which the transmission and use of valuable experiential knowledge is optimized; a society in which the information of those with experiential knowledge is used together with that of scientific and technical experts to inform decision-making.

Land Cover The physical coverage of land, usually expressed in terms of vegetation cover or lack of it. Influenced by but not synonymous with *land use*.

Land Degradation The reduction in the capability of the land to produce benefits from a particular land use under a specific form of land management.

Landscape An area of land that contains a mosaic of ecosystems, including human-dominated ecosystems. The term *cultural landscape* is often used when referring to landscapes containing significant human populations.

Land Tenure The relationship, whether legally or customarily defined, among people, as individuals or groups, with respect to land and associated natural resources (water, trees, minerals, wildlife, and so on).

Rules of tenure define how property rights in land are to be allocated within societies. Land tenure systems determine who can use what resources for how long, and under what conditions.

Land Use The human utilization of a piece of land for a certain purpose (such as irrigated agriculture or recreation). Land use is influenced by, but not synonymous with, land cover.

Leguminous Cultivated or spontaneous plants that fix atmospheric nitrogen.

Malnutrition Failure to achieve nutrient requirements, which can impair physical and/or mental health. It may result from consuming too little food or a shortage or imbalance of key nutrients (e.g., micronutrient deficiencies or excess consumption of refined sugar and fat).

Marginal Rates of Return Calculates the returns to the last dollar invested on a certain activity. It is usually estimated through econometric estimation.

Marker Assisted Selection (MAS) The use of DNA markers to improve response to selection in a population. The markers will be closely linked to one or more target loci, which may often be quantitative trait loci.

Minimum Tillage The least amount possible of cultivation or soil disturbance done to prepare a suitable seedbed. The main purposes of minimum tillage are to reduce tillage energy consumption, to conserve moisture, and to retain plant cover to minimize erosion.

Model A simplified representation of reality used to simulate a process, understand a situation, predict an outcome or analyze a problem. A model can be viewed as a selective approximation, which by elimination of incidental detail, allows hypothesized or quantified aspects of the real world to appear manipulated or tested.

Multifunctionality In IAASTD, multifunctionality is used solely to express the inescapable interconnectedness of agriculture's different roles and functions. The concept of multifunctionality recognizes agriculture as a multi-output activity producing not only commodities (food, feed, fibers, agrofuels, medicinal products and ornamentals), but also non-commodity outputs such as environmental services, landscape amenities and cultural heritages. (See Global SDM Text Box)

Natural Resources Management Includes all functions and services of nature that are directly or indirectly significant to humankind, i.e., economic functions, as well as other cultural and ecological functions or social services that are not taken into account in economic models or not entirely known.

Nanotechnology The engineering of functional systems at the atomic or molecular scale.

Net Present Value (NPV) Net present value is used to analyze the profitability of an investment or project, representing the difference between the discounted present value of benefits and the discounted present value of costs. If NPV of a prospective project is positive, then the project should be accepted. The analysis of NPV is sensitive to the reliability of future cash inflows that an investment or project will yield.

No-Till Planting without tillage. In most systems, planter-mounted coulters till a narrow seedbed assisting in the placement of fertilizer and seed. The tillage effect on weed control is replaced by herbicide use.

Obesity A chronic physical condition characterized by too much body fat, which results in higher risk for health problems such as high blood pressure, high blood cholesterol, diabetes, heart disease and stroke. Commonly it is defined as a Body Mass Index (BMI) equal to or more than 30, while overweight is equal to or more than 25. The BMI is an idex of weight-for-height and is defined as the weight in kilograms divided by the square of the height in meters (kg/m^2).

Organic Agriculture An ecological production management system that promotes and enhances biological cycles and soil biological activity. It is based on minimal use of off-farm inputs and on management practices that restore, maintain and enhance ecological harmony.

Organization Organizations can be formal or informal. Examples of organizations are government agencies (e.g., police force, ministries, etc.), administrative bodies (e.g., local government), nongovernmental organizations, associations (e.g., farmers' associations) and private companies (firms). Cf. with Institutions.

Orphan Crops Crops such as teff, finger millet, yam, roots and tubers that tend to be regionally or locally important for income and nutrition, but which are not traded globally and receive minimal attention by research networks.

Participatory Development A process that involves people (population groups, organizations, associations, political parties) actively and significantly in all decisions affecting their lives.

Participatory Domestication The process of domestication that involves agriculturalists and other community members actively and significantly in making decisions, taking action and sharing benefits.

Participatory Plant Breeding (PPB) Involvement of a range of actors, including scientists, farmers, consumers, extension agents, vendors, processors and other industry stakeholders, as well as farmer and community-based organizations and nongovernment organization (NGOs) in plant breeding research and development.

Participatory Varietal Selection (PVS) A process by which farmers and other stakeholders along the food chain are involved with researchers in the selection of varieties from formal and farmer-based collections and trials, to determine which are best suited to their own agroecosystems' needs, uses and preferences, and which should go ahead for finishing, wider release and dissemination. The information gathered may in turn be fed back into formal-led breeding programs.

Pesticide A toxic chemical or biological product that kills organisms (e.g., insecticides, fungicides, weedicides, rodenticides).

Poverty There are many definitions of poverty.

Absolute Poverty: According to a UN declaration that resulted from the World Summit on Social Development in 1995, absolute poverty is a condition characterized by severe deprivation of basic human needs, including food, safe drinking water, sanitation facilities, health, shelter, education and information. It depends not only on income but also on access to services.

Dimensions of Poverty: The individual and social characteristics of poverty such as lack of access to health and education, powerlessness or lack of dignity. Such aspects of deprivation experienced by the individual or group are not captured by measures of income or expenditure.

Extreme Poverty: Persons who fall below the defined poverty line of US$1 income per day. The measure is converted into local currencies using purchasing power parity (PPP) exchange rates. Other definitions of this concept have identified minimum subsistence requirements, the denial of basic human rights or the experience of exclusion.

Poverty Line: A minimum requirement of welfare, usually defined in relation to income or expenditure, used to identify the poor. Individuals or households with incomes or expenditure below the poverty line are poor. Those with incomes or expenditure equal to or above the line are not poor. It is common practice to draw more than one poverty line to distinguish different categories of poor, for example, the extreme poor.

Private Rate of Return The gain in net revenue to the private firm/business divided by the cost of an investment expressed in percentage.

Processes A series of actions, motions, occurrences, a method, mode, or operation, whereby a result or effect is produced.

Production Technology All methods that farmers, market agents and consumers use to cultivate, harvest, store, process, handle, transport and prepare food crops, cash crops, livestock, etc., for consumption.

Protected Area A geographically defined area which is designated or regulated and managed to achieve specific conservation objectives as defined by society.

Public Goods A good or service in which the benefit received by any one party does not diminish the availability of the benefits to others, and/or where access to the good cannot be restricted. Public goods have the properties of non-rivalry in consumption and non-excludability.

Public R&D Investment Includes R&D investments done by government agencies, nonprofit institutions, and higher-education agencies. It excludes the private for-profit enterprises.

Research and Development (R&D) Organizational strategies and methods used by research and extension program to conduct their work including scientific procedures, organizational modes, institutional strategies, interdisciplinary team research, etc.

Scenario A plausible and often simplified description of how the future may develop based on explicit and coherent and internally consistent set of assumptions about key driving forces (e.g., rate of technology change, prices) and relationships. Scenarios are neither predictions nor projections and sometimes may be based on a "narrative storyline". Scenarios may be derived from projections but are often based on additional information from other sources.

Science, Technology and Innovation Includes all forms of useful knowledge (codified and tacit) derived from diverse branches of learning and practice, ranging from basic scientific research to engineering to local knowledge. It also includes the policies used to promote scientific advance, technology development, and the commercialization of products, as well as the associated institutional innovations. *Science* refers to both basic and applied sciences. *Technology* refers to the application of science, engineering, and other fields, such as medicine. *Innovation* includes all of the processes, including business activities that bring a technology to market.

Shifting Cultivation Found mainly in the tropics, especially in humid and subhumid regions. There are different kinds; for example, in some cases a settlement is permanent, but certain fields are fallowed and cropped alternately ("rotational agriculture"). In other cases, new land is cleared when the old is no longer productive.

Slash and Burn Agriculture A pattern of agriculture in which existing vegetation is cleared and burned to provide space and nutrients for cropping.

Social Rate of Return The gain to society of a project or investment in net revenue divided by cost of the investment, expressed by percentage.

Soil and Water Conservation (SWC) A combination of appropriate technology and successful approach. Technologies promote the sustainable use of agricultural soils by minimizing soil erosion, maintaining and/or enhancing soil properties, managing water, and controlling temperature. Approaches explain the ways and means which are used to realize SWC in a given ecological and socioeconomic environment.

Soil Erosion The detachment and movement of soil from the land surface by wind and water in conditions influenced by human activities.

Soil Function Any service, role, or task that a soil performs, especially: (a) sustaining biological activity, diversity, and productivity; (b) regulating and partitioning water and solute flow; (c) filtering, buffering, degrading, and detoxifying potential pollutants; (d) storing and cycling nutrients; (e) providing support for buildings and other structures and to protect archaeological treasures.

Staple Food (Crops) Food that is eaten as daily diet.

Soil Quality The capacity of a specific kind of soil to function, within natural or managed ecosystem boundaries, to sustain plant and animal productivity, maintain or enhance water and air quality, and support human health and habitation. In short, the capacity of the soil to function.

Subsidy Transfer of resources to an entity, which either reduces the operating costs or increases the revenues of such entity for the purpose of achieving some objective.

Subsistence Agriculture Agriculture carried out for the use of the individual person or their family with few or no outputs available for sale.

Sustainable Development Development that meets the needs of the present without compromising the ability of future generations to meet their own needs.

Sustainable Land Management (SLM) A system of technologies and/or planning that aims to integrate ecological with socio-economic and political principles in the management of land for agricultural and other purposes to achieve intra- and intergenerational equity.

Sustainable Use of Natural Resources Natural resource use is sustainable if specific types of use in a particular ecosystem are considered reasonable in the light of both the internal and the external perspective on natural resources. "Reasonable" in this context means that all actors agree that resource use fulfils productive, physical, and cultural functions in ways that will meet the long-term needs of the affected population.

Technology Transfer The broad set of deliberate and spontaneous processes that give rise to the exchange and dissemination of information and technologies among different stakeholders. As a generic concept, the term is used to encompass both diffusion of technologies and technological cooperation across and within countries.

Terms of Trade The *international terms* of trade measures a relationship between the prices of exports and the prices of imports, this being known strictly as the barter terms of trade. In this sense, deterioration in the terms of trade could have resulted if unit prices of exports had risen less than unit prices for imports. The *inter-sectoral terms of trade* refers to the terms of trade between sectors of the economy, e.g., rural and urban, agriculture and industry.

Total Factor Productivity A measure of the increase in total output which is not accounted for by increases in total inputs. The total factor productivity index is computed as the ratio of an index of aggregate output to an index of aggregate inputs.

Tradeoff Management choices that intentionally or otherwise change the type, magnitude, and relative mix of services provided by ecosystems.

Transgene An isolated gene sequence used to transform an organism. Often, but not always, the transgene has been derived from a different species than that of the recipient.

Transgenic An organism that has incorporated a functional foreign gene through recombinant DNA technology. The novel gene exists in all of its cells and is passed through to progeny.

Undernourishment Food intake that is continuously inadequate to meet dietary energy requirement.

Undernutrition The result of food intake that is insufficient to meet dietary energy requirements continuously, poor absorption, and/or poor biological use of nutrients consumed.

Urban and Peri-Urban Agriculture Agriculture occurring within and surrounding the boundaries of cities throughout the world and includes crop and livestock production, fisheries and forestry, as well as the ecological services they provide. Often multiple farming and gardening systems exist in and near a single city.

Value Chain A set of value-adding activities through which a product passes from the initial production or design stage to final delivery to the consumer.

Virtual Water The volume of water used to produce a commodity. The adjective "virtual" refers to the fact that most of the water used to produce a product is not contained in the product. In accounting virtual water flows we keep track of which parts of these flows refer to green, blue and grey water, respectively.

The real-water content of products is generally negligible if compared to the virtual-water content.

Waste Water "Grey" water that has been used in homes, agriculture, industries and businesses that is not for reuse unless it is treated.

Watershed The area which supplies water by surface and subsurface flow from precipitation to a given point in the drainage system.

Watershed Management Use, regulation and treatment of water and land resources of a watershed to accomplish stated objectives.

Water Productivity An efficiency term quantified as a ration of product output (goods and services) over water input.

Expressions of water productivity. Three major expressions of water productivity can be identified: 1) the amount of carbon gain per unit of water transpired by the leaf or by the canopy (photosynthetic water productivity); 2) the amount of water transpired by the crop (biomass water productivity); or 3) the yield obtained per unit amount of water transpired by the crop (yield water productivity).

Agricultural water productivity relates net benefits gained through the use of water in crop, forestry, fishery, livestock and mixed agricultural systems. In its broadest sense, it reflects the objectives of producing more food, income, livelihood and ecological benefits at less social and environmental cost per unit of water in agriculture.

Physical water productivity relates agricultural production to water use—more crop per drop. Water use is expressed either in terms of delivery to a use, or depletion by a use through evapotranspiration, pollution, or directing water to a sink where it cannot be reused. Improving physical water productivity is important to reduce future water needs in agriculture.

Economic water productivity relates the value of agricultural production to agricultural water use. A holistic assessment should account for the benefits and costs of water, including less tangible livelihood benefits, but this is rarely done. Improving economic water productivity is important for economic growth and poverty reduction.

Annex D

Acronyms, Abbreviations and Units

AARINENA Association of Agricultural Research Institutions in the Near East and North Africa
ACSAD Arab Center for Agricultural Research in the Dry Lands and Arid Zones
AIDS Acquired immune deficiency syndrome
AKIS agriculture knowledge and information system
AKST Agricultural knowledge, science and technology
AMHY Alpine and Mediterranean Hydrology
AOAD Arab Organization for Agricultural Development
ARC Agricultural Research Council
ARI agricultural research institute
AU African Union
BCE Before current era
BNF Biological nitrogen fixation
BSE Bovine spongiform encephalopathy
Bt soil bacterium *Bacillus thuringiensis* (usually refers to plants made insecticidal using a variant of various *cry* toxin genes sourced from plasmids of these bacteria)
CAC Central Asia and the Caucasus
CA-TCN/PGR Central Asia and Trans-Caucasus Network on Plant Genetic Resources
CAP Common Agricultural Policy
CBD Convention on Biological Diversity
C carbon
Ca calcium
CA Comprehensive Assessment of Water Management in Agriculture
CBD Convention on Biological Diversity
CBO Community-based organization
Cd cadmium
CDM Clean Development Mechanism
CE current era
CEDAW Convention on the Elimination of All Forms of Discrimination Against Women
CEEC Central and Eastern European Countries
CGIAR Consultative Group on International Agricultural Research
CH_4 methane
CIAT International Center for Tropical Agriculture
CIFOR Center for International Forestry Research
CIHEAM International Center for Advanced Mediterranean Agronomic Studies
CIMMYT International Maize and Wheat Improvement Center
CIP International Potato Center
CIS Commonwealth of Independent States among Central Asia and Caucasus countries
CITES Convention on International Trade in Endangered Species of Wild Fauna and Flora
CLAES Central Laboratory for Agricultural Expert Systems
C:N carbon to nitrogen ratio
CO_2 carbon dioxide
CO_2-eq carbon dioxide equivalent
$[CO_2]$ carbon dioxide concentration
COA certified organic agriculture
Codex Codex Alimentarius
CODIS Communication, documentation and information services
COMSTECH-OIC Scientific and Technical Committee of the Organization of Islamic Conference
CPB Cartagena Protocol on Biosafety
CPI corruption perception index
CS carbon sequestration
CSO civil society organization
Cu copper
CWANA Central and West Asia and North Africa
Defra UK Department of Environment, Food and Rural Affairs
DFID UK Department of International Development
DNA deoxyribonucleic acid
dS deciSiemen (measure of salinity)
EC/PGR European Cooperative Program for Plant Genetic Resources
EU European Union
FAO Food and Agriculture Organization of the United Nations

Fe	iron
FFS	farmer field school
FLO	Fair Trade Labeling Organization
FPRE	Farmer participatory research and extension
FSRE	Farming systems research and extension
FTA	Free Trade Agreement
g	gram (10^{-3} kg)
GAB	Gender Advisory Board
GAP	South Eastern Anatolian Project
GATT	General Agreement on Trade and Tariffs
GBA	Global Biodiversity Assessment
GDP	Gross domestic product
GE	genetic engineering/genetically engineered
GEF	Global Environment Facility
GEO	Global Environment Outlook
GEWEX	Global Energy and Water Experiment
Gg	gigagram (10^6 kg)
Gha	gigahectare (10^9 hectare)
GHG	greenhouse gas
GHI	Global Hunger Index
GIS	geographic information system
GISP	Global Invasive Species Program
GLASOD	Global assessment of human-induced soil degradation
GM	genetically modified/genetic modification
GMO	genetically modified organism
GMZ	gene management zone
GNP	Gross National Product
GPS	global positioning system
GR	Green Revolution
Gt	gigaton/gigatonne; 10^{19} tonnes
GURT	Genetic Use of Restriction Technologies
GWP	global warming potential
ha	hectare (10^4 m^2)
HACCP	Hazard Analysis Critical Control Point
HDI	human development index
HIV	Human immunodeficiency virus
HRD	Human resource development
HT	herbicide tolerant
HYMEX	Hydrological Cycle in Mediterranean Experiment
HYV	High yielding variety
IAASTD	International Assessment of Agricultural Knowledge, Science and Technology for Development
IARC	International Agricultural Research Center
IAS	invasive alien species
IASUS	International Actions for the Sustainable Use of Soils
IBRD	International Bank of Rural Development
ICA	International Commodity Agreement
ICAR	Indian Council of Agricultural Research
ICARDA	International Center for Agricultural Research in the Dry Areas
ICRAF	World Agroforestry Center
ICRISAT	International Crops Research Institute for Semi-arid Tropics
ICT	information and communication technologies
IDA	International Development Agency
IEA	International Energy Agency
IFAD	International Fund for Agricultural Development
IFC	International Finance Corporation
IFI	international financial institution
IFOAM	International Federation of Organic Agriculture Movements
IFPRI	International Food Policy Research Institute
IITA	International Institute for Tropical Agriculture
IK	Indigenous knowledge
ILO	International Labour Organisation
ILRI	International Livestock Research Institute
IMF	International Monetary Fund
INM	Integrated Nutrient Management
INRM	Integrated Natural Resources Management
IP	intellectual property
IPCC	Intergovernmental Panel on Climate Change
IPGRI	Bioversity International
IPM	Integrated pest management
IPPC	International Plant Protection Convention
IPR	intellectual property rights
IRR	internal rate of return
IRRI	International Rice Research Institute
IS	innovation systems
IUCN	International Union for Conservation of Nature
IWMI	International Water Management Institute
IWRM	Integrated water resources management
K	potassium
kcal	kilocalorie
kg	kilogram, 10^3 grams
km	kilometer
kWh	kilowatt hour
LAC	Latin America and the Caribbean
LDC	least developed countries
LEISA	Low-External Input Sustainable Agriculture
LIC	low income country
LINKS	Local and Indigenous Knowledge Systems
LUC	land use change
m	10^2 cm
MA	Millennium Ecosystem Assessment
MAB/S	marker assisted breeding/selection
MDG	Millennium Development Goals

Mg magnesium
mg milligram (10^{-3} grams)
MIGA Multilateral Investment Agency
MNC multinational corporation
MRL maximum residue level
N nitrogen
NAE North America and Europe
NAFTA North American Free Trade Agreement
NARI National agricultural research institute
NARS national agricultural research systems
NBF National Biosafety Frameworks
NBSAP National Biodiversity Strategy and Action Plans
ng nanogram (10^{-9} grams)
NGO nongovernmental organization
N_2O nitrous oxide
NPK nitrogen, phosphorus, potassium
NRM Natural resource management
NTFP non-timber forest product
NUE nitrogen use efficiency
O_3 ozone
OA organic agriculture
ODA overseas development assistance
OECD Organization of Economic Cooperation and Development
OH hydroxyl
OIE World Animal Health Organization
OMC Organización Mundial del Comercio
OPEC Organization of Petroleum Exporting Countries
OSS Sahel and Sahara Observatory
Pg petagram: 10^{15} grams
P phosphorus
PES Payments for environmental services
PGRFA Plant Genetic Resources for Food and Agriculture
PIPRA Public-Sector Intellectual Property Resource for Agriculture
PPB Participatory plant breeding
ppm parts per million
ppmv parts per million by volume
PPP Purchasing Power Parity
PPQ plant protection and quarantine
PVP plant variety protection
R&D research and development
RNA ribonucleic acid
ROR rates of return
ROSELT Long Term Ecological Monitoring Observatories Network
S&T science and technology
SAP Structural adjustment policies
SDG Sustainable Development Goal
SIA Strategic Impact Assessment
SPS Sanitary and Phytosanitary
SRES Special Report on Emission Scenarios
SSA sub-Saharan Africa
TCP Technical Cooperation Program
TG Technogarden scenario
TGA third-generation agriculture
TK traditional knowledge
TNC transnational corporation
tonne 10^3 kg (metric ton)
TRIPS Trade-Related Aspects of Intellectual Property Rights
T&V training and visit
TV Traditional variety
UAE United Arab Emirates
UNCBD UN Convention on Biodiversity
UNCCD UN Commission to Combat Desertification
UNCED UN Conference on Environment and Development
UNCTAD UN Conference on Trade and Development
UNDP United Nations Development Program
UNEP United Nations Environment Programme
UNESCO United Nations Educational, Scientific and Cultural Organization
UNFCCC United Nations Framework Convention on Climate Change
UNWWAP United Nations World Water Assessment Programme
UPOV International Union for the Protection of New Varieties of Plants
USAID US Agency for International Development
USDA US Department of Agriculture
VERCON Virtual Extension and Research Communication Network
WANA West Asia and North Africa
WANANET West Asia and North Africa Network on Plant Genetic Resources
WARDA Africa Rice Center
WCED World Commission on Environment and Development (Brundtland Commission)
WEDO Women's Environment and Development Organization
WHO World Health Organization
WIPO World Intellectual Property Organization
WMO World Meteorological Organization
WOCAT World Overview of Conservation Approaches and Technologies
WP water productivity
WRI World Resources Institute
WSSD World Summit on Sustainable Development
WTO World Trade Organization

WUA	Water User Association
WUE	water use efficiency
WWF	World Wildlife Fund
yr	year
Zn	zinc

Annex E

Steering Committee for Consultative Process and Advisory Bureau for Assessment

Steering Committee

The Steering Committee was established to oversee the consultative process and recommend whether an international assessment was needed, and if so, what was the goal, the scope, the expected outputs and outcomes, governance and management structure, location of the secretariat and funding strategy.

Co-chairs

Louise Fresco, Assistant Director General for Agriculture, FAO
Seyfu Ketema, Executive Secretary, Association for Strengthening Agricultural Research in East and Central Africa (ASARECA)
Claudia Martinez Zuleta, Former Deputy Minister of the Environment, Colombia
Rita Sharma, Principal Secretary and Rural Infrastructure Commissioner, Government of Uttar Pradesh, India
Robert T. Watson, Chief Scientist, The World Bank

Nongovernmental Organizations

Benny Haerlin, Advisor, Greenpeace International
Marcia Ishii-Eiteman, Senior Scientist, Pesticide Action Network North America Regional Center (PANNA)
Monica Kapiriri, Regional Program Officer for NGO Enhancement and Rural Development, Aga Khan
Raymond C. Offenheiser, President, Oxfam America
Daniel Rodriguez, International Technology Development Group (ITDG), Latin America Regional Office, Peru

UN Bodies

Ivar Baste, Chief, Environment Assessment Branch, UN Environment Programme
Wim van Eck, Senior Advisor, Sustainable Development and Healthy Environments, World Health Organization
Joke Waller-Hunter, Executive Secretary, UN Framework Convention on Climate Change
Hamdallah Zedan, Executive Secretary, UN Convention on Biological Diversity

At-large Scientists

Adrienne Clarke, Laureate Professor, School of Botany, University of Melbourne, Australia
Denis Lucey, Professor of Food Economics, Dept. of Food Business & Development, University College Cork, Ireland, and Vice-President NATURA
Vo-tong Xuan, Rector, Angiang University, Vietnam

Private Sector

Momtaz Faruki Chowdhury, Director, Agribusiness Center for Competitiveness and Enterprise Development, Bangladesh
Sam Dryden, Managing Director, Emergent Genetics
David Evans, Former Head of Research and Technology, Syngenta International
Steve Parry, Sustainable Agriculture Research and Development Program Leader, Unilever
Mumeka M. Wright, Director, Bimzi Ltd., Zambia

Consumer Groups

Michael Hansen, Consumers International
Greg Jaffe, Director, Biotechnology Project, Center for Science in the Public Interest
Samuel Ochieng, Chief Executive, Consumer Information Network

Producer Groups

Mercy Karanja, Chief Executive Officer, Kenya National Farmers' Union
Prabha Mahale, World Board, International Federation Organic Agriculture Movements (IFOAM)
Tsakani Ngomane, Director Agricultural Extension Services, Department of Agriculture, Limpopo Province, Republic of South Africa
Armando Paredes, Presidente, Consejo Nacional Agropecuario (CNA)

Scientific Organizations

Jorge Ardila Vásquez, Director Area of Technology and Innovation, Inter-American Institute for Cooperation on Agriculture (IICA)
Samuel Bruce-Oliver, NARS Senior Fellow, Global Forum for Agricultural Research Secretariat
Adel El-Beltagy, Chair, Center Directors Committee, Consultative Group on International Agricultural Research (CGIAR)
Carl Greenidge, Director, Center for Rural and Technical Cooperation, Netherlands
Mohamed Hassan, Executive Director, Third World Academy of Sciences (TWAS)
Mark Holderness, Head Crop and Pest Management, CAB International
Charlotte Johnson-Welch, Public Health and Gender Specialist and Nata Duvvury, Director Social Conflict and Transformation Team, International Center for Research on Women (ICRW)
Thomas Rosswall, Executive Director, International Council for Science (ICSU)

Judi Wakhungu, Executive Director, African Center for Technology Studies

Governments

Australia: Peter Core, Director, Australian Centre for International Agricultural Research
China: Keming Qian, Director General Inst. Agricultural Economics, Dept. of International Cooperation, Chinese Academy of Agricultural Science
Finland: Tiina Huvio, Senior Advisor, Agriculture and Rural Development, Ministry of Foreign Affairs
France: Alain Derevier, Senior Advisor, Research for Sustainable Development, Ministry of Foreign Affairs
Germany: Hans-Jochen de Haas, Head, Agricultural and Rural Development, Federal Ministry of Economic Cooperation and Development (BMZ)
Hungary: Zoltan Bedo, Director, Agricultural Research Institute, Hungarian Academy of Sciences
Ireland: Aidan O'Driscoll, Assistant Secretary General, Department of Agriculture and Food
Morocco: Hamid Narjisse, Director General, INRA
Russia: Eugenia Serova, Head, Agrarian Policy Division, Institute for Economy in Transition
Uganda: Grace Akello, Minister of State for Northern Uganda Rehabilitation
United Kingdom: Paul Spray, Head of Research, DFID
United States: Rodney Brown, Deputy Under Secretary of Agriculture and Hans Klemm, Director of the Office of Agriculture, Biotechnology and Textile Trade Affairs, Department of State

Foundations and Unions

Susan Sechler, Senior Advisor on Biotechnology Policy, Rockefeller Foundation
Achim Steiner, Director General, The World Conservation Union (IUCN)
Eugene Terry, Director, African Agricultural Technology Foundation

Advisory Bureau

Nongovernment Representatives

Consumer Groups

Jaime Delgado • Asociación Peruana de Consumidores y Usuarios
Greg Jaffe • Center for Science in the Public Interest
Catherine Rutivi • Consumers International
Indrani Thuraisingham • Southeast Asia Council for Food Security and Trade
Jose Vargas Niello • Consumers International Chile

International organizations

Nata Duvvury • International Center for Research on Women
Emile Frison • CGIAR
Mohamed Hassan • Third World Academy of Sciences
Mark Holderness • GFAR
Jeffrey McNeely • World Conservation Union (IUCN)
Dennis Rangi • CAB International
John Stewart • International Council of Science (ICSU)

NGOs

Kevin Akoyi • Vredeseilanden
Hedia Baccar • Association pour la Protection de l'Environment de Kairouan
Benedikt Haerlin • Greenpeace International
Juan Lopez • Friends of the Earth International
Khadouja Mellouli • Women for Sustainable Development
Patrick Mulvaney • Practical Action
Romeo Quihano • Pesticide Action Network
Maryam Rahmaniam • CENESTA
Daniel Rodriguez • International Technology Development Group

Private Sector

Momtaz Chowdhury • Agrobased Technology and Industry Development
Giselle L. D'Almeida • Interface
Eva Maria Erisgen • BASF
Armando Paredes • Consejo Nacional Agropecuario
Steve Parry • Unilever
Harry Swaine • Syngenta (resigned)

Producer Groups

Shoaib Aziz • Sustainable Agriculture Action Group of Pakistan
Philip Kiriro • East African Farmers Federation
Kristie Knoll • Knoll Farms
Prabha Mahale • International Federation of Organic Agriculture Movements
Anita Morales • Apit Tako
Nizam Selim • Pioneer Hatchery

Government Representatives

Central and West Asia and North Africa

Egypt • Ahlam Al Naggar
Iran • Hossein Askari
Kyrgyz Republic • Djamin Akimaliev
Saudi Arabia • Abdu Al Assiri, Taqi Elldeen Adar, Khalid Al Ghamedi
Turkey • Yalcin Kaya, Mesut Keser

East and South Asia and the Pacific

Australia • Simon Hearn
China • Puyun Yang
India • PK Joshi
Japan • Ryuko Inoue
Philippines • William Medrano

Latin America and Caribbean

Brazil • Sebastiao Barbosa, Alexandre Cardoso, Paulo Roberto Galerani, Rubens Nodari
Dominican Republic • Rafael Perez Duvergé
Honduras • Arturo Galo, Roberto Villeda Toledo
Uruguay • Mario Allegri

North America and Europe

Austria • Hedwig Woegerbauer
Canada • Iain MacGillivray
Finland • Marja-Liisa Tapio-Bistrom
France • Michel Dodet
Ireland • Aidan O'Driscoll , Tony Smith
Russia • Eugenia Serova, Sergey Alexanian
United Kingdom • Jim Harvey, David Howlett, John Barret
United States • Christian Foster

Sub-Saharan Africa

Benin • Jean Claude Codjia
Gambia • Sulayman Trawally
Kenya • Evans Mwangi
Mozambique • Alsácia Atanásio, Júlio Mchola
Namibia • Gillian Maggs-Kölling
Senegal • Ibrahim Diouck

Annex F
Secretariat and Cosponsor Focal Points

Secretariat

World Bank

Marianne Cabraal, Leonila Castillo, Jodi Horton, Betsi Isay, Pekka Jamsen, Pedro Marques, Beverly McIntyre, Wubi Mekonnen, June Remy

UNEP

Marcus Lee, Nalini Sharma, Anna Stabrawa

UNESCO

Guillen Calvo

With special thanks to the publications team: Audrey Ringler (logo design), Pedro Marques (proofing and graphics), Ketill Berger and Eric Fuller (graphic design)

Regional Institutes

Sub-Saharan Africa – African Centre for Technology Studies (ACTS)

Ronald Ajengo, Elvin Nyukuri, Judi Wakhungu

Central and West Asia and North Africa – International Center for Agricultural Research in the Dry Areas (ICARDA)

Mustapha Guellouz, Lamis Makhoul, Caroline Msrieh-Seropian, Ahmed Sidahmed, Cathy Farnworth

Latin America and the Caribbean – Inter-American Institute for Cooperation on Agriculture (IICA)

Enrique Alarcon, Jorge Ardila Vásquez, Viviana Chacon, Johana Rodríguez, Gustavo Sain

East and South Asia and the Pacific – WorldFish Center

Karen Khoo, Siew Hua Koh, Li Ping Ng, Jamie Oliver, Prem Chandran Venugopalan

Cosponsor Focal Points

GEF	Mark Zimsky
UNDP	Philip Dobie
UNEP	Ivar Baste
UNESCO	Salvatore Arico, Walter Erdelen
WHO	Jorgen Schlundt
World Bank	Mark Cackler, Kevin Cleaver, Eija Pehu, Juergen Voegele

Index

Page references followed by *f*, *t*, and *b* indicate figures, tables, and boxes, respectively.

About Island Press

Since 1984, the nonprofit Island Press has been stimulating, shaping, and communicating the ideas that are essential for solving environmental problems worldwide. With more than 800 titles in print and some 40 new releases each year, we are the nation's leading publisher on environmental issues. We identify innovative thinkers and emerging trends in the environmental field. We work with world-renowned experts and authors to develop cross-disciplinary solutions to environmental challenges.

Island Press designs and implements coordinated book publication campaigns in order to communicate our critical messages in print, in person, and online using the latest technologies, programs, and the media. Our goal: to reach targeted audiences—scientists, policymakers, environmental advocates, the media, and concerned citizens—who can and will take action to protect the plants and animals that enrich our world, the ecosystems we need to survive, the water we drink, and the air we breathe.

Island Press gratefully acknowledges the support of its work by the Agua Fund, Inc., Annenberg Foundation, The Christensen Fund, The Nathan Cummings Foundation, The Geraldine R. Dodge Foundation, Doris Duke Charitable Foundation, The Educational Foundation of America, Betsy and Jesse Fink Foundation, The William and Flora Hewlett Foundation, The Kendeda Fund, The Andrew W. Mellon Foundation, The Curtis and Edith Munson Foundation, Oak Foundation, The Overbrook Foundation, the David and Lucile Packard Foundation, The Summit Fund of Washington, Trust for Architectural Easements, Wallace Global Fund, The Winslow Foundation, and other generous donors.